PRACTICAL CELESTIAL NAVIGATION

PRACTICAL CELESTIAL NAVIGATION

BY TOM BOTTOMLEY

TAB TAB BOOKS Inc.

BLUE RIDGE SUMMIT, PA. 17214

FIRST EDITION

FIRST PRINTING

Copyright © 1983 by TAB BOOKS Inc.

Printed in the United States of America

Library of Congress Cataloging in Publication Data

Bottomley, Tom.
    Practical celestial navigation.

    Includes index.
    1. Navigation. 2. Celestial astronomy.
I. Title.
VK555.B727   1983   623.89   82-19370
ISBN 0-8306-1386-2 (pbk.)

Cover photo courtesy of *Motor Boating and Sailing* magazine.

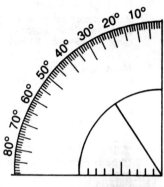

Contents

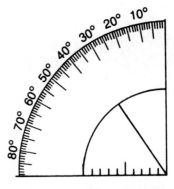

Foreword

Navigation is often described as an art and a science. From either viewpoint, it makes a fascinating topic for study. I would not hesitate to say that a majority of those who read navigation books do so for the love of learning a new and interesting subject rather than for their own use at sea. Celestial navigation is especially interesting because it combines the lore of the skies with mathematics and the thrill—no other word will do—of using a sextant to determine one's position.

Books on celestial navigation span the wide range from the encyclopedic *Bowditch* and comprehensive *Dutton's* to the many brief (too brief, actually), "cookbooks" that tell the reader to do "Step 1," "Step 2," etc., to achieve an answer—but with no knowledge of, or feeling for, the process followed. Tom Bottomley's book strikes a happy middle course. In *Practical Celestial Navigation,* he has fully described a basic procedure with enough theory and supplemental material to keep the reader in the picture.

Modern electronic calculator use is also covered but not so emphasized that the romance of celestial navigation is lost to a mechanical procedure. The text includes numerous problems (with answers) worked through in detail to show just how a solution is reached.

This book contains all that the reader needs to become proficient and knowledgeable in celestial navigation; all, that is, except many hours of practice with a sextant, accurate time, and some heavenly bodies. *Practical Celestial Navigation* truly lives up to its name, and it will be a welcome addition to any boatman's library.

Elbert S. Maloney

Elbert S. Maloney has authored editions of Chapman's *Piloting, Seamanship and Small Boat Handling* and Dutton's *Navigation and Piloting.*

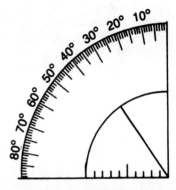

Acknowledgments

The following illustrations are reproduced from *The Nautical Almanac* (No. 008-054-00079-7) courtesy of the Nautical Almanac Office, United States Naval Observatory:

Tables 3-1 through 3-8
Tables 5-1 and 5-2
Tables 6-6 through 6-8
Tables 9-1, 9-2, and 9-5 through 9-9
Table 11-1
Figures 12-1 and 12-2
Table 13-1
Table 15-1

The following illustrations are reproduced from *Sight Reduction Tables for Marine Navigation Latitudes* 30 - 40 (Pub. No. 229, vol. 3) inclusive published by the Defense Mapping Agency Hydrographic Center:

Tables 6-1 through 6-10
Tables 9-3, 9-4, 9-7, and 9-10
Tables 13-1 and 13-3 through 13-6
Tables 17-1 through 17-3

The following illustrations are reproduced from *The Air Almanac* 1980 July-December (No. 1980 0-236-662) courtesy of the United States Naval Observatory:

Tables 3-9 through 3-12
Figure 12-3

The following illustrations are reproduced from *Sight Reduction Tables for Air Navigation* (Pub. 249, Volumes 1-3) published by the Defense Mapping Agency Hydrographic Center:

Tables 14-4 through 14-10

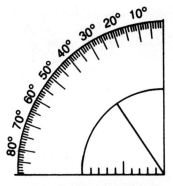

Introduction

My interest in celestial navigation began with the sextant I inherited from my father. That might make it sound like I am carrying on a family seafaring tradition, but my father's voyaging experience was limited to troopship crossings of the Atlantic to France and back during World War I. The sextant, purchased in a Chicago junk shop in 1944, served as the base for a lamp that he constructed. The sight tube, mirrors, and vernier scale magnifier were missing, as well as a number of miscellaneous screws. It wasn't a fit instrument for practical use.

When the sextant came into my possession, I discarded the lamp and had the sextant restored by Sam Mellor, a retired sea captain with skills in both optics and fine machine work. Made by Wilfred O. White in about 1916, according to records now at the Danforth Co., it carries a vernier scale that reads to 10 seconds of arc. It is a difficult scale to read even with magnification and more so now that the gradations are worn and pitted. After the United States Power Squadrons navigation courses and a lot of practice, I find no difficulty in using the sextant for daylight shots. In morning or evening twilight I much prefer the modern instrument with its micrometer drum scale.

But why bother with celestial navigation at all, now that the wonders of modern marine electronics engineering provides us with instruments that not only tell us our exact position almost anywhere on earth, but also provide course and distance to our destination—and then sound an alert as we near it? The answer, of course, is that not all boat owners can afford to lay out thousands of dollars for this sophisticated equipment. Even where a Loran-C or Omega is carried, there is the possibility of an electrical failure that could put it out of commission. Certainly the skipper who is forced to take to his life raft will have to rely on his sextant.

This book is designed to show you how to use the sextant to get accurate position information based on the sun, moon, planets, and stars. The investment is modest; even an inexpensive sextant will provide reasonably acceptable results. The process involved is not difficult.

The navigation method described here is based on use of the *Nautical Almanac* (published jointly by the U.S. Naval Observatory Office and the U.K.'s Her Majesty's Stationary Office) and the *Publication No. 229 Sight Reduction Tables* (by the Defence Mapping Agency Hydrographic Center) because these provide the most accurate results with the least amount of work and confusion—in my opinion. Other methods are discussed briefly, where appropriate, with a little extra emphasis given to use of the *Air Almanac* (by the U.S. Naval Observatory) because this is in fairly widespread use.

There are few skills that a boater can acquire that provide more satisfaction than the ability to determine position or the course to steer from observations of the heavenly bodies.

Chapter 1

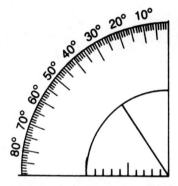

The Earth and the Heavens Above

To practice celestial navigation, you need to accept a few misconceptions. Discard from your mind some of the scientific discoveries of the past 2,000 years and assume that the Earth is a perfect sphere. The sun, moon, planets, and stars are all equidistant from the Earth. The stars are fixed in place and the sun, moon, and planets move in their appointed orbits on the surface of the celestial sphere to which the stars are plastered. Assume that the Earth is stationary and that the celestial sphere revolves about the Earth. It's also helpful to assume that no matter where you are on the face of the Earth, your bit of geography is at the top when you take a sight. That makes sense. If you were at the bottom, you'd fall off—wouldn't you?

THE EARTH COORDINATES

For the convenience of navigators, the sphere of the earth is divided into coordinates of *longitude* and *latitude*. *Meridians* of longitude are great circles that pass through the North and South Poles. The plane of a great circle passes through the center of a sphere. The earth is divided into 360 degrees of longitude. This is measured as 180 degrees west from the 0-degree line that passes through the former Naval Observatory at Greenwich, England (the *Prime Meridian*) and 180 degrees east from this line. See Fig. 1.

Parallels of latitude are small circles that are parallel to the plane of the equator (Fig. 1-2). The equator is a great circle because its plane passes through the equator. The earth is divided into 90 degrees of latitude north of the equator, 90 degrees south of the equator, the 0-degree line is the equator, and the 90-degree marks are at the respective poles.

1

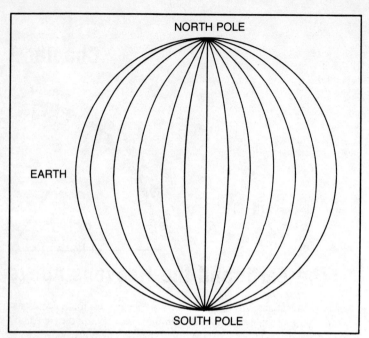

Fig. 1-1. Meridians of longitude.

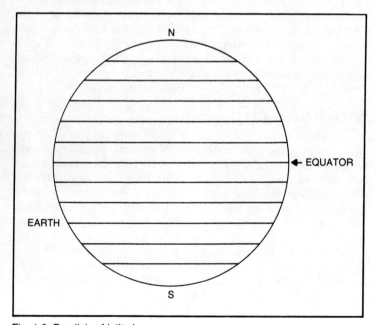

Fig. 1-2. Parallels of latitude.

One degree of longitude or latitude is equal to 60 minutes of *arc;* each minute of arc equals 60 seconds of arc. As a convenience in making calculations, minutes and seconds of arc are usually shown as minutes and tenths of arc in the relevant publications. This is because the micrometer drum of the modern sextant is calibrated in decimal fractions of a minute of arc (0.5′, 0.2′, or 0.1′). One-tenth of a minute of arc equals six seconds of arc.

Any position on the earth's surface can be defined by its navigation coordinates. For example, the food concession pavillion at Parking Lot No. 2, Robert Moses State Park (Fire Island, New York), is at 73°16.7′ west longitude, and 40°37.3′ north latitude. When doing computations, these cordinates would be written as: Lo73°16.7′W, L40°37.3′N. (Some navigators use the lower case Greek lambda (λ) to signify longitude.)

Another help to the navigator is the fact that 1 minute of arc of latitude is exactly equal to 1 nautical mile in distance. Hence 1 minute of arc of any great circle is equal to 60 nautical miles. Remember that parallels of latitude run east and west, parallel to the equator, but latitude is measured north and south. If the angle is measured from the center of the earth between any two points on its surface, this angle can be converted to the distance between the points by multiplying whole degrees of arc by 60 and adding the remainder.

Here is an example. A great circle angle of 33°22.8′ of arc is measured between points A and B. Distance between the points is $33° \times 60 = 1980 + 22.8′ = 2002.8$ nautical miles.

Caution: Do not try to convert the designated angle between meridians of longitude to distance unless the angle is measured along the plane of the equator. For example, Point A on the equator is at 0°00.0′ (the prime meridian); Point B is at 45°00.0′W, also on the equator, and the center of the earth is Point C. This forms a great circle triangle ACB and the distance from A to B is $45° \times 60 = 2,300$ nautical miles.

If Points A and B maintain their positions on their meridians of longitude, but are at a latitude of 45° north, the angle of longitude is still 45°—angle A′C′B′, see Fig. 1-3. But as the meridians of longitude converge toward the poles, distance between A′ and B′ is only 1,680 nautical miles. This is reflected in a great circle angle of A′C′B′ that is 28° of arc ($28° \times 60 = 1,680$ nautical miles).

CELESTIAL COORDINATES

The celestial sphere, with the stars plastered to it and the other bodies running along their appointed tracks, also is divided into coordinates as an aid to navigation. To make it clear that the celestial sphere is involved in a calculation, and not the earth, the north-south equivalent of meridians of longitude are *Greenwich hour angles* (GHA), see Fig. 1-4. The equivalent of parallels of latitude are degrees of declination (Fig. 1-5).

Because the earth is centered exactly within this sphere, the axis running from its north pole to its south pole, if extended in each direction, will touch the north and south poles of the celestial sphere. The plane of the

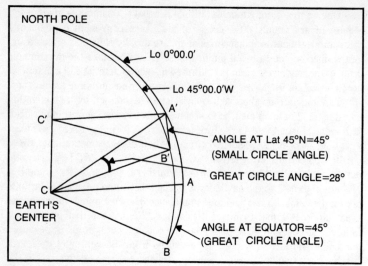

NORTH POLE

Lo 0°00.0'

Lo 45°00.0'W

A'

C'

ANGLE AT Lat 45°N=45°
(SMALL CIRCLE ANGLE)

B'

GREAT CIRCLE ANGLE=28°

A

C

EARTH'S
CENTER

ANGLE AT EQUATOR=45°
(GREAT CIRCLE ANGLE)

B

Fig. 1-3. Distance between meridians of longitude decreases with distance north
or south of the equator.

equator, if extended into space, will pass exactly through the celestial
equator.

Like the earth, the celestial shere is divided into 360 degrees with
GHA angles up to 180 degrees west of the Greenwich 0° GHA point, and
180 degrees east of this point. Declination runs from 0° at the celestial
equator to 90° north and 90° south at the respective poles.

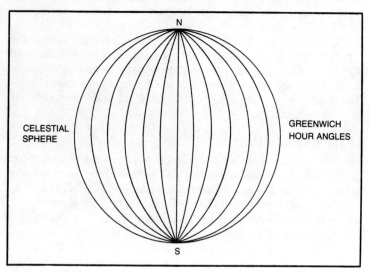

N

CELESTIAL
SPHERE

GREENWICH
HOUR ANGLES

S

Fig. 1-4. Greenwich hour angles.

GEOGRAPHICAL POSITION

In coastal piloting, if you determine your distance from an object that you can see—a navigational aid or landmark shown on a chart—and you also determine the bearing of the object from your boat, you can quickly plot your position on the chart. The bearing gives you a *line of position* (LOP). With your dividers, you mark off the appropriate distance from the sighted object along this LOP—and there you are.

In celestial navigation, you do essentially the same thing. You determine the geographical position (GP) of a celestial object and then determine the bearing and distance to this GP from your boat. At any given moment, a line drawn from the center of the earth to any body on the celestial sphere passes through the surface of the earth at the longitude and latitude that correspond to the GHA and declination of the body at that instant. This point on the earth's surface is the geographical position of the body at that time (see Fig. 1-6). The *Nautical Almanac* and *Air Almanac* both provide the means of establishing the GP at any instant for all the celestial bodies used for navigation.

When you take a sight, you note the exact time. Then you look up the body in the *Almanac* and determine its GP for that instant. The angle measured by the sextant is converted to distance from the GP, and sight reduction tables or calculations give you the bearing to this GP. If you had a chart large enough, you could plot your position as the intersection of a distance-off measurement on the bearing LOP.

In practice, the sight reduction tables are used to provide distance and bearing (*azimuth* in celestial navigation) from a known position to the GP of

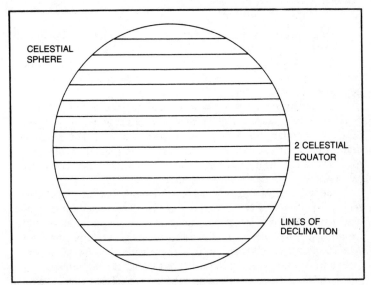

Fig. 1-5. Lines of declination.

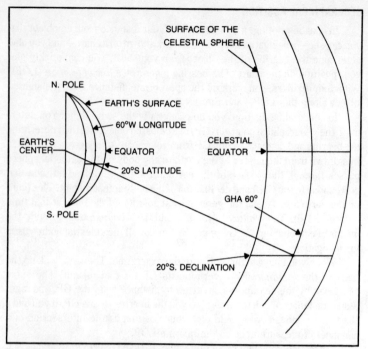

Fig. 1-6. Latitude and longitude of the geographical position (GP) of a body is exactly equal to its declination and Greenwich Hour Angle (GHA) at any given moment.

the sighted object. The known position may be your *dead reckoning* (DR) position at the time of the sight or an *assumed position* (AP), depending on the sight reduction method used. The known position is always so close to your actual position at the time of the sight that the calculated azimuth can be considered actual azimuth to the GP of the body.

The difference between the calculated distance to the GP and the distance actually measured by the sextant is then plotted from your DR or AP on the chart (see Fig. 1-7). Here the dashed line (called the *intercept*) is drawn from a plotted AP in the direction of the azimuth to the GP (its *reciprocal* is discussed later) for a length equal to the difference between calculated and observed (measured) distance to the GP.

The line at right angles is called a line of position although it represents a small segment of a circle that has the GP of the body at its center. Your actual position is somewhere along this LOP. In the absence of other information, it is plotted as an *estimated position* (EP) at the point closest to your DR position at the time of the sight. Figure 1-8 shows an intercept plotted from a DR position; the EP is at the end of the intercept.

If sights are taken on two or more bodies at or close to the same time, your position is a fix at the intersection of the LOPs that result.

6

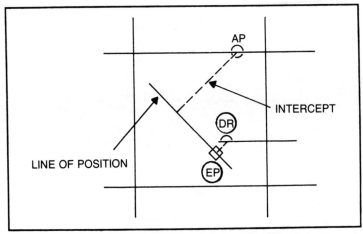

Fig. 1-7. In some sight reduction methods, the line of position that represents distance from the GP of a celestial body is based on an assumed position used for the calculations.

THE SEXTANT ANGLE

To relate the sextant angle that you measured to the GP of the sighted body, you have to assume that this angle, measured from your position on the earth's surface between the body and the horizon, is equal to an angle measured from the center of the earth to the GP of the body and a horizon that cuts through the center of the earth (the celestial horizon). In Fig. 1-9, Angle A equals Angle B.

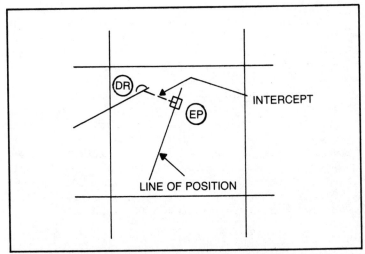

Fig. 1-8. In other sight reduction methods, the line of position is based on the vessel's dead reckoning (DR) position at the time of the sight.

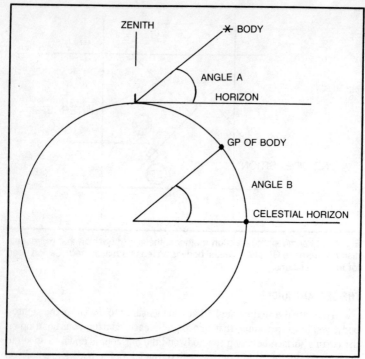

Fig. 1-9. The angle that a sighted body makes with the horizon, after certain corrections have been made, is the same as the angle as measured from the center of the earth to the GP of the body.

In the case of stars and the planets, which are at vast distances from the earth, there really is no difference in the angles (or at least the differences are so small that they are insignificant). In the case of the sun and moon, corrections (supplied in the *Nautical* and *Air Almanacs*) are applied to the sextant reading to "equalize" the angles.

The point of the celestial sphere that is directly overhead when you take a sight is the *zenith* (Fig. 1-10). Your position at the "top" of the earth is P, directly below the zenith. Your measured angle, from the center of the earth between the body and the celestial horizon, in Fig. 1-10, is 40 degrees. Subtract this angle from 90 degrees to get *zenith distance;* this is the angle between the zenith and the body. Because it is also the angle between your position and the GP of the body, the zenith distance—in this case 50 degrees—represents distance from your position to the GP of the body at the time of the sight: $50 \times 60 = 3,000$ nautical miles.

THE HORIZON

On a clear day, with calm seas, you look out and see a fairly distinct line where sky and sea appear to meet. This is the *horizon*, and when using a sextant, you measure the angle between this line and a celestial body somewhere above it.

The horizon you see is called the *visible horizon,* but due to the refraction of light passing through the earth's atmosphere, it isn't where you think it is. A line from your eye, tangent to the physical surface of the earth, would locate it properly in the plane of the *geometrical horizon.* You think you see a horizon directly out in front of you at eye level and, because this seems to make sense, this is the sensible horizon.

When you take a sight you try to establish the angle from your position between the body and a plane tangent to the surface of the earth at your position—the *geoidal horizon*—because this will be the same as the angle as measured from the center of the earth between the GP of the body and the plane of the *celestial horizon*.

Confused? Figure 1-11 should help. The distinctions between various types of horizon are given here so you'll have a better understanding of what you see and why sextant corrections are necessary for each sight you take. The *Nautical* and *Air Almanac* give you these corrections in convenient tables for each type of celestial body (sun, moon, planets, stars) based on the sextant angle you actually measure.

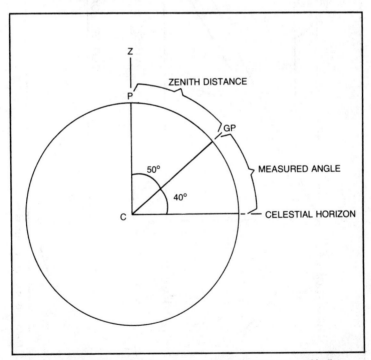

Fig. 1-10. Subtract the measured angle from 90 degrees to get zenith distance, which can be converted to distance in nautical miles from your position to the GP of the body.

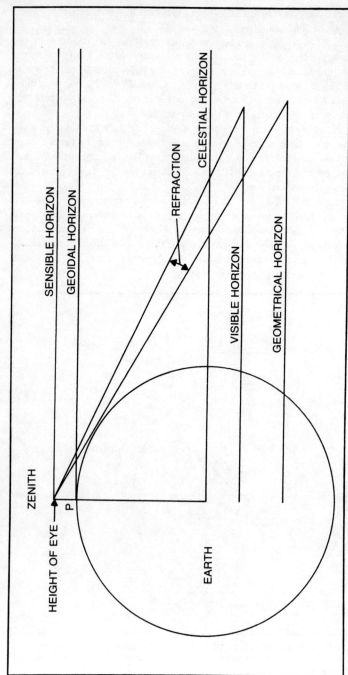

Fig. 1-11. Names of the various forms of horizon.

REFRACTION AND PARALLAX

As noted above, the earth's atmosphere actually bends light passing through it just as light rays are bent (refracted) as they pass through a lens. The closer to the horizon the origin of the rays the more they are bent. Figure 1-12 shows this in exaggerated form. Note that a body that is below the horizon might actually appear to be above it! As can be seen, this bending of light decreases with altitude to the point of no refraction for bodies at or near the zenith.

The angle you measure with the sextant is not necessarily the angle you need unless the body is at the zenith, directly overhead. The main correction given in each of the *Almanacs* includes the refraction factor.

For the sun and moon, there are additional problems. These are so close to the earth that the angle measured from the surface of the earth, corrected for refraction, is not the same as the angle would be if it were measured from the center of the earth. The difference here is called

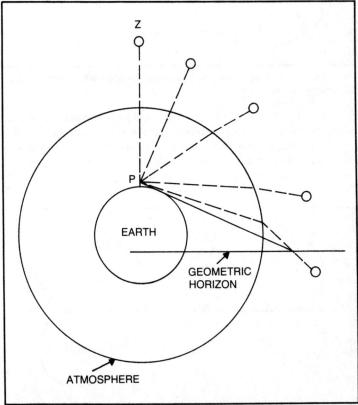

Fig. 1-12. The earth's atmosphere bends light rays, with maximum distortion when a body is near the horizon and least when it is overhead.

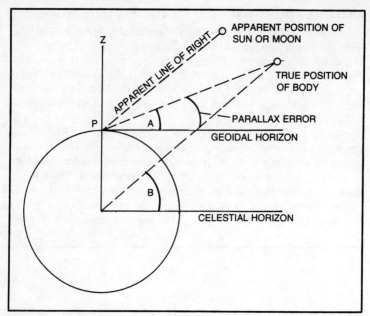

Fig. 1-13. Because the sun and moon are so close to the earth, a correction must be made for parallax.

parallax. In Fig. 1-13, Angle A is not equal to Angle B. Fortunately, the main correction provided in the *Almanacs* includes parallax along with refraction.

Also, for greatest accuracy in taking a sun or moon sight, wait until either the bottom edge or top edge (lower or upper limb) just touches the horizon. The altitude of the body is measured to its center. Therefore, for the sun and moon, *semidiameter* corrections are also needed. The *Nautical*

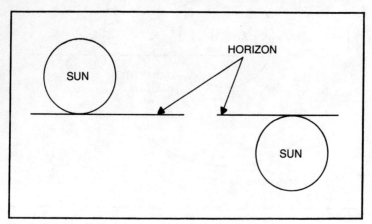

Fig. 1-14. Lower (left) and upper (right) limb sights of the sun.

Almanac incorporates these in the main correction along with refraction and parallax; *Air Almanac* gives daily semidiameter (SD) figures that must be added or subtracted as necessary (Fig. 1-14).

The moon has another correction necessary, for *horizontal parallax* (HP) as it passes through its phases. Only when it is full does it appear as a complete circle. As it waxes or wanes, circumference of the visible portion is not an arc of a perfect circle. A separate HP correction, given in the *Almanacs,* is used to adjust for this.

Chapter 2

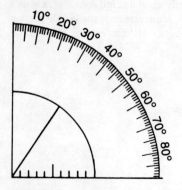

The Sextant

The sextant sight is the key to celestial navigation. It provides the information that leads to establishing the geographical position of a celestial body at a given time, and from this GP the navigator can determine his location. The accuracy of a sight can be no greater than the accuracy of the instrument itself, but when a sight is taken from the pitching deck of a small boat it is difficult to take precise readings with even the most accurate of instruments. Because with practical celestial navigation, consider the use of a good plastic sextant unless you have the money for the more expensive metal type. A plastic sextant designed for marine *use*—not a toy or practice type—is precise enough for the small boat skipper.

Even an inexpensive sextant, however, must be treated as the precision instrument that it is. Every caution should be observed to keep it clean and in proper adjustment. Lens tissues and alcohol lens cleaner, available from camera stores, should be used to clean lenses, mirrors, and sun screens. Don't use a handkerchief or similar cloth for this purpose because sensitive optic surfaces can be scratched. Be sure that the sextant is kept in its case when not in use and that the case is well secured. Handle the instrument with respect and it will provide years of faithful service.

A sextant is so named because on the original models the index arm moved through an arc of 60 degrees—one sixth of a circle. Earlier instruments, octants, covered one eighth of a circle—45 degrees. The index arm on the sextant moves only half the distance of the angle that is measured so the index scale on a 60-degree index arc reads from 0 to 120 degrees. Actually, most modern sextants have a larger index arc with scales that read from 0 to 130 degrees or as much as 150 degrees (see Fig. 2-1).

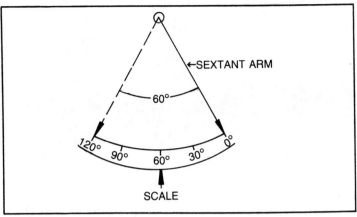

Fig. 2-1. Each degree of movement of a sextant arm covers two degrees on the sextant's index scale.

SEXTANT NOMENCLATURE

A marine sextant is made up of a triangular frame with the bottom of the triangle in the form of an arc. This, with its scale, is the *index arc*. An *eyepiece ring* is attached to the rear leg of the sextant frame in line with a *horizon glass* at the front of the mirror. The horizon glass is silvered on the half next to the frame. This portion of the glass is the *horizon mirror*.

An *index arm* is pivoted to the top of the frame and the *index mirror* is mounted at the top of this arm. The bottom end of the arm has a clamping mechanism that holds it to the index arc, plus a *micrometer drum*. The clamp is released so the arm can be swung freely along the arc to bring a sighted object close to the horizon. The clamp is then set and the micrometer drum is turned until the body is brought exactly to the horizon.

Because you can't look directly at the sun without damage to the eyes, shades in the form of polarizing filters are fitted to the sextants. One is fitted so that it can be swung in front of the index mirror and adjusted to provide the necessary density so that a sun sight can be taken without discomfort. A second filter swings in front of the horizon glass to reduce glare along the horizon. In older sextants, dark red or green filters were used singly or in combination to make sun sights possible.

The eyepiece ring on the sextant accepts a *single tube* that in effect is a telescope without lenses. Some navigators find this useful when taking sun or moon sights. A telescope (sometimes more than one) of modest power—usually 2× to 4×—may be provided. The telescope is not needed so much to magnify the sighted body, but because its greater light-gathering capability makes a dim star easier to observe. Most of the time, it isn't necessary to use either the sight tube or scope. Just observe the horizon glass through the eyepiece ring.

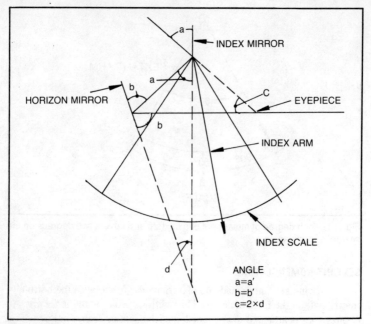

Fig. 2-2. Optical principles of a sextant.

SEXTANT OPTICS

Light from a celestial body striking the index mirror is reflected at an angle equal to the angle at which it strikes the mirror (the angle of reflection is equal to the angle of incidence). The index arm is swung until this reflected light shows up as the image of the body on the horizon mirror. The sextant is adjusted until this image appears to be touching the horizon. Again, the angle of the light striking the horizon mirror is equal to the angle that it is reflected back through the eyepiece ring. Because the eyepiece ring and horizon mirror are both immovable, these angles are constant.

The angle formed by the planes of the index mirror and horizon mirror surfaces is exactly half the size of the angle measured by the instrument from the horizon to the sighted body. Therefore, swinging the index arm only half a degree provides a full degree of change in the measured angle (see Fig. 2-2).

USING THE SEXTANT

Ideally, two persons are involved in taking a sight: the navigator, with the sextant, and an assistant with a timepiece. The navigator sets the index arm to the zero mark on the scale. This way the image reflected by the index mirror to the horizon is the same as the real thing seen through the horizon glass. If aimed at the horizon, for example, the horizon appears as an unbroken line across the horizon glass and mirror.

If aimed at a celestial body, two of the body will be seen; the real thing through the horizon glass, and the reflected image alongside it in the horizon mirror. Many sextants are adjusted so that the reflected image appears to be superimposed directly over the real object. In most cases, however, the index arm is set at zero and the celestial body is sighted as above. The clamp is released and the index arm is slid slowly forward along the index arc as the front of the instrument is gently lowered. The body seen through the horizon glass disappears as the front of the sextant moves down, but the image of the sighted body is held on the horizon mirror.

When the image of the sighted body is almost at the horizon, the clamp is reset by releasing the levers at the bottom of the arm. The image may jump up or down slightly as the clamp sets. The micrometer drum is then turned for the fine adjustment that brings the body down until it just touches the horizon. It is a good practice to rock the sextant slightly from side to side at this point to make sure the reading is correct (see Fig. 2-3) with the instrument exactly vertical.

Note that the rise and fall of a boat in waves will not affect the reading as long as the horizon is in sight through the horizon glass. Try taking practice sights from a stable platform. You'll see that tilting the sextant slightly up and down will not change the position of the image in relation to the horizon. This is true as long as the sextant doesn't lean to one side or the other.

Timing the Sight

As soon as you are sure you have the body exactly touching the line of the horizon, call out "mark." Your assistant then notes the time. Usually, as

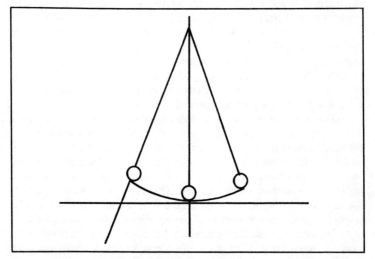

Fig. 2-3. A sextant should be rocked slightly from side to side to make sure the image of the body just touches the horizon at the lowest part of the arc.

you get the body close to the horizon, you give the alert by saying "stand by" or "ready." If you don't have an assistant, quickly shift your eyes from the sextant to your timepiece. With practice, this should take no more than one second (you can deduct it from the recorded time).

In recording the time, note the seconds first, then minutes, and finally the hour. Once you have the seconds established, it makes a quick recheck of the minutes possible. Be especially careful when the second hand of a watch is between the 45 seconds and 60 seconds marks. It is easy to record the up-coming minute rather than the previous one and this could introduce a serious error into your computations.

As soon as the time is noted, put the sextant reading alongside it. If you're working with an assistant, call out the reading so he or she can write it down. If you don't have an assistant note it down yourself. It is good practice to take at least three sights in quick succession (within one minute of each other), noting the time and reading for each. In some cases, you'll want a run of at least five sights in order to come up with an accurate position.

The Sextant Reading

The index arc of a sextant is divided into a scale of whole minutes of arc. If the micrometer drum is set at zero and the arm clamp is set, the sextant is reading a whole degree of arc. The micrometer drum is calibrated in a scale of 60 minutes that reads against an index line or a vernier scale. If there's just a single index mark, the readout is in whole minutes of arc. If a vernier scale is provided, the readout may be to 0.5', 0.2', or even 0.1' of arc. The nearest whole minute is read at the zero end of the vernier scale. The decimal percentage of a minute is read on the vernier scale where one of its marks is aligned exactly with a minute mark on the micrometer drum scale.

It is not a difficult sequence to follow. Read whole degrees on the index arc. Read whole minutes on the micrometer drum scale against the index mark or zero on the vernier scale. Then read the decimal portion of a minute on the vernier scale where one of its lines is in alignment with a line on the micrometer drum minute scale. If you can practice from a known point, you can check the accuracy of your readings very quickly.

Older sextants, such as mine, do not have a micrometer drum. This is replaced by a vernier scale on the bottom of the index arm that reads against the index scale on the index arc itself. The sighted object is brought as close as possible to the horizon by swinging the index arm. The clamp is locked and then a fine adjustment screw, located at the front of the arm, is turned until the object just touches the horizon. The arc index scale is read against the zero mark on the vernier scale and the vernier scale is read at the point where one of its lines is in exact alignment with a scale mark on the arc.

The arc index scale can be in whole degrees of arc, with the vernier in whole minutes, but smaller gradations are generally the rule. My index scale is to 10 minutes on the index arc and 10 seconds on the vernier scale (just under 0.2').

Sextant Corrections

As noted in the previous chapter, the sextant reading you record is not the true angle of the sighted body. Errors exist due to refraction, in the case of stars and planets, and refraction plus parallax in the case of the sun and moon. There's also index error, an error based on height of eye above sea level, and there can be operator error as well.

Index error is common; don't worry about it. Metal and plastic sextants can expand or contract slightly as temperatures change. This subtly changes the relationships between index and horizon mirrors. The error that results is instrument error and its extent—and direction—can be quickly found using the following procedure.

With the index arm set at zero, sight a celestial body or the horizon. Adjust the instrument until the real and reflected images are in exact alignment in the horizon glass and mirror. Note the sextant reading at this point. If you're lucky, it is exactly 0°00.0′, but this is seldom the case. The reading might be slightly above or below this. If it is above (on the scale), the error is *positive*. The index correction is *negative* and is subtracted from the reading when a sight is taken. If the reading is less than zero (off the scale), the error is *negative* and the correction, being positive, is added to your sight readings.

A good practice before taking any run of sights is to check the index error and note the index correction (IC) with the sight information. Organizations such as the United States Power Squadrons recommend use of fairly elaborate "sight logs," with this and other data recorded for each sight taken. Some modern sextants have a quick adjustment that can be made to correct the reading to zero before you take sights, thus eliminating this particular error.

There might also be slight irregularities in the scale markings or a slight warp in materials during manufacture. This will produce small errors because the index arm is swung along its arc, even if there is zero index error. A table of such errors, if they exist, is supplied by the manufacturer on a card pasted inside the instrument case. These instrument errors also must be corrected as necessary and they are combined with the index error in the calculations.

If you are standing on the bridge of a container ship, 60 feet or so above the surface of the sea, the angle that you measure between a body and the horizon is not the same as the angle would be if measured from the deck of your pleasure boat with your eye about 10 feet above the sea (Fig. 2-4). This height-of-eye error is called *dip*. The higher you are the greater the dip error. And because this always provides a reading greater than the angle would be if measured from the sea surface, its correction is *always negative*. Subtract dip from sextant readings. The relationship between height of eye and the amount of error is exact so handy tables are provided in the *Almanacs* that show just what the correction should be.

If you are with a group of other navigators (as in a school program) all taking sights about the same time, and you find your readings are consis-

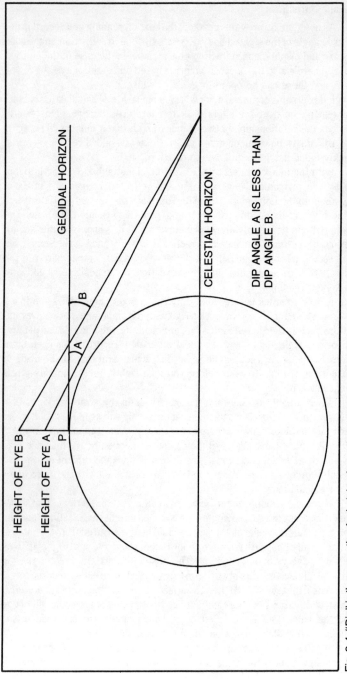

Fig. 2-4. "Dip" is the correction for height of eye above sea level; the greater the eye height, the greater the correction.

20

tently slightly greater or less than these of the others, you might be bringing the body slightly above or below the horizon each time you record the sight. This is *operator error*. If you know how much it is, and in which direction, you can correct for it. In general practice, it is assumed not to exist.

ALTITUDE DESIGNATIONS

The angle measured to a celestial body is called *altitude* and the uppercase letter H is used to designate altitude. The raw sextant sight that you record, before any corrections, is *sextant altitude* (Hs). After index and dip corrections are made, you have *apparent altitude* (Ha). After corrections are made for refraction and parallax, if necessary, the result is the *observed altitude* (Ho). This represents the actual angle between your position and the GP of the body at the time of the sight—if the sight was accurate to begin with. The altitude determined from sight reduction tables, based on your DR position or an assumed position (AP), is *calculated altitude* (Hc). It is the difference between Ho and Hc that gives you an intercept distance that you can plot.

Many sights are taken at dawn or dusk—morning or evening twilight—when it's possible to see stars and planets as well as the horizon. At these times, a small flashlight is needed to read sextant scales (unless the instrument is fitted with its own light). Some people use red bulbs or lenses on such lights, but I don't think this is necessary. There's usually enough light in the sky so that the small amount of light you use doesn't leave you temporarily "night-blind."

Chapter 3

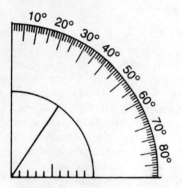

The Nautical Almanac

A sextant sight determines direction and distance from the geographical position of a celestial body. Unless you know the GP of the body for the time of the sight, the sextant information by itself will be worthless. Fortunately, the GP of the sun, moon, and the navigational stars and planets can be established quickly by using the *Nautical Almanac*, published jointly by the U.S. Naval Observatory in this country, and Her Majesty's Stationery Office in Great Britain.

DIP CORRECTION DATA

As noted in the previous chapter, the raw sextant reading you record is sextant altitude (Hs), and this must be corrected first to apparent altitude (Ha) with index error and height-of-eye correction factors. Index error is determined before a run of sights is taken and its correction is noted on your sight log. The height-of-eye (dip) correction is taken from the *Almanac*. The inside front cover (Table 3-1) carries tables that provide dip corrections from eye heights ranging from two feet to 155 feet above sea level in U.S. units and from one meter to 48 meters above sea level in metric units.

For most small boat skippers, the left-hand column under the dip heading will be used. Simply look up the figures that bracket your actual or estimated height of eye and then take the correction factor that is between these. For example, you figure your height of eye is 12.9 feet above sea level; heights of 12.6 feet and 13.3 feet are shown in the table and the correction factor between them is −3.4′ seconds. The correction is in minutes and tenths of minutes of arc, and it is always subtracted from the sextant reading.

If you are using the metric system, and your eye height is 2.7m above

Table 3-1. Altitude Correction Tables from the *Nautical Almanac*.

A2 ALTITUDE CORRECTION TABLES 10°-90°—SUN, STARS, PLANETS

SUN

OCT.–MAR. App. Alt.	Lower Limb	Upper Limb	APR.–SEPT. App. Alt.	Lower Limb	Upper Limb
9 34	+10.8	−21.5	9 39	+10.6	−21.2
9 45	+10.9	−21.4	9 51	+10.7	−21.1
9 56	+11.0	−21.3	10 03	+10.8	−21.0
10 08	+11.1	−21.2	10 15	+10.9	−20.9
10 21	+11.2	−21.1	10 27	+11.0	−20.8
10 34	+11.3	−21.0	10 40	+11.1	−20.7
10 47	+11.4	−20.9	10 54	+11.2	−20.6
11 01	+11.5	−20.8	11 08	+11.3	−20.5
11 15	+11.6	−20.7	11 23	+11.4	−20.4
11 30	+11.7	−20.6	11 38	+11.5	−20.3
11 46	+11.8	−20.5	11 54	+11.6	−20.2
12 02	+11.9	−20.4	12 10	+11.7	−20.1
12 19	+12.0	−20.3	12 28	+11.8	−20.0
12 37	+12.1	−20.2	12 46	+11.9	−19.9
12 55	+12.2	−20.1	13 05	+12.0	−19.8
13 14	+12.3	−20.0	13 24	+12.1	−19.7
13 35	+12.4	−19.9	13 45	+12.2	−19.6
13 56	+12.5	−19.8	14 07	+12.3	−19.5
14 18	+12.6	−19.7	14 30	+12.4	−19.4
14 42	+12.7	−19.6	14 54	+12.5	−19.3
15 06	+12.8	−19.5	15 19	+12.6	−19.2
15 32	+12.9	−19.4	15 46	+12.7	−19.1
15 59	+13.0	−19.3	16 14	+12.8	−19.0
16 28	+13.1	−19.2	16 44	+12.9	−18.9
16 59	+13.2	−19.1	17 15	+13.0	−18.8
17 32	+13.3	−19.0	17 48	+13.1	−18.7
18 06	+13.4	−18.9	18 24	+13.2	−18.6
18 42	+13.5	−18.8	19 01	+13.3	−18.5
19 21	+13.6	−18.7	19 42	+13.4	−18.4
20 03	+13.7	−18.6	20 25	+13.5	−18.3
20 48	+13.8	−18.5	21 11	+13.6	−18.2
21 35	+13.9	−18.4	22 00	+13.7	−18.1
22 26	+14.0	−18.3	22 54	+13.8	−18.0
23 22	+14.1	−18.2	23 51	+13.9	−17.9
24 21	+14.2	−18.1	24 53	+14.0	−17.8
25 26	+14.3	−18.0	26 00	+14.1	−17.7
26 36	+14.4	−17.9	27 13	+14.2	−17.6
27 52	+14.5	−17.8	28 33	+14.3	−17.5
29 15	+14.6	−17.7	30 00	+14.4	−17.4
30 46	+14.7	−17.6	31 35	+14.5	−17.3
32 26	+14.8	−17.5	33 20	+14.6	−17.2
34 17	+14.9	−17.4	35 17	+14.7	−17.1
36 20	+15.0	−17.3	37 26	+14.8	−17.0
38 36	+15.1	−17.2	39 50	+14.9	−16.9
41 08	+15.2	−17.1	42 31	+15.0	−16.8
43 59	+15.3	−17.0	45 31	+15.1	−16.7
47 10	+15.4	−16.9	48 55	+15.2	−16.6
50 46	+15.5	−16.8	52 44	+15.3	−16.5
54 49	+15.6	−16.7	57 02	+15.4	−16.4
59 23	+15.7	−16.6	61 51	+15.5	−16.3
64 30	+15.8	−16.5	67 17	+15.6	−16.2
70 12	+15.9	−16.4	73 16	+15.7	−16.1
76 26	+16.0	−16.3	79 43	+15.8	−16.0
83 05	+16.1	−16.2	86 32	+15.9	−15.9
90 00			90 00		

STARS AND PLANETS

App. Alt.	Corrn
9 56	−5.3
10 08	−5.2
10 20	−5.1
10 33	−5.0
10 46	−4.9
11 00	−4.8
11 14	−4.7
11 29	−4.6
11 45	−4.5
12 01	−4.4
12 18	−4.3
12 35	−4.2
12 54	−4.1
13 13	−4.0
13 33	−3.9
13 54	−3.8
14 16	−3.7
14 40	−3.6
15 04	−3.5
15 30	−3.4
15 57	−3.3
16 26	−3.2
16 56	−3.1
17 28	−3.0
18 02	−2.9
18 38	−2.8
19 17	−2.7
19 58	−2.6
20 42	−2.5
21 28	−2.4
22 19	−2.3
23 13	−2.2
24 11	−2.1
25 14	−2.0
26 22	−1.9
27 36	−1.8
28 56	−1.7
30 24	−1.6
32 00	−1.5
33 45	−1.4
35 40	−1.3
37 48	−1.2
40 08	−1.1
42 44	−1.0
45 36	−0.9
48 47	−0.8
52 18	−0.7
56 11	−0.6
60 28	−0.5
65 08	−0.4
70 11	−0.3
75 34	−0.2
81 13	−0.1
87 03	0.0
90 00	

App. Alt. — Additional Corrn

1980

VENUS

Jan. 1-Feb. 26
0
42 + 0.1

Feb. 27-Apr. 13
0
47 + 0.2

Apr. 14-May 9
0
46 + 0.3

May 10-May 25
0
11 + 0.4
41 + 0.5

May 26-June 3
0
6 + 0.5
20 + 0.7
31

June 4-June 26
0
4 + 0.6
12 + 0.7
22 + 0.8

June 27-July 6
0
6 + 0.5
20 + 0.6
31 + 0.7

July 7-July 21
0
11 + 0.4
41 + 0.5

July 22-Aug. 17
0
46 + 0.3

Aug. 18-Oct. 2
0
47 + 0.2

Oct. 3-Dec. 31
0
42 + 0.1

MARS

Jan. 1-Apr. 28
0
41 + 0.2
75 + 0.1

Apr. 29-Dec. 31
0
60 + 0.1

DIP

Ht. of Eye (m)	Corrn	Ht. of Eye (ft.)	Corrn	Ht. of Eye	Corrn
2.4	−2.8	8.0	−1.8		
2.6	−2.9	8.6	−2.2		
2.8	−3.0	9.2	−2.5		
3.0	−3.1	9.8	−2.8		
3.2	−3.2	10.5	−3.0		
3.4	−3.2	11.2	See table		
3.6	−3.3	11.9	←		
3.8	−3.4	12.6		**m**	
4.0	−3.5	13.3		20 −	7.9
4.3	−3.6	14.1		22 −	8.3
4.5	−3.7	14.9		24 −	8.6
4.7	−3.8	15.7		26 −	9.0
5.0	−3.9	16.5		28 −	9.3
5.2	−4.0	17.4			
5.5	−4.1	18.3		30 −	9.6
5.8	−4.2	19.1		32 −	10.0
6.1	−4.3	20.1		34 −	10.3
6.3	−4.4	21.0		36 −	10.6
6.6	−4.5	22.0		38 −	10.8
6.9	−4.6	22.9			
7.2	−4.7	23.9		40 −	11.1
7.5	−4.8	24.9		42 −	11.4
7.9	−4.9	26.0		44 −	11.7
8.2	−5.0	27.1		46 −	11.9
8.5	−5.1	28.1		48 −	12.2
8.8	−5.2	29.2			
9.2	−5.3	30.4		**ft.**	
9.5	−5.4	31.5		2 −	1.4
9.9	−5.5	32.7		4 −	1.9
10.3	−5.6	33.9		6 −	2.4
10.6	−5.7	35.1		8 −	2.7
11.0	−5.8	36.3		10 −	3.1
11.4	−5.9	37.6		See table	
11.8	−6.0	38.9		←	
12.2	−6.1	40.1		**ft.**	
12.6	−6.2	41.5		70 −	8.1
13.0	−6.3	42.8		75 −	8.4
13.4	−6.4	44.2		80 −	8.7
13.8	−6.5	45.5		85 −	8.9
14.2	−6.6	46.9		90 −	9.2
14.7	−6.7	48.4		95 −	9.5
15.1	−6.8	49.8			
15.5	−6.9	51.3		100 −	9.7
16.0	−7.0	52.8		105 −	9.9
16.5	−7.1	54.3		110 −	10.2
16.9	−7.2	55.8		115 −	10.4
17.4	−7.3	57.4		120 −	10.6
17.9	−7.4	58.9		125 −	10.8
18.4	−7.5	60.5			
18.8	−7.6	62.1		130 −	11.1
19.3	−7.7	63.8		135 −	11.3
19.8	−7.8	65.4		140 −	11.5
20.4	−7.9	67.1		145 −	11.7
20.9	−8.0	68.8		150 −	11.9
21.4	−8.1	70.5		155 −	12.1

App. Alt. = Apparent altitude = Sextant altitude corrected for index error and dip.

sea level, the bracketing figures are 2.6m and 2.8m, with a correction factor between these of −2.9′.

If your eye height is between 2 and 10 feet, or between 70 and 155 feet above sea level, you would use the table at the right under the dip heading.

It is necessary to interpolate in order to get correction for heights between the printed figures. Say your eye height is 5.3 feet above sea level. The table shows:

Height	Corr.
4.0 feet	−1.9′
6.0 feet	−2.4′

There's a 2.0-foot difference in the height figures and a −0.5′ difference in the correction factors. Your height of eye is 1.3 feet more than the 4.0-foot figure in the height table, so:

$$\frac{1.3' \times -0.5'}{2} = -0.325$$

Thus you find there's a −0.3′ increase in the correction factor for the 1.3-foot increase in eye height from 4.0 to 5.3 feet. Add this −0.3′ to the −1.9′ factor for 4.0 feet and your dip correction is −2.2′. The right-hand column also is used for metric heights from 1.0m to 3.0m, and from 20m to 48m above sea level.

MAIN CORRECTION

Once the apparent altitude of the sight has been established, it can be converted to the observed altitude (Ho). This is the figure used to give you your actual line of position. To get this, a *main correction* is applied to the apparent altitude. This correction may just account for refraction in the case of planets and stars, or refraction, semidiameter, and parallax in the case of the sun and moon.

Main corrections for the sun and for the navigational planets and stars are given on the inside front cover of the *Almanac*, to the left of the dip tables. Note that the different tables are provided for sun sights taken from October through March and for those taken from April through September. Because in most cases the bottom rim (lower limb) of the sun just touches the horizon when a sight is recorded, the lower limb corrections are shown in large type. If a sight is recorded when the top edge (upper limb) of the sun touches the horizon, with the body itself below the horizon, the small figures are used to obtain the main correction.

Note that all lower limb corrections are *positive* and are added to the apparent altitude (Ha) to get observed altitude (Ho). All upper limb corrections are *negative* and must be subtracted from the apparent altitude.

To establish a sun sight main correction, look up the Ha figures in the appropriate seasonal column that bracket the Ha you have determined. Use the correction factor that lies between these figures: to the right for upper limb sights, and to the left for lower limb sights.

For example, on 26 June 1980, you have recorded a lower limb sun

sight with the sextant reading 48°26.7'. Your sextant er
Almanac dip table shows a correction of −6.3' for your
feet. Here's the computation for observed altitude:

$$
\begin{array}{lll}
& 48°26.7' & \text{(Hs)} \\
\text{IC} & +\ 0.7' & \text{(index correction)} \\
\text{Dip} & -\ 6.3' & \text{(height of eye correction)} \\
\hline
& 48°21.1' & \text{(Ha)}
\end{array}
$$

The altitude correction table for April-September sun sights shows
that for apparent altitudes of sun sights between 45°31' and 48°55', the main
correction, lower limb, is +15.1'. This gives you:

$$
\begin{array}{ll}
48°21.1' & \text{(Ha)} \\
+\ 15.1' & \text{(main correction)} \\
\hline
48°36.2' & \text{(Ho)}
\end{array}
$$

The observed altitude for your sight is 48°36.2'. Here's another example, set out as it should be shown on a sight reduction worksheet: On 3
November 1980, you take an upper limb sun sight, and get a reading of
39°42.7' on the sextant, with an index error of +0.8'. Your height of eye is
23.0 feet. Check Table 3-1 and you should find the dip and main corrections
that are used here.

$$
\begin{array}{ll}
\text{Hs} & 39°42.7' \\
\text{IC} & -0.8' \\
\text{Dip} & -2.3' \\
\hline
\text{Ha} & 39°39.6' \\
\text{Main} & -17.2' \\
\hline
\text{Ho} & 39°22.4'
\end{array}
$$

WRONG DATA

The main correction for the stars and planets is used in the same
manner as the sun main correction columns; except that there is no upper or
lower limb to worry about. Except for Mars and Venus, the figures are good
for an entire year. For these two planets, there is an additional correction
that must be added to obtain Ho. For example, on 14 July 1980, you record a
Venus sight with a sextant reading of 28°16.6'. Index correction is +2.2' dip
correction is −4.2':

$$
\begin{array}{ll}
\text{Hs} & 28°16.6' \\
\text{IC} & +2.2' \\
\text{Dip} & -4.2' \\
\hline
\text{Ha} & 28°14.6' \\
\text{Main} & -1.8' \\
\text{Add'l} & c. +0.5' \\
\hline
\text{Ho} & 28°13.3'
\end{array}
$$

The moon requires special treatment because, as it goes through its phases, there is an additional error factor called *horizontal parallax*. The

Table 3-2. Information Taken from Tables to Provide Main and H.P. Corrections for a Moon Sight; See Text for the Problem.

ALTITUDE CORRECTION TABLES 0°–35°—MOON

App. Alt.	0°–4° Corr^n	5°–9° Corr^n	10°–14° Corr^n	15°–19° Corr^n	20°–24° Corr^n	25°–29° Corr^n	30°–34° Corr^n	App. Alt.
00	0 33.8	5 58.2	10 62.1	15 62.8	20 62.2	25 60.8	30 58.9	00
10	35.9	58.5	62.2	62.8	62.1	60.8	58.8	10
20	37.8	58.7	62.2	62.8	62.1	60.7	58.8	20
30	39.6	58.9	62.3	62.8	62.1	60.7	58.7	30
40	41.2	59.1	62.3	62.8	62.0	60.6	58.6	40
50	42.6	59.3	62.4	62.7	62.0	60.6	58.5	50
00	1 44.0	6 59.5	11 62.4	16 62.7	21 62.0	26 60.5	31 58.5	00
10	45.2	59.7	62.4	62.7	61.9	60.4	58.4	10
20	46.3	59.9	62.5	62.7	61.9	60.4	58.3	20
30	47.3	60.0	62.5	62.7	61.9	60.3	58.2	30
40	48.3	60.2	62.5	62.7	61.8	60.3	58.2	40
50	49.2	60.3	62.6	62.7	61.8	60.2	58.1	50
00	2 50.0	7 60.5	12 62.6	17 62.7	22 61.7	27 60.1	32 58.0	00
10	50.8	60.6	62.6	62.6	61.7	60.1	57.9	10
20	51.4	60.7	62.6	62.6	61.6	60.0	57.8	20
30	52.1	60.9	62.7	62.6	61.6	59.9	57.8	30
40	52.7	61.0	62.7	62.6	61.5	59.9	57.7	40
50	53.3	61.1	62.7	62.6	61.5	59.8	57.6	50
00	3 53.8	8 61.2	13 62.7	18 62.5	23 61.5	28 59.7	33 57.5	00
10	54.3	61.3	62.7	62.5	61.4	59.7	57.4	10
20	54.8	61.4	62.7	62.5	61.4	59.6	57.4	20
30	55.2	61.5	62.8	62.4	61.3	59.6	57.3	30
40	55.6	61.6	62.8	62.4	61.3	59.5	57.2	40
50	56.0	61.6	62.8	62.4	61.2	59.4	57.1	50
00	4 56.4	9 61.7	14 62.8	19 62.4	24 61.2	29 59.3	34 57.0	00
10	56.7	61.8	62.8	62.3	61.1	59.3	56.9	10
20	57.1	61.9	62.8	62.3	61.1	59.2	56.9	20
30	57.4	61.9	62.8	62.3	61.0	59.1	56.8	30
40	57.7	62.0	62.8	62.2	60.9	59.1	56.7	40
50	57.9	62.1	62.8	62.2	60.9	59.0	56.6	50

H.P.	0°–4° L U	5°–9° L U	10°–14° L U	15°–19° L U	20°–24° L U	25°–29° L U	30°–34° L U	H.P.
54.0	0.3 0.9	0.3 0.9	0.4 1.0	0.5 1.1	0.6 1.2	0.7 1.3	0.9 1.5	54.0
54.3	0.7 1.1	0.7 1.2	0.7 1.2	0.8 1.3	0.9 1.4	1.1 1.5	1.2 1.7	54.3
54.6	1.1 1.4	1.1 1.4	1.1 1.4	1.2 1.5	1.3 1.6	1.4 1.7	1.5 1.8	54.6
54.9	1.4 1.6	1.5 1.6	1.5 1.6	1.6 1.7	1.6 1.8	1.8 1.9	1.9 2.0	54.9
55.2	1.8 1.8	1.8 1.8	1.9 1.9	1.9 1.9	2.0 2.0	2.1 2.1	2.2 2.2	55.2
55.5	2.2 2.0	2.2 2.0	2.3 2.1	2.3 2.1	2.4 2.2	2.4 2.3	2.5 2.4	55.5
55.8	2.6 2.2	2.6 2.2	2.6 2.3	2.7 2.3	2.7 2.4	2.8 2.4	2.9 2.5	55.8
56.1	3.0 2.4	3.0 2.5	3.0 2.5	3.0 2.5	3.1 2.6	3.1 2.6	3.2 2.7	56.1
56.4	3.4 2.7	3.4 2.7	3.4 2.7	3.4 2.7	3.4 2.8	3.5 2.8	3.5 2.9	56.4
56.7	3.7 2.9	3.7 2.9	3.8 2.9	3.8 2.9	3.8 3.0	3.8 3.0	3.9 3.0	56.7
57.0	4.1 3.1	4.1 3.1	4.1 3.1	4.1 3.1	4.2 3.1	4.2 3.2	4.2 3.2	57.0
57.3	4.5 3.3	4.5 3.3	4.5 3.3	4.5 3.3	4.5 3.3	4.5 3.4	4.6 3.4	57.3
57.6	4.9 3.5	4.9 3.5	4.9 3.5	4.9 3.5	4.9 3.5	4.9 3.5	4.9 3.6	57.6
57.9	5.3 3.8	5.3 3.8	5.2 3.8	5.2 3.7	5.2 3.7	5.2 3.7	5.2 3.7	57.9
58.2	5.6 4.0	5.6 4.0	5.6 4.0	5.6 4.0	5.6 3.9	5.6 3.9	5.6 3.9	58.2
58.5	6.0 4.2	6.0 4.2	6.0 4.2	6.0 4.2	6.0 4.1	5.9 4.1	5.9 4.1	58.5
58.8	6.4 4.4	6.4 4.4	6.4 4.4	6.3 4.4	6.3 4.3	6.3 4.3	6.2 4.2	58.8
59.1	6.8 4.6	6.8 4.6	6.7 4.6	6.7 4.6	6.7 4.5	6.6 4.5	6.6 4.4	59.1
59.4	7.2 4.8	7.1 4.8	7.1 4.8	7.1 4.8	7.0 4.7	7.0 4.7	6.9 4.6	59.4
59.7	7.5 5.1	7.5 5.0	7.5 5.0	7.5 5.0	7.4 4.9	7.3 4.8	7.2 4.7	59.7
60.0	7.9 5.3	7.9 5.3	7.9 5.2	7.8 5.2	7.8 5.1	7.7 5.0	7.6 4.9	60.0
60.3	8.3 5.5	8.3 5.5	8.2 5.4	8.2 5.4	8.1 5.3	8.0 5.2	7.9 5.1	60.3
60.6	8.7 5.7	8.7 5.7	8.6 5.7	8.6 5.6	8.5 5.5	8.4 5.4	8.2 5.3	60.6
60.9	9.1 5.9	9.0 5.9	9.0 5.9	8.9 5.8	8.8 5.7	8.7 5.6	8.6 5.4	60.9
61.2	9.5 6.2	9.4 6.1	9.4 6.1	9.3 6.0	9.2 5.9	9.1 5.8	8.9 5.6	61.2
61.5	9.8 6.4	9.8 6.3	9.7 6.3	9.7 6.2	9.5 6.1	9.4 5.9	9.2 5.8	61.5

DIP

Ht. of Eye (m)	Corr^n	Ht. of Eye (ft)	Ht. of Eye (m)	Corr^n	Ht. of Eye (ft)
2.4	−2.8	8.0	9.5	−5.5	31.5
2.6	−2.9	8.6	9.9	−5.6	32.7
2.8	−3.0	9.2	10.3	−5.7	33.9
3.0	−3.1	9.8	10.6	−5.8	35.1
3.2	−3.2	10.5	11.0	−5.9	36.3
3.4	−3.3	11.2	11.4	−6.0	37.6
3.6	−3.4	11.9	11.8	−6.1	38.9
3.8	−3.5	12.6	12.2	−6.2	40.1
4.0	−3.6	13.3	12.6	−6.3	41.5
4.3	−3.7	14.1	13.0	−6.4	42.8
4.5	−3.8	14.9	13.4	−6.5	44.2
4.7	−3.9	15.7	13.8	−6.6	45.5
5.0	−4.0	16.5	14.2	−6.7	46.9
5.2	−4.1	17.4	14.7	−6.8	48.4
5.5	−4.2	18.3	15.1	−6.9	49.8
5.8	−4.3	19.1	15.5	−7.0	51.3
6.1	−4.4	20.1	16.0	−7.1	52.8
6.3	−4.5	21.0	16.5	−7.2	54.3
6.6	−4.6	22.0	16.9	−7.3	55.8
6.9	−4.7	22.9	17.4	−7.4	57.4
7.2	−4.8	23.9	17.9	−7.5	58.9
7.5	−4.9	24.9	18.4	−7.6	60.5
7.9	−5.0	26.0	18.8	−7.7	62.1
8.2	−5.1	27.1	19.3	−7.8	63.8
8.5	−5.2	28.1	19.8	−7.9	65.4
8.8	−5.3	29.2	20.4	−8.0	67.1
9.2	−5.4	30.4	20.9	−8.1	68.8
9.5		31.5	21.4		70.5

MOON CORRECTION TABLE

The correction is in two parts; the first correction is taken from the upper part of the table with argument apparent altitude, and the second from the lower part, with argument H.P., in the same column as that from which the first correction was taken. Separate corrections are given in the lower part for lower (L) and upper (U) limbs. All corrections are to be added to apparent altitude, but 30' is to be subtracted from the altitude of the upper limb.

For corrections for pressure and temperature see page A4.

For bubble sextant observations ignore dip, take the mean of upper and lower limb corrections and subtract 15' from the altitude.

App. Alt. = Apparent altitude = Sextant altitude corrected for index error and dip.

edge of the moon that you bring down to the horizon when taking a sight will not represent the true upper or lower limb of the body because a small portion of it is in the earth's shadow.

Main corrections for the moon include refraction, semidiameter, and normal parallax. These corrections are in tables at the top of the *Almanac* inside back cover and its facing page. Below these tables are the horizontal parallax (HP) factor corrections. Note that the main corrections do not take into account whether the sight was upper or lower limb, but the bottom tables do. The HP factors range from 54.0′ to 61.5′. The HP factor for a sight is found on the *daily page*, along with the moon's GHA and declination for the time of the sight.

Main corrections are given for apparent altitudes ranging from 0°00.0′ to 89°50.0′, at 10.0′ intervals. Because the corrections change so slowly, you can pick the tabulated figure that is closest to your calculated Ha and use the correction figure given—unless the moon is at a fairly low altitude. In that case, it's best to interpolate using the tabulated figures that bracket your calculated Ha.

When you have determined the main correction, follow its column down into the lower table until you are in line with the HP factor that you got from the daily page for the sight. Pick the HP correction for the type of sight—upper or lower limb—and add this to the main correction. This gives you your observed altitude.

In many cases, the HP factor from the daily pages falls *between* tabulated figures. In this event, it is necessary to interpolate in order to establish the actual correction. For example, your apparent altitude for a lower limb moon sight is 34°15.9′, and from the daily page for the sight you get an HP factor of 58.1′. In the right hand column of the upper table on the page facing the back cover (Table 3-2), you see that for apparent altitudes of 34°10′ to 34°20′ the main correction is 56.9′. When you follow this column down to the lower table, you see a lower limb correction of 5.2′ for an HP of 57.9′, and a correction of 5.6′ for an HP of 58.2′. While an "eyeball interpolation" might give you the correction for an HP of 58.1′, it's best to work it out mathematically.

In this case, there's an increase of 0.4′ in the correction factor for an increase of 0.3′ in the HP factor, or an increase of .133 correction for each .100 HP factor, thus:

HP	Correction	
57.9′	5.2′	
58.0′	5.2′ + .133 = 5.33′ = 5.3′	rounded to nearest tenth
58.1′	5.33′ + .133 = 5.466 = 5.5′	rounded to nearest tenth
58.2′	5.466 + .133 = 5.6′	, as tabulated

Both the moon main correction and the HP correction are always added to the apparent altitude to get observed altitude. If it is an *upper* limb sight, however, you must first subtract 30′ of arc.

Now let's work out the observed altitude of the lower limb sight selected for the above example:

Ha	34°15.9′
Main corr.	+56.9′
HP corr.	+ 5.5′
Ho	35°18.3′

Here's an example using an upper limb sight. Apparent altitude is 50°42.8′ and the HP factor, from the daily page, is 54.7′. On the inside back cover of the *Almanac* (Table 3-3) you find a main correction of 46.4′ for an Ha of 50°40′. This will do for your Ha of 50°42.8′. Follow this column down to the lower table. For an HP of 54.6′, there's a correction of 2.6′ for an upper limb sight and a correction of 2.7′ for an HP of 54.9′. Therefore:

HP	Correction
54.6′	2.6′
54.7′	2.6′ + .033 = 2.633; round to 2.6′
54.8′	2.633 + .033 = 2.666; round to 2.7′
54.9′	2.666 + .033 = 2.699; round to 2.7′ as tabulated

The correction for an HP of 54.7′ is 2.7′. These HP correction factors often can be found by "eyeball interpolation" if the exact figure is not tabulated, but it's best to work the mathematics if you want maximum precision. Now you can calculate the observed altitude for the sight:

Ha	50°42.8′
Main corr.	+46.4′
HP corr.	+ 2.7′
UL corr.	−30.0′
Ho	51°01.9′

BOOKMARK

A separate "bookmark" supplied with the *Almanac* is a loose, stiff page that on one side duplicates all the information on the inside front cover of the book. The backside of the bookmark lists the 57 stars most commonly used in navigation, first in alphabetical order and then in order of sidereal hour angle. Each list gives the star number (they are numbered in order of sidereal hour angle), magnitude, sidereal hour angle, and declination. Both sidereal hour angle (SHA) and declination are in whole degrees, a convenience in helping to locate the stars, but these figures are not the ones used in sight reduction. The angles given on the daily pages are much more precise.

It is handy to be able to slip this bookmark into the *Almanac* at the daily page you are using. This way you won't need to switch from the daily page to the inside front cover when you are working up a sight.

Table 3-3. More Moon Sight Corrections; See Text for the Problem.

ALTITUDE CORRECTION TABLES 35°-90°—MOON

App. Alt.	35°–39° Corrⁿ	40°–44° Corrⁿ	45°–49° Corrⁿ	50°–54° Corrⁿ	55°–59° Corrⁿ	60°–64° Corrⁿ	65°–69° Corrⁿ	70°–74° Corrⁿ	75°–79° Corrⁿ	80°–84° Corrⁿ	85°–89° Corrⁿ	App. Alt.
00	35 56.5	40 53.7	45 50.5	50 46.9	55 43.1	60 38.9	65 34.6	70 30.1	75 25.3	80 20.5	85 15.6	00
10	56.4	53.6	50.4	46.8	42.9	38.8	34.4	29.9	25.2	20.4	15.5	10
20	56.3	53.5	50.2	46.7	42.8	38.7	34.3	29.7	25.0	20.2	15.3	20
30	56.2	53.4	50.1	46.5	42.7	38.5	34.1	29.6	24.9	20.0	15.1	30
40	56.2	53.3	50.0	46.4	42.5	38.4	34.0	29.4	24.7	19.9	15.0	40
50	56.1	53.2	49.9	46.3	42.4	38.2	33.8	29.3	24.5	19.7	14.8	50
00	36 56.0	41 53.1	46 49.8	51 46.2	56 42.3	61 38.1	66 33.7	71 29.1	76 24.4	81 19.6	86 14.6	00
10	55.9	53.0	49.7	46.0	42.1	37.9	33.5	29.0	24.2	19.4	14.5	10
20	55.8	52.8	49.5	45.9	41.9	37.8	33.4	28.8	24.1	19.3	14.3	20
30	55.7	52.7	49.4	45.8	41.8	37.7	33.2	28.7	23.9	19.1	14.1	30
40	55.6	52.6	49.3	45.7	41.7	37.5	33.1	28.5	23.8	18.9	14.0	40
50	55.5	52.5	49.2	45.5	41.6	37.4	32.9	28.3	23.6	18.7	13.8	50
00	37 55.4	42 52.4	47 49.1	52 45.4	57 41.4	62 37.2	67 32.8	72 28.2	77 23.4	82 18.6	87 13.7	00
10	55.3	52.3	49.0	45.3	41.3	37.1	32.6	28.0	23.3	18.4	13.5	10
20	55.2	52.2	48.8	45.2	41.2	36.9	32.5	27.9	23.1	18.2	13.3	20
30	55.1	52.1	48.7	45.0	41.0	36.8	32.3	27.7	22.9	18.1	13.2	30
40	55.0	52.0	48.6	44.9	40.9	36.6	32.2	27.6	22.8	17.9	13.0	40
50	55.0	51.9	48.5	44.8	40.8	36.5	32.0	27.4	22.6	17.8	12.8	50
00	38 54.9	43 51.8	48 48.4	53 44.6	58 40.6	63 36.4	68 31.9	73 27.2	78 22.5	83 17.6	88 12.7	00
10	54.8	51.7	48.2	44.5	40.5	36.2	31.7	27.1	22.3	17.4	12.5	10
20	54.7	51.6	48.1	44.4	40.3	36.1	31.6	26.9	22.1	17.3	12.3	20
30	54.6	51.5	48.0	44.2	40.2	35.9	31.4	26.8	22.0	17.1	12.2	30
40	54.5	51.4	47.9	44.1	40.1	35.8	31.3	26.6	21.8	16.9	12.0	40
50	54.4	51.2	47.8	44.0	39.9	35.6	31.1	26.5	21.7	16.8	11.8	50
00	39 54.3	44 51.1	49 47.6	54 43.9	59 39.8	64 35.5	69 31.0	74 26.3	79 21.6	84 16.6	89 11.7	00
10	54.2	51.0	47.5	43.7	39.6	35.3	30.8	26.1	21.3	16.5	11.5	10
20	54.1	50.9	47.4	43.6	39.5	35.2	30.7	26.0	21.2	16.3	11.4	20
30	54.0	50.8	47.3	43.5	39.4	35.0	30.5	25.8	21.0	16.1	11.2	30
40	53.9	50.7	47.2	43.3	39.2	34.9	30.4	25.7	20.9	16.0	11.0	40
50	53.8	50.6	47.0	43.2	39.1	34.7	30.2	25.5	20.7	15.8	10.9	50

H.P.	L U	L U	L U	L U	L U	L U	L U	L U	L U	L U	L U	H.P.
54.0	1.1 1.7	1.3 1.9	1.5 2.1	1.7 2.4	2.0 2.6	2.3 2.9	2.6 3.2	2.9 3.5	3.2 3.8	3.5 4.1	3.8 4.5	54.0
54.3	1.4 1.8	1.6 2.0	1.8 2.2	2.0 2.5	2.3 2.7	2.5 3.0	2.8 3.2	3.0 3.5	3.3 3.8	3.6 4.1	3.9 4.4	54.3
54.6	1.7 2.0	1.9 2.2	2.1 2.4	2.3 2.6	2.5 2.8	2.7 3.0	3.0 3.3	3.2 3.5	3.5 3.8	3.7 4.1	4.0 4.3	54.6
54.9	2.0 2.2	2.2 2.3	2.3 2.5	2.5 2.7	2.7 2.9	2.9 3.1	3.2 3.3	3.4 3.5	3.6 3.8	3.9 4.0	4.1 4.3	54.9
55.2	2.3 2.3	2.5 2.4	2.6 2.6	2.8 2.8	3.0 2.9	3.2 3.1	3.4 3.3	3.6 3.5	3.8 3.7	4.0 4.0	4.2 4.2	55.2
55.5	2.7 2.5	2.8 2.6	2.9 2.7	3.1 2.9	3.2 3.0	3.4 3.2	3.6 3.4	3.7 3.5	3.9 3.7	4.1 3.9	4.3 4.1	55.5
55.8	3.0 2.6	3.1 2.7	3.2 2.8	3.3 3.0	3.5 3.1	3.6 3.3	3.8 3.4	3.9 3.6	4.1 3.7	4.2 3.9	4.4 4.0	55.8
56.1	3.3 2.8	3.4 2.9	3.5 3.0	3.6 3.1	3.7 3.2	3.9 3.3	4.0 3.5	4.1 3.6	4.2 3.7	4.4 3.8	4.5 4.0	56.1
56.4	3.6 2.9	3.7 3.0	3.8 3.1	3.9 3.2	3.9 3.3	4.0 3.4	4.1 3.5	4.3 3.6	4.4 3.7	4.5 3.8	4.6 3.9	56.4
56.7	3.9 3.1	4.0 3.1	4.1 3.2	4.1 3.3	4.2 3.3	4.3 3.4	4.3 3.5	4.4 3.6	4.5 3.7	4.6 3.8	4.7 3.8	56.7
57.0	4.3 3.2	4.3 3.3	4.3 3.3	4.4 3.4	4.4 3.4	4.5 3.5	4.5 3.5	4.6 3.6	4.7 3.6	4.7 3.7	4.8 3.8	57.0
57.3	4.6 3.4	4.6 3.4	4.6 3.4	4.6 3.5	4.7 3.5	4.7 3.6	4.7 3.6	4.8 3.6	4.8 3.6	4.8 3.7	4.9 3.7	57.3
57.6	4.9 3.6	4.9 3.6	4.9 3.6	4.9 3.6	4.9 3.6	4.9 3.6	4.9 3.6	5.0 3.6	5.0 3.6	5.0 3.6	5.0 3.6	57.6
57.9	5.2 3.7	5.2 3.7	5.2 3.7	5.2 3.7	5.2 3.7	5.1 3.6	5.1 3.6	5.1 3.6	5.1 3.6	5.1 3.6	5.1 3.6	57.9
58.2	5.5 3.9	5.5 3.8	5.5 3.8	5.4 3.8	5.4 3.7	5.4 3.7	5.3 3.7	5.3 3.6	5.2 3.6	5.2 3.5	5.2 3.5	58.2
58.5	5.9 4.0	5.8 4.0	5.8 3.9	5.7 3.9	5.6 3.8	5.6 3.8	5.5 3.7	5.5 3.6	5.4 3.6	5.3 3.5	5.3 3.4	58.5
58.8	6.2 4.2	6.1 4.1	6.0 4.1	6.0 4.0	5.9 3.9	5.8 3.8	5.7 3.7	5.6 3.6	5.5 3.5	5.5 3.4	5.3 3.4	58.8
59.1	6.5 4.3	6.4 4.3	6.3 4.2	6.2 4.1	6.1 4.0	6.0 3.9	5.9 3.8	5.8 3.6	5.7 3.5	5.6 3.4	5.4 3.3	59.1
59.4	6.8 4.5	6.7 4.4	6.6 4.3	6.5 4.2	6.4 4.1	6.2 3.9	6.1 3.8	6.0 3.7	5.8 3.5	5.7 3.4	5.5 3.2	59.4
59.7	7.1 4.6	7.0 4.5	6.9 4.4	6.8 4.3	6.6 4.1	6.5 4.0	6.3 3.8	6.2 3.7	6.0 3.5	5.8 3.3	5.6 3.2	59.7
60.0	7.5 4.8	7.3 4.7	7.2 4.5	7.0 4.4	6.9 4.2	6.7 4.0	6.5 3.9	6.3 3.7	6.1 3.5	5.9 3.3	5.7 3.1	60.0
60.3	7.8 5.0	7.6 4.8	7.5 4.7	7.3 4.5	7.1 4.3	6.9 4.1	6.7 3.9	6.5 3.7	6.3 3.5	6.0 3.2	5.8 3.0	60.3
60.6	8.1 5.1	7.9 5.0	7.7 4.8	7.6 4.6	7.3 4.4	7.1 4.2	6.9 3.9	6.7 3.7	6.4 3.4	6.2 3.2	5.9 2.9	60.6
60.9	8.4 5.3	8.2 5.1	8.0 4.9	7.8 4.7	7.6 4.5	7.3 4.2	7.1 4.0	6.8 3.7	6.6 3.4	6.3 3.2	6.0 2.9	60.9
61.2	8.7 5.4	8.5 5.2	8.3 5.0	8.1 4.8	7.8 4.5	7.6 4.3	7.3 4.0	7.0 3.7	6.7 3.4	6.4 3.1	6.1 2.8	61.2
61.5	9.1 5.6	8.8 5.4	8.6 5.1	8.3 4.9	8.1 4.6	7.8 4.3	7.5 4.0	7.2 3.7	6.9 3.4	6.5 3.1	6.2 2.7	61.5

THE DAILY PAGES

The main body of the *Nautical Almanac* is made up of "daily pages," each of which actually covers a three-day period. Facing pages for each three-day period provide GHA and declination data for all the celestial bodies used in navigation, plus additional information that is of use to the navigator.

On the left-hand pages, the GHA is given in Greenwich mean time (GMT) for Aries, Venus, Jupiter, and Saturn. Aries is an arbitrary point on the celestial sphere that is presumed to lie on the celestial equator. When the sun, moving north, crosses the celestial equator the GHA of Aries and the sun are exactly the same—very close to 0°00.0′ at noon, GMT on 21 March.

Aries is important to navigators because it helps to establish the GHA of the navigational stars at any given time. Each star maintains a constant SHA from Aries. If it were necessary to tabulate the GHA for each of the 57 stars for each hour of the year, the *Almanac* would be enormous in size and unwieldy in use. By listing the SHA for each star and giving the GHA for Aries, the book is kept small and easy to use. The GHA of a star at any given time is simply the sun of its SHA plus the GHA of Aries. The 57 stars are listed in the right-hand column of the page (Table 3-4), in alphabetical order, with the SHA and declination given for each.

At the head of each planet column, a figure is given alongside the planet name. This is its brightness factor. At some point, astronomers decided that a particular star had a brightness arbitrarily designated as 1.0. All other stars were either brighter or dimmer and the planets were included in the ratings. Stars that are dimmer have higher positive numbers. While those that are brighter have higher negative numbers. A star of −2.0 is considered to be twice as bright as the 1.0 star. The 2.0 (positive) star is twice as dim. Planets generally are well on the bright side of the mark, particularly Venus and Jupiter, although their brightness does vary as they orbit about the sun.

The column for each planet carries the hourly GHA and declination for each of the three days covered by the page. Note that the complete figure for declination is shown at the head of each six-hour grouping and only minutes and tenths of minutes are shown for the other five hours. The entire figure must be used in your computations.

At the bottom of each planet column, *v* and *d* figures are shown. These are factors used in calculating the exact GHA and declination for times between the whole hours. Also at the bottom of the page is the meridian passage of Aries (you don't need to bother with this) and the sidereal hour angles of the planets.

Daily Sun and Moon Information

On the right-hand daily page (Table 3-5) is all the information concerning the sun and the moon. GHA and declination for both bodies are given on an hourly basis, and again all times are GMT. The moon columns also include v and d factors and the HP factor for each hour. A semidiameter (S.D.) factor is given for the sun and for the moon at the bottom of their

Table 3-4. *Nautical Almanac* Daily Page Information For Aries, Planets, and Stars.

1980 AUGUST 16, 17, 18 (SAT., SUN., MON.)

G.M.T.	ARIES GHA	VENUS −4.0 GHA	VENUS Dec	MARS +1.4 GHA	MARS Dec	JUPITER −1.2 GHA	JUPITER Dec	SATURN +1.3 GHA	SATURN Dec
16 00	324 32.3	226 14.7 N19 29.6		124 41.6 S 8 33.5		158 14.5 N 6 58.1		147 42.9 N 3 38.4	
01	339 34.8	241 14.9	29.7	139 42.6	34.1	173 16.5	59.1	162 45.2	38.3
02	354 37.2	256 15.0	29.7	154 43.7	34.7	188 18.5	58.9	177 47.4	38.2
03	9 39.7	271 15.1	29.8	169 44.7	35.3	203 20.5	58.7	192 49.6	38.1
04	24 42.1	286 15.2	29.9	184 45.7	35.9	218 22.5	58.5	207 51.8	38.0
05	39 44.6	301 15.3	29.9	199 46.7	36.6	233 24.4	58.3	222 54.0	37.8
06	54 47.1	316 15.4 N19 30.0		214 47.7 S 8 37.2		248 26.4 N 6 58.1		237 56.2 N 3 37.7	
07	69 49.5	331 15.5	30.0	229 48.7	37.8	263 28.4	57.8	252 58.4	37.6
S 08	84 52.0	346 15.6	30.1	244 49.7	38.4	278 30.4	57.6	268 00.6	37.5
A 09	99 54.5	1 15.7	30.1	259 50.7	39.0	293 32.4	57.4	283 02.8	37.4
T 10	114 56.9	16 15.8	30.2	274 51.7	39.7	308 34.3	57.2	298 05.0	37.3
U 11	129 59.4	31 15.9	30.2	289 52.7	40.3	323 36.3	57.0	313 07.2	37.2
R 12	145 01.9	46 16.1 N19 30.3		304 53.7 S 8 40.9		338 38.3 N 6 56.8		328 09.5 N 3 37.1	
D 13	160 04.3	61 16.2	30.4	319 54.7	41.5	353 40.3	56.6	343 11.7	36.9
A 14	175 06.8	76 16.3	30.4	334 55.7	42.1	8 42.3	56.4	358 13.9	36.8
Y 15	190 09.3	91 16.4	30.5	349 56.7	42.7	23 44.2	56.2	13 16.1	36.7
16	205 11.7	106 16.5	30.5	4 57.7	43.4	38 46.2	56.0	28 18.3	36.6
17	220 14.2	121 16.5	30.6	19 58.7	44.0	53 48.2	55.8	43 20.5	36.5
18	235 16.6	136 16.6 N19 30.6		34 59.7 S 8 44.6		68 50.2 N 6 55.6		58 22.7 N 3 36.4	
19	250 19.1	151 16.7	30.7	50 00.7	45.2	83 52.2	55.4	73 24.9	36.3
20	265 21.6	166 16.8	30.7	65 01.7	45.8	98 54.1	55.2	88 27.1	36.1
21	280 24.0	181 16.9	30.8	80 02.7	46.5	113 56.1	55.0	103 29.3	36.0
22	295 26.5	196 17.0	30.8	95 03.7	47.1	128 58.1	54.8	118 31.5	35.9
23	310 29.0	211 17.1	30.8	110 04.7	47.7	144 00.1	54.6	133 33.7	35.8
17 00	325 31.4	226 17.2 N19 30.9		125 05.7 S 8 48.3		159 02.1 N 6 54.4		148 36.0 N 3 35.7	
01	340 33.9	241 17.3	30.9	140 06.7	48.9	174 04.0	54.2	163 38.2	35.6
02	355 36.4	256 17.4	31.0	155 07.7	49.5	189 06.0	54.0	178 40.4	35.5
03	10 38.8	271 17.5	31.0	170 08.7	50.2	204 08.0	53.8	193 42.6	35.3
04	25 41.3	286 17.5	31.1	185 09.7	50.8	219 10.0	53.6	208 44.8	35.2
05	40 43.7	301 17.6	31.1	200 10.7	51.4	234 11.9	53.4	223 47.0	35.1
06	55 46.2	316 17.7 N19 31.2		215 11.7 S 8 52.0		249 13.9 N 6 53.2		238 49.2 N 3 35.0	
07	70 48.7	331 17.8	31.2	230 12.7	52.6	264 15.9	53.0	253 51.4	34.9
08	85 51.1	346 17.9	31.2	245 13.7	53.2	279 17.9	52.8	268 53.6	34.8
S 09	100 53.6	1 18.0	31.3	260 14.7	53.9	294 19.9	52.6	283 55.8	34.7
U 10	115 56.1	16 18.0	31.3	275 15.7	54.5	309 21.8	52.4	298 58.0	34.6
N 11	130 58.5	31 18.1	31.4	290 16.7	55.1	324 23.8	52.2	314 00.2	34.4
D 12	146 01.0	46 18.2 N19 31.4		305 17.6 S 8 55.7		339 25.8 N 6 52.0		329 02.4 N 3 34.3	
A 13	161 03.5	61 18.3	31.4	320 18.6	56.3	354 27.8	51.8	344 04.7	34.2
Y 14	176 05.9	76 18.3	31.5	335 19.6	56.9	9 29.8	51.6	359 06.9	34.1
15	191 08.4	91 18.4	31.5	350 20.6	57.6	24 31.7	51.4	14 09.1	34.0
16	206 10.9	106 18.5	31.6	5 21.6	58.2	39 33.7	51.2	29 11.3	33.9
17	221 13.3	121 18.6	31.6	20 22.6	58.8	54 35.7	51.0	44 13.5	33.8
18	236 15.8	136 18.6 N19 31.6		35 23.6 S 8 59.4		69 37.7 N 6 50.8		59 15.7 N 3 33.6	
19	251 18.2	151 18.7	31.7	50 24.6	9 00.0	84 39.7	50.6	74 17.9	33.5
20	266 20.7	166 18.8	31.7	65 25.6	00.6	99 41.6	50.3	89 20.1	33.4
21	281 23.2	181 18.8	31.8	80 26.6	01.3	114 43.6	50.1	104 22.3	33.3
22	296 25.6	196 18.9	31.8	95 27.6	01.9	129 45.6	49.9	119 24.5	33.2
23	311 28.1	211 19.0	31.8	110 28.6	02.5	144 47.6	49.7	134 26.7	33.1
18 00	326 30.6	226 19.0 N19 31.8		125 29.6 S 9 03.1		159 49.5 N 6 49.5		149 28.9 N 3 33.0	
01	341 33.0	241 19.1	31.9	140 30.6	03.7	174 51.5	49.3	164 31.1	32.8
02	356 35.5	256 19.2	31.9	155 31.6	04.3	189 53.5	49.1	179 33.3	32.7
03	11 38.0	271 19.2	31.9	170 32.6	05.0	204 55.5	48.9	194 35.5	32.6
04	26 40.4	286 19.3	32.0	185 33.5	05.6	219 57.5	48.7	209 37.8	32.5
05	41 42.9	301 19.3	32.0	200 34.5	06.2	234 59.4	48.5	224 40.0	32.4
06	56 45.4	316 19.4 N19 32.0		215 35.5 S 9 06.8		250 01.4 N 6 48.3		239 42.2 N 3 32.3	
07	71 47.8	331 19.4	32.0	230 36.5	07.4	265 03.4	48.1	254 44.4	32.2
08	86 50.3	346 19.5	32.1	245 37.5	08.0	280 05.4	47.9	269 46.6	32.0
M 09	101 52.7	1 19.6	32.1	260 38.5	08.7	295 07.3	47.7	284 48.8	31.9
O 10	116 55.2	16 19.6	32.1	275 39.5	09.3	310 09.3	47.5	299 51.0	31.8
N 11	131 57.7	31 19.7	32.1	290 40.5	09.9	325 11.3	47.3	314 53.2	31.7
D 12	147 00.1	46 19.7 N19 32.2		305 41.5 S 9 10.5		340 13.3 N 6 47.1		329 55.4 N 3 31.6	
A 13	162 02.6	61 19.8	32.2	320 42.5	11.1	355 15.3	46.9	344 57.6	31.5
Y 14	177 05.1	76 19.8	32.2	335 43.5	11.7	10 17.2	46.7	359 59.8	31.4
15	192 07.5	91 19.9	32.2	350 44.4	12.4	25 19.2	46.5	15 02.0	31.2
16	207 10.0	106 19.9	32.3	5 45.4	13.0	40 21.2	46.3	30 04.2	31.1
17	222 12.5	121 20.0	32.3	20 46.4	13.6	55 23.2	46.1	45 06.4	31.0
18	237 14.9	136 20.0 N19 32.3		35 47.4 S 9 14.2		70 25.1 N 6 45.9		60 08.6 N 3 30.9	
19	252 17.4	151 20.0	32.3	50 48.4	14.8	85 27.1	45.7	75 10.8	30.8
20	267 19.8	166 20.1	32.4	65 49.4	15.4	100 29.1	45.5	90 13.0	30.7
21	282 22.3	181 20.1	32.4	80 50.4	16.0	115 31.1	45.3	105 15.2	30.5
22	297 24.8	196 20.2	32.4	95 51.4	16.7	130 33.1	45.1	120 17.5	30.4
23	312 27.2	211 20.2	32.4	110 52.3	17.3	145 35.0	44.8	135 19.7	30.3
Mer. Pass.	2 17.5	v 0.1 d 0.0		v 1.0 d 0.6		v 2.0 d 0.2		v 2.2 d 0.1	

STARS

Name	S.H.A.	Dec
Acamar	315 37.2	S40 22.7
Achernar	335 45.1	S57 19.9
Acrux	173 37.7	S62 59.6
Adhara	255 32.4	S28 56.6
Aldebaran	291 18.2	N16 28.2
Alioth	166 43.0	N56 04.2
Alkaid	153 18.8	N49 25.0
Al Na'ir	28 14.6	S47 03.2
Alnilam	276 11.9	S 1 12.8
Alphard	218 20.9	S 8 34.4
Alphecca	126 32.2	N26 47.1
Alpheratz	358 09.1	N28 58.9
Altair	62 32.3	N 8 49.2
Ankaa	353 40.0	S42 24.5
Antares	112 56.8	S26 23.3
Arcturus	146 18.6	N19 17.3
Atria	108 20.9	S68 59.8
Avior	234 28.8	S59 26.7
Bellatrix	278 59.0	N 6 19.9
Betelgeuse	271 28.5	N 7 24.2
Canopus	264 07.6	S52 40.9
Capella	281 11.6	N45 58.5
Deneb	49 48.0	N45 12.8
Denebola	182 59.4	N14 41.0
Diphda	349 20.7	S18 05.5
Dubhe	194 22.9	N61 51.5
Elnath	278 44.4	N28 35.4
Eltanin	90 57.5	N51 29.8
Enif	34 11.3	N 9 47.3
Fomalhaut	15 51.2	S29 43.4
Gacrux	172 29.2	S57 00.3
Gienah	176 18.3	S17 25.9
Hadar	149 23.6	S60 16.9
Hamal	328 28.8	N23 22.1
Kaus Aust.	84 16.7	S34 23.6
Kochab	137 19.3	N74 14.5
Markab	14 02.9	N15 06.1
Menkar	314 41.2	N 4 00.8
Menkent	148 37.2	S36 16.5
Miaplacidus	221 45.8	S69 38.2
Mirfak	309 16.1	N49 47.3
Nunki	76 29.0	S26 19.2
Peacock	53 57.5	S56 47.5
Pollux	243 58.6	N28 04.4
Procyon	245 26.1	N 5 16.6
Rasalhague	96 29.5	N12 34.7
Regulus	208 10.4	N12 03.9
Rigel	281 36.2	S 8 13.4
Rigil Kent.	140 26.0	S60 45.4
Sabik	102 41.1	S15 42.0
Schedar	350 08.6	N56 25.7
Shaula	96 55.6	S37 05.4
Sirius	258 56.0	S16 41.3
Spica	158 57.8	S11 03.5
Suhail	223 12.2	S43 21.2
Vega	80 55.6	N38 46.2
Zuben'ubi	137 33.2	S15 57.6

	S H A	Mer. Pass
Venus	260 45.8	8 55
Mars	159 34.3	15 39
Jupiter	193 30.6	13 22
Saturn	183 04.5	14 04

respective columns. You can ignore these because semidiameter is included in the main corrections (as shown earlier).

At the right of the page, there are data covering times of sunrise, sunset, moonrise, moonset, and periods of twilight. This information is used to predetermine the best time to take morning or evening sights. At the bottom right portion of the page, there is information on the *equation of time* and the sun's meridian passage. The information on the moon, at the bottom of the page, is seldom needed or used by pleasure-boat skippers.

Increments and Corrections

Daily pages provide GHA on an hourly basis for each of the celestial bodies used in navigation. To bring GHA to the instant of a sight—the exact minute and second—the Increment and Correction tables printed on buff paper at the back of the book are used. These provide correction figures for every minute and second from 0 minutes, 1 second through 59 minutes, 59 seconds. One column for each minute has the increments for the sun and planets, a second column is for Aries, and a third is for the moon.

The correction figures for v and d factors from the daily pages are given

Table 3-5. Daily Page Information for a Moon Shot at 0100 on 18 August 1980.

1980 AUGUST 16, 17, 18 (SAT., SUN., MON.)

G.M.T.	SUN G.H.A.	SUN Dec.	MOON G.H.A.	v	Dec.	d	H.P.
16 00	178 55.8	N13 45.7	123 24.7	15.5	S 3 54.5	9.8	54.1
01	193 55.9	44.9	137 59.2	15.5	4 04.3	9.7	54.1
02	208 56.1	44.1	152 33.7	15.5	4 14.0	9.7	54.1
03	223 56.2	.. 43.3	167 08.2	15.5	4 23.7	9.7	54.1
04	238 56.3	42.5	181 42.7	15.5	4 33.4	9.7	54.1
05	253 56.5	41.7	196 17.2	15.4	4 43.1	9.6	54.1
06	268 56.6	N13 40.9	210 51.6	15.5	S 4 52.7	9.7	54.1
07	283 56.7	40.1	225 26.1	15.4	5 02.4	9.6	54.1
S 08	298 56.8	39.3	240 00.5	15.4	5 12.0	9.6	54.1
A 09	313 57.0	.. 38.5	254 34.9	15.4	5 21.6	9.6	54.1
T 10	328 57.1	37.8	269 09.3	15.3	5 31.2	9.5	54.1
U 11	343 57.2	37.0	283 43.6	15.4	5 40.7	9.6	54.1
R 12	358 57.4	N13 36.2	298 18.0	15.3	S 5 50.3	9.5	54.2
D 13	13 57.5	35.4	312 52.3	15.4	5 59.8	9.5	54.2
A 14	28 57.6	34.6	327 26.7	15.3	6 09.3	9.4	54.2
Y 15	43 57.7	.. 33.8	342 01.0	15.2	6 18.7	9.5	54.2
16	58 57.9	33.0	356 35.2	15.3	6 28.2	9.4	54.2
17	73 58.0	32.2	11 09.5	15.2	6 37.6	9.4	54.2
18	88 58.1	N13 31.4	25 43.7	15.2	S 6 47.0	9.3	54.2
19	103 58.3	30.6	40 17.9	15.2	6 56.3	9.4	54.2
20	118 58.4	29.8	54 52.1	15.2	7 05.7	9.3	54.2
21	133 58.5	.. 29.0	69 26.3	15.1	7 15.0	9.3	54.2
22	148 58.7	28.2	84 00.4	15.1	7 24.3	9.2	54.2
23	163 58.8	27.4	98 34.5	15.1	7 33.5	9.3	54.2
17 00	178 58.9	N13 26.6	113 08.6	15.0	S 7 42.8	9.2	54.2
01	193 59.1	25.8	127 42.6	15.1	7 52.0	9.2	54.2
02	208 59.2	25.0	142 16.7	15.0	8 01.2	9.1	54.2
03	223 59.3	.. 24.2	156 50.7	14.9	8 10.3	9.1	54.3
04	238 59.5	23.4	171 24.6	15.0	8 19.4	9.1	54.3
05	253 59.6	22.6	185 58.6	14.9	8 28.5	9.0	54.3
06	268 59.7	N13 21.8	200 32.5	14.8	S 8 37.5	9.1	54.3
07	283 59.9	21.0	215 06.3	14.9	8 46.6	8.9	54.3
08	299 00.0	20.2	229 40.2	14.8	8 55.5	9.0	54.3
S 09	314 00.1	.. 19.4	244 14.0	14.8	9 04.5	8.9	54.3
U 10	329 00.3	18.6	258 47.8	14.7	9 13.4	8.9	54.3
N 11	344 00.4	17.8	273 21.5	14.7	9 22.3	8.8	54.3
D 12	359 00.5	N13 17.0	287 55.2	14.7	S 9 31.1	8.8	54.3
A 13	14 00.7	16.2	302 28.9	14.6	9 39.9	8.8	54.4
Y 14	29 00.8	15.4	317 02.5	14.6	9 48.7	8.8	54.4
15	44 00.9	.. 14.6	331 36.1	14.6	9 57.5	8.7	54.4
16	59 01.1	13.8	346 09.7	14.5	10 06.2	8.6	54.4
17	74 01.2	13.0	0 43.2	14.5	10 14.8	8.6	54.4
18	89 01.3	N13 12.2	15 16.7	14.4	S10 23.4	8.6	54.4
19	104 01.5	11.4	29 50.1	14.4	10 32.0	8.6	54.4
20	119 01.6	10.6	44 23.5	14.4	10 40.6	8.5	54.4
21	134 01.8	.. 09.8	58 56.9	14.3	10 49.1	8.4	54.5
22	149 01.9	09.0	73 30.2	14.3	10 57.5	8.4	54.5
23	164 02.0	08.2	88 03.5	14.2	11 05.9	8.4	54.5
18 00	179 02.2	N12 07.4	102 36.7	14.2	S11 14.3	8.3	54.5
01	194 02.3	06.6	117 09.9	14.1	11 22.6	8.3	54.5
02	209 02.4	05.8	131 43.0	14.1	11 30.9	8.3	54.5
03	224 02.6	.. 04.9	146 16.1	14.1	11 39.2	8.2	54.6
04	239 02.7	04.1	160 49.2	14.0	11 47.4	8.1	54.6
05	254 02.9	03.3	175 22.2	14.0	11 55.5	8.1	54.6
06	269 03.0	N12 02.5	189 55.2	13.9	S12 03.6	8.1	54.6
07	284 03.1	01.7	204 28.1	13.8	12 11.7	8.0	54.6
08	299 03.3	00.9	219 00.9	13.9	12 19.7	8.0	54.6
M 09	314 03.4	13 00.1	233 33.8	13.7	12 27.7	7.9	54.7
O 10	329 03.6	12 59.3	248 06.5	13.8	12 35.6	7.9	54.7
N 11	344 03.7	58.5	262 39.3	13.6	12 43.5	7.8	54.7
D 12	359 03.8	N12 57.7	277 11.9	13.6	S12 51.3	7.7	54.7
A 13	14 04.0	56.9	291 44.5	13.6	12 59.0	7.7	54.7
Y 14	29 04.1	.. 56.2	306 17.1	13.5	13 06.7	7.7	54.7
15	44 04.3	55.4	320 49.6	13.5	13 14.4	7.6	54.8
16	59 04.4	54.4	335 22.1	13.4	13 22.0	7.6	54.8
17	74 04.5	53.6	349 54.5	13.4	13 29.6	7.5	54.8
18	89 04.7	N12 52.8	4 26.9	13.3	S13 37.1	7.4	54.8
19	104 04.8	52.0	18 59.2	13.3	13 44.5	7.4	54.8
20	119 05.0	51.2	33 31.4	13.2	13 51.9	7.3	54.9
21	134 05.1	.. 50.4	48 03.6	13.1	13 59.2	7.3	54.9
22	149 05.3	49.5	62 35.7	13.1	14 06.5	7.2	54.9
23	164 05.4	48.7	77 07.8	13.1	14 13.7	7.2	54.9
	S.D. 15.8	d 08	S.D. 14.8		14.8		14.9

Lat.	Twilight Naut.	Civil	Sunrise	Moonrise 16	17	18	19
N 72	////	////	02 38	11 19	12 58	14 43	16 41
N 70	////	01 00	03 07	11 10	12 42	14 17	15 56
68	////	01 59	03 28	11 03	12 29	13 57	15 26
66	////	02 32	03 45	10 57	12 19	13 41	15 04
64	00 53	02 55	03 58	10 53	12 10	13 28	14 47
62	01 45	03 14	04 09	10 48	12 03	13 18	14 32
60	02 15	03 29	04 19	10 45	11 56	13 08	14 20
N 58	02 37	03 41	04 27	10 42	11 51	13 00	14 10
56	02 55	03 52	04 35	10 39	11 46	12 53	14 01
54	03 09	04 01	04 41	10 36	11 42	12 47	13 53
52	03 21	04 10	04 47	10 34	11 38	12 42	13 46
50	03 32	04 17	04 53	10 32	11 34	12 36	13 39
45	03 53	04 32	05 04	10 27	11 26	12 26	13 25
N 40	04 10	04 45	05 13	10 23	11 20	12 17	13 14
35	04 23	04 55	05 21	10 20	11 14	12 09	13 04
30	04 34	05 03	05 28	10 17	11 09	12 02	12 56
20	04 51	05 18	05 41	10 12	11 01	11 50	12 41
N 10	05 05	05 29	05 51	10 08	10 53	11 40	12 29
0	05 15	05 39	06 01	10 04	10 47	11 31	12 17
S 10	05 24	05 49	06 10	10 00	10 40	11 21	12 05
20	05 32	05 58	06 20	09 56	10 32	11 11	11 53
30	05 39	06 07	06 32	09 51	10 24	11 00	11 39
35	05 42	06 12	06 38	09 48	10 20	10 53	11 30
40	05 46	06 18	06 45	09 45	10 14	10 46	11 21
45	05 49	06 24	06 54	09 41	10 08	10 37	11 10
S 50	05 53	06 31	07 04	09 37	10 01	10 27	10 57
52	05 54	06 34	07 09	09 35	09 57	10 22	10 51
54	05 56	06 37	07 14	09 33	09 54	10 17	10 44
56	05 57	06 41	07 20	09 31	09 50	10 11	10 37
58	05 59	06 45	07 26	09 28	09 45	10 05	10 28
S 60	06 00	06 49	07 34	09 25	09 40	09 57	10 19

Lat.	Sunset	Twilight Civil	Naut.	Moonset 16	17	18	19
N 72	21 25	////	////	20 49	20 39	20 26	20 05
N 70	20 57	22 55	////	20 59	20 56	20 53	20 51
68	20 37	22 03	////	21 08	21 10	21 14	21 21
66	20 21	21 32	////	21 15	21 22	21 31	21 44
64	20 08	21 10	23 03	21 21	21 31	21 44	22 02
62	19 57	20 52	22 17	21 26	21 39	21 56	22 17
60	19 47	20 37	21 49	21 31	21 47	22 06	22 30
N 58	19 39	20 25	21 28	21 35	21 53	22 14	22 40
56	19 32	20 14	21 11	21 39	21 59	22 22	22 50
54	19 25	20 05	20 57	21 42	22 04	22 28	22 58
52	19 20	19 57	20 45	21 45	22 08	22 35	23 06
50	19 14	19 50	20 35	21 48	22 12	22 40	23 13
45	19 03	19 35	20 14	21 53	22 21	22 52	23 27
N 40	18 54	19 23	19 57	21 58	22 29	23 02	23 39
35	18 46	19 13	19 44	22 03	22 35	23 10	23 49
30	18 38	19 04	19 34	22 06	22 41	23 18	23 58
20	18 27	18 50	19 17	22 13	22 51	23 31	24 14
N 10	18 17	18 38	19 04	22 19	22 59	23 42	24 27
0	18 07	18 29	18 53	22 24	23 08	23 53	24 40
S 10	17 58	18 19	18 44	22 30	23 16	24 03	00 03
20	17 48	18 11	18 37	22 36	23 25	24 15	00 15
30	17 37	18 01	18 29	22 42	23 35	24 28	00 28
35	17 30	17 56	18 26	22 46	23 40	24 35	00 35
40	17 23	17 51	18 23	22 50	23 47	24 44	00 44
45	17 15	17 45	18 19	22 56	23 55	24 54	00 54
S 50	17 04	17 38	18 16	23 02	24 04	00 04	01 06
52	17 00	17 35	18 15	23 04	24 08	00 08	01 12
54	16 55	17 31	18 14	23 07	24 13	00 13	01 18
56	16 49	17 28	18 12	23 11	24 18	00 18	01 26
58	16 43	17 24	18 10	23 15	24 24	00 24	01 33
S 60	16 35	17 20	18 09	23 19	24 31	00 31	01 42

Day	SUN Eqn. of Time 00h	12h	Mer. Pass.	MOON Mer. Pass. Upper	Lower	Age	Phase
16	04 17	04 11	12 04	16 14	03 53	06	
17	04 05	03 58	12 04	16 57	04 35	07	
18	03 52	03 45	12 04	17 42	05 19	08	

to the right of the minute to which they apply. The v corrections are always added to the GHA except those for Venus which are sometimes subtracted. In this case, the v figure on the daily page will be preceded by a minus sign (−). The d correction may be added or subtracted from the declination, depending on whether declination is increasing or decreasing arithmetically. This can be determined from inspection of the information on the daily page for the sight.

Here's an example: a moon sight for 18 August 1980, with a GMT of 01-18-42. For 0100 on 18 August, the *Almanac* shows on the daily page: GHA 117°09.9′, v factor 14.1′, declination S 11°22.6′ and d factor 8.3′. Because declination is S 11°14.3′ at 0000 and S 11°30.9′ at 0200, it is increasing mathematically although the moon is actually lower on the celestial sphere (see Table 3-5).

On page xi of the buff pages, the increment for 18 minutes, 42 seconds is found as 4°27.7′ in the moon column. To the right on this page, the v or d correction for 14.1′ is 4.3′ and the v or d correction for a factor of 8.3′ is 2.6′ (Table 3-6).

For the GHA of the moon at 01-18-42 on 18 August, we then have:

01h	117°09.9′
18m 42s	+ 4°27.7′
v corr.	+ 4.3′
GHA moon	117°41.9′

For the declination, the calculation is:

01h	S 11°22.6′
d corr.	+ 2.6′
Dec.	S 11°25.2′

At 01-18-42 on 18 August 1980, the geographical position of the moon is GHA 117°41.9′ and declination S 11°25.2′.

Here's an example using Venus at a time when the v correction is negative. On 8 January 1980, we need to know GHA and declination for GMT 21-41-31. The daily page shows at 2100: GHA 98°31.3′ and Dec. S 16°07.1′. At the bottom of the column, the v factor is − 0.6′ and the d factor is 1.0′. Inspection of the declination column shows declination is decreasing arithmetically. Therefore, the d correction also will be negative. See Table 3-7.

Page xxii of the buff pages at the back of the *Almanac* shows an increment of 10°22.8′ for 41 minutes, 31 seconds of time (Table 3-8). The correction for a v factor of 0.6′ is 0.4′ and the correction for a d factor of 1.0′ is 0.7′. Both these factors will be subtracted in the calculations. For GHA we have:

21h	98°31.3′
41m 31s	+10°22.8′
v corr.	− 0.4′
GHA Venus	108°53.7′

Table 3-6. GHA Increment for 18m 42s for a Moon Sight, and the Correction for a *v* of 8.3′ and *d* of 14.1′ Taken from the Daily Page (Table 3-5).

18	SUN PLANETS	ARIES	MOON	v or Corr d		v or Corr d		v or Corr d	
s	° ′	° ′	° ′	′	′	′	′	′	′
00	4 30.0	4 30.7	4 17.7	0.0	0.0	6.0	1.9	12.0	3.7
01	4 30.3	4 31.0	4 17.9	0.1	0.0	6.1	1.9	12.1	3.7
02	4 30.5	4 31.2	4 18.2	0.2	0.1	6.2	1.9	12.2	3.8
03	4 30.8	4 31.5	4 18.4	0.3	0.1	6.3	1.9	12.3	3.8
04	4 31.0	4 31.7	4 18.7	0.4	0.1	6.4	2.0	12.4	3.8
05	4 31.3	4 32.0	4 18.9	0.5	0.2	6.5	2.0	12.5	3.9
06	4 31.5	4 32.2	4 19.1	0.6	0.2	6.6	2.0	12.6	3.9
07	4 31.8	4 32.5	4 19.4	0.7	0.2	6.7	2.1	12.7	3.9
08	4 32.0	4 32.7	4 19.6	0.8	0.2	6.8	2.1	12.8	3.9
09	4 32.3	4 33.0	4 19.8	0.9	0.3	6.9	2.1	12.9	4.0
10	4 32.5	4 33.2	4 20.1	1.0	0.3	7.0	2.2	13.0	4.0
11	4 32.8	4 33.5	4 20.3	1.1	0.3	7.1	2.2	13.1	4.0
12	4 33.0	4 33.7	4 20.6	1.2	0.4	7.2	2.2	13.2	4.1
13	4 33.3	4 34.0	4 20.8	1.3	0.4	7.3	2.3	13.3	4.1
14	4 33.5	4 34.2	4 21.0	1.4	0.4	7.4	2.3	13.4	4.1
15	4 33.8	4 34.5	4 21.3	1.5	0.5	7.5	2.3	13.5	4.2
16	4 34.0	4 34.8	4 21.5	1.6	0.5	7.6	2.3	13.6	4.2
17	4 34.3	4 35.0	4 21.8	1.7	0.5	7.7	2.4	13.7	4.2
18	4 34.5	4 35.3	4 22.0	1.8	0.6	7.8	2.4	13.8	4.3
19	4 34.8	4 35.5	4 22.2	1.9	0.6	7.9	2.4	13.9	4.3
20	4 35.0	4 35.8	4 22.5	2.0	0.6	8.0	2.5	14.0	4.3
21	4 35.3	4 36.0	4 22.7	2.1	0.6	8.1	2.5	14.1	4.3
22	4 35.5	4 36.3	4 22.9	2.2	0.7	8.2	2.5	14.2	4.4
23	4 35.8	4 36.5	4 23.2	2.3	0.7	8.3	2.6	14.3	4.4
24	4 36.0	4 36.8	4 23.4	2.4	0.7	8.4	2.6	14.4	4.4
25	4 36.3	4 37.0	4 23.7	2.5	0.8	8.5	2.6	14.5	4.5
26	4 36.5	4 37.3	4 23.9	2.6	0.8	8.6	2.7	14.6	4.5
27	4 36.8	4 37.5	4 24.1	2.7	0.8	8.7	2.7	14.7	4.5
28	4 37.0	4 37.8	4 24.4	2.8	0.9	8.8	2.7	14.8	4.6
29	4 37.3	4 38.0	4 24.6	2.9	0.9	8.9	2.7	14.9	4.6
30	4 37.5	4 38.3	4 24.9	3.0	0.9	9.0	2.8	15.0	4.6
31	4 37.8	4 38.5	4 25.1	3.1	1.0	9.1	2.8	15.1	4.7
32	4 38.0	4 38.8	4 25.3	3.2	1.0	9.2	2.8	15.2	4.7
33	4 38.3	4 39.0	4 25.6	3.3	1.0	9.3	2.9	15.3	4.7
34	4 38.5	4 39.3	4 25.8	3.4	1.0	9.4	2.9	15.4	4.7
35	4 38.8	4 39.5	4 26.1	3.5	1.1	9.5	2.9	15.5	4.8
36	4 39.0	4 39.8	4 26.3	3.6	1.1	9.6	3.0	15.6	4.8
37	4 39.3	4 40.0	4 26.5	3.7	1.1	9.7	3.0	15.7	4.8
38	4 39.5	4 40.3	4 26.8	3.8	1.2	9.8	3.0	15.8	4.9
39	4 39.8	4 40.5	4 27.0	3.9	1.2	9.9	3.1	15.9	4.9
40	4 40.0	4 40.8	4 27.2	4.0	1.2	10.0	3.1	16.0	4.9
41	4 40.3	4 41.0	4 27.5	4.1	1.3	10.1	3.1	16.1	5.0
42	4 40.5	4 41.3	4 27.7	4.2	1.3	10.2	3.1	16.2	5.0
43	4 40.8	4 41.5	4 28.0	4.3	1.3	10.3	3.2	16.3	5.0
44	4 41.0	4 41.8	4 28.2	4.4	1.4	10.4	3.2	16.4	5.1
45	4 41.3	4 42.0	4 28.4	4.5	1.4	10.5	3.2	16.5	5.1
46	4 41.5	4 42.3	4 28.7	4.6	1.4	10.6	3.3	16.6	5.1
47	4 41.8	4 42.8	4 28.9	4.7	1.4	10.7	3.3	16.7	5.1
48	4 42.0	4 42.8	4 29.2	4.8	1.5	10.8	3.3	16.8	5.2
49	4 42.3	4 43.0	4 29.4	4.9	1.5	10.9	3.4	16.9	5.2
50	4 42.5	4 43.3	4 29.6	5.0	1.5	11.0	3.4	17.0	5.2
51	4 42.8	4 43.5	4 29.9	5.1	1.6	11.1	3.4	17.1	5.3
52	4 43.0	4 43.8	4 30.1	5.2	1.6	11.2	3.5	17.2	5.3
53	4 43.3	4 44.0	4 30.3	5.3	1.6	11.3	3.5	17.3	5.3
54	4 43.5	4 44.3	4 30.6	5.4	1.7	11.4	3.5	17.4	5.4
55	4 43.8	4 44.5	4 30.8	5.5	1.7	11.5	3.5	17.5	5.4
56	4 44.0	4 44.8	4 31.1	5.6	1.7	11.6	3.6	17.6	5.4
57	4 44.3	4 45.0	4 31.3	5.7	1.8	11.7	3.6	17.7	5.5
58	4 44.5	4 45.3	4 31.5	5.8	1.8	11.8	3.6	17.8	5.5
59	4 44.8	4 45.5	4 31.8	5.9	1.8	11.9	3.7	17.9	5.5
60	4 45.0	4 45.8	4 32.0	6.0	1.9	12.0	3.7	18.0	5.6

19	SUN PLANETS	ARIES	MOON	v or Corr d		v or Corr d		v or Corr d	
s	° ′	° ′	° ′	′	′	′	′	′	′
00	4 45.0	4 45.8	4 32.0	0.0	0.0	6.0	2.0	12.0	3.9
01	4 45.3	4 46.0	4 32.3	0.1	0.0	6.1	2.0	12.1	3.9
02	4 45.5	4 46.3	4 32.5	0.2	0.1	6.2	2.0	12.2	4.0
03	4 45.8	4 46.5	4 32.7	0.3	0.1	6.3	2.0	12.3	4.0
04	4 46.0	4 46.8	4 33.0	0.4	0.1	6.4	2.1	12.4	4.0
05	4 46.3	4 47.0	4 33.2	0.5	0.2	6.5	2.1	12.5	4.1
06	4 46.5	4 47.3	4 33.4	0.6	0.2	6.6	2.1	12.6	4.1
07	4 46.8	4 47.5	4 33.7	0.7	0.2	6.7	2.2	12.7	4.1
08	4 47.0	4 47.8	4 33.9	0.8	0.3	6.8	2.2	12.8	4.2
09	4 47.3	4 48.0	4 34.2	0.9	0.3	6.9	2.2	12.9	4.2
10	4 47.5	4 48.3	4 34.4	1.0	0.3	7.0	2.3	13.0	4.2
11	4 47.8	4 48.5	4 34.6	1.1	0.4	7.1	2.3	13.1	4.3
12	4 48.0	4 48.8	4 34.9	1.2	0.4	7.2	2.3	13.2	4.3
13	4 48.3	4 49.0	4 35.1	1.3	0.4	7.3	2.4	13.3	4.3
14	4 48.5	4 49.3	4 35.4	1.4	0.5	7.4	2.4	13.4	4.4
15	4 48.8	4 49.5	4 35.6	1.5	0.5	7.5	2.4	13.5	4.4
16	4 49.0	4 49.8	4 35.8	1.6	0.5	7.6	2.5	13.6	4.4
17	4 49.3	4 50.0	4 36.1	1.7	0.6	7.7	2.5	13.7	4.5
18	4 49.5	4 50.3	4 36.3	1.8	0.6	7.8	2.5	13.8	4.5
19	4 49.8	4 50.5	4 36.6	1.9	0.6	7.9	2.6	13.9	4.5
20	4 50.0	4 50.8	4 36.8	2.0	0.7	8.0	2.6	14.0	4.6
21	4 50.3	4 51.0	4 37.0	2.1	0.7	8.1	2.6	14.1	4.6
22	4 50.5	4 51.3	4 37.3	2.2	0.7	8.2	2.7	14.2	4.6
23	4 50.8	4 51.5	4 37.5	2.3	0.7	8.3	2.7	14.3	4.6
24	4 51.0	4 51.8	4 37.7	2.4	0.8	8.4	2.7	14.4	4.7
25	4 51.3	4 52.0	4 38.0	2.5	0.8	8.5	2.8	14.5	4.7
26	4 51.5	4 52.3	4 38.2	2.6	0.8	8.6	2.8	14.6	4.7
27	4 51.8	4 52.5	4 38.5	2.7	0.9	8.7	2.8	14.7	4.8
28	4 52.0	4 52.8	4 38.7	2.8	0.9	8.8	2.9	14.8	4.8
29	4 52.3	4 53.1	4 38.9	2.9	0.9	8.9	2.9	14.9	4.8
30	4 52.5	4 53.3	4 39.2	3.0	1.0	9.0	2.9	15.0	4.9
31	4 52.8	4 53.6	4 39.4	3.1	1.0	9.1	3.0	15.1	4.9
32	4 53.0	4 53.8	4 39.7	3.2	1.0	9.2	3.0	15.2	4.9
33	4 53.3	4 54.1	4 39.9	3.3	1.1	9.3	3.0	15.3	5.0
34	4 53.5	4 54.3	4 40.1	3.4	1.1	9.4	3.1	15.4	5.0
35	4 53.8	4 54.6	4 40.4	3.5	1.1	9.5	3.1	15.5	5.0
36	4 54.0	4 54.8	4 40.6	3.6	1.2	9.6	3.1	15.6	5.1
37	4 54.3	4 55.1	4 40.8	3.7	1.2	9.7	3.2	15.7	5.1
38	4 54.5	4 55.3	4 41.1	3.8	1.2	9.8	3.2	15.8	5.1
39	4 54.8	4 55.6	4 41.3	3.9	1.3	9.9	3.2	15.9	5.2
40	4 55.0	4 55.8	4 41.6	4.0	1.3	10.0	3.3	16.0	5.2
41	4 55.3	4 56.1	4 41.8	4.1	1.3	10.1	3.3	16.1	5.2
42	4 55.5	4 56.3	4 42.0	4.2	1.4	10.2	3.3	16.2	5.3
43	4 55.8	4 56.6	4 42.3	4.3	1.4	10.3	3.4	16.3	5.3
44	4 56.0	4 56.8	4 42.5	4.4	1.4	10.4	3.4	16.4	5.3
45	4 56.3	4 57.1	4 42.8	4.5	1.5	10.5	3.4	16.5	5.4
46	4 56.5	4 57.3	4 43.0	4.6	1.5	10.6	3.5	16.6	5.4
47	4 56.8	4 57.6	4 43.2	4.7	1.5	10.7	3.5	16.7	5.4
48	4 57.0	4 57.8	4 43.5	4.8	1.6	10.8	3.5	16.8	5.5
49	4 57.3	4 58.1	4 43.7	4.9	1.6	10.9	3.6	16.9	5.5
50	4 57.5	4 58.3	4 43.9	5.0	1.6	11.0	3.6	17.0	5.5
51	4 57.8	4 58.6	4 44.2	5.1	1.7	11.1	3.6	17.1	5.6
52	4 58.0	4 58.8	4 44.4	5.2	1.7	11.2	3.6	17.2	5.6
53	4 58.3	4 59.1	4 44.7	5.3	1.7	11.3	3.7	17.3	5.6
54	4 58.5	4 59.3	4 44.9	5.4	1.8	11.4	3.7	17.4	5.7
55	4 58.8	4 59.6	4 45.1	5.5	1.8	11.5	3.7	17.5	5.7
56	4 59.0	4 59.8	4 45.4	5.6	1.8	11.6	3.8	17.6	5.7
57	4 59.3	5 00.1	4 45.6	5.7	1.9	11.7	3.8	17.7	5.8
58	4 59.5	5 00.3	4 45.9	5.8	1.9	11.8	3.8	17.8	5.8
59	4 59.8	5 00.6	4 46.1	5.9	1.9	11.9	3.9	17.9	5.8
60	5 00.0	5 00.8	4 46.3	6.0	2.0	12.0	3.9	18.0	5.9

The declination calculation shows:

21h	S 16°07.1′
d corr.	− 0.7′
Dec.	S 16°06.4′

ADDITIONAL ALMANAC INFORMATION

Following the daily pages, the *Almanac* has information on how to use the book, and corrections to GMT for standard time in all nations of the world as well as individual regions of the United States. Star charts for both the Northern and Southern Hemispheres are provided, and there's a listing

Table 3-7. Daily Page Information for a Venus Sight at 2100 on 8 January 1980. Note That Whole Degrees of Declination are Shown at Six-Hour Intervals, and the *v* and *d* Values are Given at the Bottom of the Column.

14 — 1980 JANUARY 7, 8, 9 (MON., TUES., WED.)

G.M.T.	ARIES G.H.A.	VENUS −3.4 G.H.A.	Dec.	MARS 0.0 G.H.A.	Dec.	JUPITER −1.9 G.H.A.	Dec.	SATURN +1.1 G.H.A.	Dec.	STARS Name	S.H.A.	Dec.
7 00	105 43.5	143 56.5	S16 52.2	298 24.1	N 9 02.4	303 43.0	N 8 52.5	287 35.4	N 3 11.6	Acamar	315 37.4	S40 23.4
01	120 46.0	158 56.0	51.2	313 26.3	02.4	318 45.5	52.5	302 37.8	11.7	Achernar	335 45.6	S57 20.7
02	135 48.4	173 55.4	50.2	328 28.5	02.4	333 48.1	52.6	317 40.3	11.7	Acrux	173 37.5	S62 59.0
03	150 50.9	188 54.8 ··	49.2	343 30.7 ··	02.3	348 50.6 ··	52.6	332 42.8 ··	11.7	Adhara	255 32.1	S28 56.8
04	165 53.4	203 54.2	48.2	358 32.9	02.3	ˈ3 53.2	52.7	347 45.2	11.7	Aldebaran	291 18.3	N16 28.1
05	180 55.8	218 53.7	47.3	13 35.0	02.3	18 55.7	52.7	2 47.7	11.7			
06	195 58.3	233 53.1	S16 46.3	28 37.2	N 9 02.3	33 58.3	N 8 52.8	17 50.1	N 3 11.7	Alioth	166 43.0	N56 03.9
07	211 00.8	248 52.5	45.3	43 39.4	02.2	49 00.8	52.8	32 52.6	11.7	Alkaid	153 19.0	N49 24.6
08	226 03.2	263 52.0	44.3	58 41.6	02.2	64 03.4	52.8	47 55.1	11.7	Al Na'ir	28 15.9	S47 03.7
M 09	241 05.7	278 51.4 ··	43.3	73 43.8 ··	02.2	79 05.9 ··	52.9	62 57.5 ··	11.8	Alnilam	276 11.8	S 1 13.0
O 10	256 08.2	293 50.8	42.3	88 46.0	02.1	94 08.5	52.9	78 00.0	11.8	Alphard	218 20.7	S 8 34.3
N 11	271 10.6	308 50.2	41.3	103 48.1	02.1	109 11.0	53.0	93 02.4	11.8			
D 12	286 13.1	323 49.7	S16 40.3	118 50.3	N 9 02.1	124 13.6	N 8 53.0	108 04.9	N 3 11.8	Alphecca	126 32.8	N26 46.9
A 13	301 15.6	338 49.1	39.3	133 52.5	02.1	139 16.1	53.1	123 07.4	11.8	Alpheratz	358 09.9	N28 58.9
Y 14	316 18.0	353 48.5	38.3	148 54.7	02.0	154 18.7	53.1	138 09.8	11.8	Altair	62 33.3	N 8 49.0
15	331 20.5	8 48.0 ··	37.3	163 56.9 ··	02.0	169 21.2 ··	53.2	153 12.3 ··	11.8	Ankaa	353 40.8	S42 25.2
16	346 22.9	23 47.4	36.3	178 59.1	02.0	184 23.8	53.2	168 14.7	11.8	Antares	112 57.7	S26 23.1
17	1 25.4	38 46.8	35.3	194 01.3	02.0	199 26.3	53.3	183 17.2	11.8			
18	16 27.9	53 46.3	S16 34.3	209 03.5	N 9 01.9	214 28.9	N 8 53.3	198 19.7	N 3 11.9	Arcturus	146 19.0	N19 17.1
19	31 30.3	68 45.7	33.3	224 05.7	01.9	229 31.4	53.4	213 22.1	11.9	Atria	108 22.8	S68 59.3
20	46 32.8	83 45.1	32.3	239 07.9	01.9	244 34.0	53.4	228 24.6	11.9	Avior	234 27.7	S59 26.7
21	61 35.3	98 44.6 ··	31.3	254 10.1 ··	01.9	259 36.5 ··	53.5	243 27.1 ··	11.9	Bellatrix	278 58.9	N 6 19.8
22	76 37.7	113 44.0	30.3	269 12.3	01.8	274 39.1	53.5	258 29.5	11.9	Betelgeuse	271 28.5	N 7 24.1
23	91 40.2	128 43.5	29.3	284 14.5	01.8	289 41.6	53.5	273 32.0	11.9			
8 00	106 42.7	143 42.9	S16 28.3	299 16.7	N 9 01.8	304 44.2	N 8 53.6	288 34.5	N 3 11.9	Canopus	264 06.9	S52 41.3
01	121 45.1	158 42.3	27.3	314 18.9	01.8	319 46.7	53.6	303 36.9	11.9	Capella	281 11.6	N45 58.7
02	136 47.6	173 41.8	26.3	329 21.1	01.8	334 49.3	53.7	318 39.4	12.0	Deneb	49 49.2	N45 12.7
03	151 50.1	188 41.2 ··	25.3	344 23.3 ··	01.7	349 51.9 ··	53.7	333 41.8 ··	12.0	Denebola	182 59.4	N14 40.9
04	166 52.5	203 40.7	24.3	359 25.5	01.7	4 54.4	53.8	348 44.3	12.0	Diphda	349 21.4	S18 06.0
05	181 55.0	218 40.1	23.3	14 27.7	01.7	19 57.0	53.8	3 46.8	12.0			
06	196 57.4	233 39.5	S16 22.3	29 29.9	N 9 01.7	34 59.5	N 8 53.9	18 49.2	N 3 12.0	Dubhe	194 22.5	N61 51.3
07	211 59.9	248 39.0	21.3	44 32.1	01.7	50 02.1	53.9	33 51.7	12.0	Elnath	278 44.4	N28 35.4
08	227 02.4	263 38.4	20.3	59 34.3	01.7	65 04.6	54.0	48 54.2	12.0	Eltanin	90 58.5	N51 29.5
T 09	242 04.8	278 37.9 ··	19.2	74 36.5 ··	01.6	80 07.2 ··	54.0	63 56.6 ··	12.1	Enif	34 12.3	N 9 47.0
U 10	257 07.3	293 37.3	18.2	89 38.7	01.6	95 09.7	54.1	78 59.1	12.1	Fomalhaut	15 52.2	S29 43.9
E 11	272 09.8	308 36.8	17.2	104 40.9	01.6	110 12.3	54.1	94 01.6	12.1			
S 12	287 12.2	323 36.2	S16 16.2	119 43.2	N 9 01.6	125 14.8	N 8 54.2	109 04.0	N 3 12.1	Gacrux	172 29.1	S56 59.8
D 13	302 14.7	338 35.7	15.2	134 45.4	01.6	140 17.4	54.2	124 06.5	12.1	Gienah	176 18.4	S17 25.7
A 14	317 17.2	353 35.1	14.2	149 47.6	01.6	155 20.0	54.3	139 09.0	12.1	Hadar	149 24.1	S60 16.3
Y 15	332 19.6	8 34.6 ··	13.2	164 49.8 ··	01.5	170 22.5 ··	54.3	154 11.4 ··	12.1	Hamal	328 29.3	N23 22.1
16	347 22.1	23 34.0	12.2	179 52.0	01.5	185 25.1	54.4	169 13.9	12.1	Kaus Aust.	84 17.9	S34 23.6
17	2 24.5	38 33.5	11.1	194 54.2	01.5	200 27.6	54.4	184 16.4	12.2			
18	17 27.0	53 32.9	S16 10.1	209 56.5	N 9 01.5	215 30.2	N 8 54.5	199 18.8	N 3 12.2	Kochab	137 19.7	N74 14.1
19	32 29.5	68 32.4	09.1	224 58.7	01.5	230 32.8	54.5	214 21.3	12.2	Markab	14 03.8	N15 05.9
20	47 31.9	83 31.8	08.1	240 00.9	01.5	245 35.3	54.6	229 23.8	12.2	Menkar	314 41.5	N 4 00.6
21	62 34.4	98 31.3 ··	07.1	255 03.1 ··	01.5	260 37.9 ··	54.6	244 26.2 ··	12.2	Menkent	148 37.6	S36 16.1
22	77 36.9	113 30.7	06.0	270 05.4	01.4	275 40.4	54.7	259 28.7	12.2	Miaplacidus	221 44.1	S69 38.0
23	92 39.3	128 30.2	05.0	285 07.6	01.4	290 43.0	54.7	274 31.2	12.2			
9 00	107 41.8	143 29.6	S16 04.0	300 09.8	N 9 01.4	305 45.6	N 8 54.8	289 33.6	N 3 12.3	Mirfak	309 16.4	N49 47.5
01	122 44.3	158 29.1	03.0	315 12.0	01.4	320 48.1	54.8	304 36.1	12.3	Nunki	76 30.1	S26 19.3
02	137 46.7	173 28.5	01.9	330 14.3	01.4	335 50.7	54.9	319 38.6	12.3	Peacock	53 59.8	S56 48.1
03	152 49.2	188 28.0	16 00.9	345 16.5 ··	01.4	350 53.2 ··	54.9	334 41.0 ··	12.3	Pollux	243 58.4	N28 04.4
04	167 51.7	203 27.4	15 59.9	0 18.7	01.4	5 55.8	55.0	349 43.5	12.3	Procyon	245 25.9	N 5 16.5
05	182 54.1	218 26.9	58.9	15 21.0	01.4	20 58.4	55.0	4 46.0	12.3			
06	197 56.6	233 26.3	S15 57.8	30 23.2	N 9 01.4	36 00.9	N 8 55.1	19 48.5	N 3 12.3	Rasalhague	96 30.4	N12 34.5
W 07	212 59.0	248 25.8	56.8	45 25.4	01.4	51 03.5	55.1	34 50.9	12.4	Regulus	208 10.3	N12 03.8
E 08	228 01.5	263 25.3	55.8	60 27.7	01.4	66 06.0	55.2	49 53.4	12.4	Rigel	281 36.2	S 8 13.6
D 09	243 04.0	278 24.7 ··	54.8	75 29.9 ··	01.4	81 08.6 ··	55.2	64 55.9 ··	12.4	Rigil Kent.	140 26.7	S60 44.8
N 10	258 06.4	293 24.2	53.7	90 32.1	01.4	96 11.2	55.3	79 58.3	12.4	Sabik	102 42.0	S15 41.9
E 11	273 08.9	308 23.6	52.7	105 34.4	01.4	111 13.7	55.4	95 00.8	12.4			
S 12	288 11.4	323 23.1	S15 51.7	120 36.6	N 9 01.4	126 16.3	N 8 55.4	110 03.3	N 3 12.4	Schedar	350 09.5	N56 25.9
D 13	303 13.8	338 22.6	50.6	135 38.9	01.4	141 18.9	55.5	125 05.7	12.5	Shaula	96 56.8	S37 05.2
A 14	318 16.3	353 22.0	49.6	150 41.1	01.3	156 21.4	55.5	140 08.2	12.5	Sirius	258 55.8	S16 41.5
Y 15	333 18.8	8 21.5 ··	48.6	165 43.3 ··	01.3	171 24.0 ··	55.6	155 10.7 ··	12.5	Spica	158 58.1	S11 03.3
16	348 21.2	23 21.0	47.5	180 45.6	01.3	186 26.6	55.6	170 13.2	12.5	Suhail	223 10.7	S43 21.1
17	3 23.7	38 20.4	46.5	195 47.8	01.3	201 29.1	55.7	185 15.6	12.5			
18	18 26.2	53 19.9	S15 45.5	210 50.1	N 9 01.3	216 31.7	N 8 55.7	200 18.1	N 3 12.6	Vega	80 56.6	N38 45.9
19	33 28.6	68 19.3	44.4	225 52.3	01.3	231 34.3	55.8	215 20.6	12.6	Zuben'ubi	137 33.7	S15 57.4
20	48 31.1	83 18.8	43.4	240 54.6	01.3	246 36.8	55.8	230 23.0	12.6			
21	63 33.5	98 18.3 ··	42.4	255 56.8 ··	01.3	261 39.4 ··	55.9	245 25.5 ··	12.6		S.H.A.	Mer. Pass.
22	78 36.0	113 17.7	41.3	270 59.1	01.3	276 42.0	55.9	260 28.0	12.6	Venus	200 31.2	14 26
23	93 38.5	128 17.2	40.3	286 01.3	01.3	291 44.5	56.0	275 30.5	12.6	Mars	192 34.0	4 02
Mer. Pass. 16 50.4		*v* −0.6 *d* 1.0		*v* 2.2 *d* 0.0		*v* 2.6 *d* 0.0		*v* 2.5 *d* 0.0		Jupiter	198 01.5	3 40
										Saturn	181 51.8	4 45

Table 3-8. Increments and Corrections for the Venus Shot for 40m 31s.

40ᵐ   **41ᵐ**

40ᵐ	SUN PLANETS	ARIES	MOON	v or Corrⁿ d	v or Corrⁿ d	v or Corrⁿ d
00	10 00·0	10 01·6	9 32·7	0·0 0·0	6·0 4·1	12·0 8·1
01	10 00·3	10 01·9	9 32·9	0·1 0·1	6·1 4·1	12·1 8·2
02	10 00·5	10 02·1	9 33·1	0·2 0·1	6·2 4·2	12·2 8·2
03	10 00·8	10 02·4	9 33·4	0·3 0·2	6·3 4·3	12·3 8·3
04	10 01·0	10 02·6	9 33·6	0·4 0·3	6·4 4·3	12·4 8·4
05	10 01·3	10 02·9	9 33·9	0·5 0·3	6·5 4·4	12·5 8·4
06	10 01·5	10 03·1	9 34·1	0·6 0·4	6·6 4·5	12·6 8·5
07	10 01·8	10 03·4	9 34·3	0·7 0·5	6·7 4·5	12·7 8·6
08	10 02·0	10 03·6	9 34·6	0·8 0·5	6·8 4·6	12·8 8·6
09	10 02·3	10 03·9	9 34·8	0·9 0·6	6·9 4·7	12·9 8·7
10	10 02·5	10 04·1	9 35·1	1·0 0·7	7·0 4·7	13·0 8·8
11	10 02·8	10 04·4	9 35·3	1·1 0·7	7·1 4·8	13·1 8·8
12	10 03·0	10 04·7	9 35·5	1·2 0·8	7·2 4·9	13·2 8·9
13	10 03·3	10 04·9	9 35·8	1·3 0·9	7·3 4·9	13·3 9·0
14	10 03·5	10 05·2	9 36·0	1·4 0·9	7·4 5·0	13·4 9·0
15	10 03·8	10 05·4	9 36·2	1·5 1·0	7·5 5·1	13·5 9·1
16	10 04·0	10 05·7	9 36·5	1·6 1·1	7·6 5·1	13·6 9·2
17	10 04·3	10 05·9	9 36·7	1·7 1·1	7·7 5·2	13·7 9·2
18	10 04·5	10 06·2	9 37·0	1·8 1·2	7·8 5·3	13·8 9·3
19	10 04·8	10 06·4	9 37·2	1·9 1·3	7·9 5·3	13·9 9·4
20	10 05·0	10 06·7	9 37·4	2·0 1·4	8·0 5·4	14·0 9·5
21	10 05·3	10 06·9	9 37·7	2·1 1·4	8·1 5·5	14·1 9·5
22	10 05·5	10 07·2	9 37·9	2·2 1·5	8·2 5·5	14·2 9·6
23	10 05·8	10 07·4	9 38·2	2·3 1·6	8·3 5·6	14·3 9·7
24	10 06·0	10 07·7	9 38·4	2·4 1·6	8·4 5·7	14·4 9·7
25	10 06·3	10 07·9	9 38·6	2·5 1·7	8·5 5·7	14·5 9·8
26	10 06·5	10 08·2	9 38·9	2·6 1·8	8·6 5·8	14·6 9·9
27	10 06·8	10 08·4	9 39·1	2·7 1·8	8·7 5·9	14·7 9·9
28	10 07·0	10 08·7	9 39·3	2·8 1·9	8·8 5·9	14·8 10·0
29	10 07·3	10 08·9	9 39·6	2·9 2·0	8·9 6·0	14·9 10·1
30	10 07·5	10 09·2	9 39·8	3·0 2·0	9·0 6·1	15·0 10·1
31	10 07·8	10 09·4	9 40·1	3·1 2·1	9·1 6·1	15·1 10·2
32	10 08·0	10 09·7	9 40·3	3·2 2·2	9·2 6·2	15·2 10·3
33	10 08·3	10 09·9	9 40·5	3·3 2·2	9·3 6·3	15·3 10·3
34	10 08·5	10 10·2	9 40·8	3·4 2·3	9·4 6·3	15·4 10·4
35	10 08·8	10 10·4	9 41·0	3·5 2·4	9·5 6·4	15·5 10·5
36	10 09·0	10 10·7	9 41·3	3·6 2·4	9·6 6·5	15·6 10·5
37	10 09·3	10 10·9	9 41·5	3·7 2·5	9·7 6·5	15·7 10·6
38	10 09·5	10 11·2	9 41·7	3·8 2·6	9·8 6·6	15·8 10·7
39	10 09·8	10 11·4	9 42·0	3·9 2·6	9·9 6·7	15·9 10·7
40	10 10·0	10 11·7	9 42·2	4·0 2·7	10·0 6·8	16·0 10·8
41	10 10·3	10 11·9	9 42·4	4·1 2·8	10·1 6·8	16·1 10·9
42	10 10·5	10 12·2	9 42·7	4·2 2·8	10·2 6·9	16·2 10·9
43	10 10·8	10 12·4	9 42·9	4·3 2·9	10·3 7·0	16·3 11·0
44	10 11·0	10 12·7	9 43·2	4·4 3·0	10·4 7·0	16·4 11·1
45	10 11·3	10 12·9	9 43·4	4·5 3·0	10·5 7·1	16·5 11·1
46	10 11·5	10 13·2	9 43·6	4·6 3·1	10·6 7·2	16·6 11·2
47	10 11·8	10 13·4	9 43·9	4·7 3·2	10·7 7·2	16·7 11·3
48	10 12·0	10 13·7	9 44·1	4·8 3·2	10·8 7·3	16·8 11·3
49	10 12·3	10 13·9	9 44·4	4·9 3·3	10·9 7·4	16·9 11·4
50	10 12·5	10 14·2	9 44·6	5·0 3·4	11·0 7·4	17·0 11·5
51	10 12·8	10 14·4	9 44·8	5·1 3·4	11·1 7·5	17·1 11·5
52	10 13·0	10 14·7	9 45·1	5·2 3·5	11·2 7·6	17·2 11·6
53	10 13·3	10 14·9	9 45·3	5·3 3·6	11·3 7·6	17·3 11·7
54	10 13·5	10 15·2	9 45·6	5·4 3·6	11·4 7·7	17·4 11·7
55	10 13·8	10 15·4	9 45·8	5·5 3·7	11·5 7·8	17·5 11·8
56	10 14·0	10 15·7	9 46·0	5·6 3·8	11·6 7·8	17·6 11·9
57	10 14·3	10 15·9	9 46·3	5·7 3·8	11·7 7·9	17·7 11·9
58	10 14·5	10 16·2	9 46·5	5·8 3·9	11·8 8·0	17·8 12·0
59	10 14·8	10 16·4	9 46·7	5·9 4·0	11·9 8·0	17·9 12·1
60	10 15·0	10 16·7	9 47·0	6·0 4·1	12·0 8·1	18·0 12·2

41ᵐ	SUN PLANETS	ARIES	MOON	v or Corrⁿ d	v or Corrⁿ d	v or Corrⁿ d
00	10 15·0	10 16·7	9 47·0	0·0 0·0	6·0 4·2	12·0 8·3
01	10 15·3	10 16·9	9 47·2	0·1 0·1	6·1 4·2	12·1 8·4
02	10 15·5	10 17·2	9 47·5	0·2 0·1	6·2 4·3	12·2 8·4
03	10 15·8	10 17·4	9 47·7	0·3 0·2	6·3 4·4	12·3 8·5
04	10 16·0	10 17·7	9 47·9	0·4 0·3	6·4 4·4	12·4 8·6
05	10 16·3	10 17·9	9 48·2	0·5 0·3	6·5 4·5	12·5 8·6
06	10 16·5	10 18·2	9 48·4	0·6 0·4	6·6 4·6	12·6 8·7
07	10 16·8	10 18·4	9 48·7	0·7 0·5	6·7 4·6	12·7 8·8
08	10 17·0	10 18·7	9 48·9	0·8 0·6	6·8 4·7	12·8 8·9
09	10 17·3	10 18·9	9 49·1	0·9 0·6	6·9 4·8	12·9 8·9
10	10 17·5	10 19·2	9 49·4	1·0 0·7	7·0 4·8	13·0 9·0
11	10 17·8	10 19·4	9 49·6	1·1 0·8	7·1 4·9	13·1 9·1
12	10 18·0	10 19·7	9 49·8	1·2 0·8	7·2 5·0	13·2 9·1
13	10 18·3	10 19·9	9 50·1	1·3 0·9	7·3 5·0	13·3 9·2
14	10 18·5	10 20·2	9 50·3	1·4 1·0	7·4 5·1	13·4 9·3
15	10 18·8	10 20·4	9 50·6	1·5 1·0	7·5 5·2	13·5 9·3
16	10 19·0	10 20·7	9 50·8	1·6 1·1	7·6 5·3	13·6 9·4
17	10 19·3	10 20·9	9 51·0	1·7 1·2	7·7 5·3	13·7 9·5
18	10 19·5	10 21·2	9 51·3	1·8 1·2	7·8 5·4	13·8 9·5
19	10 19·8	10 21·4	9 51·5	1·9 1·3	7·9 5·5	13·9 9·6
20	10 20·0	10 21·7	9 51·8	2·0 1·4	8·0 5·5	14·0 9·7
21	10 20·3	10 21·9	9 52·0	2·1 1·5	8·1 5·6	14·1 9·8
22	10 20·5	10 22·2	9 52·2	2·2 1·5	8·2 5·7	14·2 9·8
23	10 20·8	10 22·5	9 52·5	2·3 1·6	8·3 5·7	14·3 9·9
24	10 21·0	10 22·7	9 52·7	2·4 1·7	8·4 5·8	14·4 10·0
25	10 21·3	10 22·9	9 52·9	2·5 1·7	8·5 5·9	14·5 10·0
26	10 21·5	10 23·2	9 53·2	2·6 1·8	8·6 5·9	14·6 10·1
27	10 21·8	10 23·5	9 53·4	2·7 1·9	8·7 6·0	14·7 10·2
28	10 22·0	10 23·7	9 53·7	2·8 1·9	8·8 6·1	14·8 10·2
29	10 22·3	10 24·0	9 53·9	2·9 2·0	8·9 6·2	14·9 10·3
30	10 22·5	10 24·2	9 54·1	3·0 2·1	9·0 6·2	15·0 10·4
31	10 22·8	10 24·5	9 54·4	3·1 2·1	9·1 6·3	15·1 10·4
32	10 23·0	10 24·7	9 54·6	3·2 2·2	9·2 6·4	15·2 10·5
33	10 23·3	10 25·0	9 54·9	3·3 2·3	9·3 6·4	15·3 10·6
34	10 23·5	10 25·2	9 55·1	3·4 2·4	9·4 6·5	15·4 10·7
35	10 23·8	10 25·5	9 55·3	3·5 2·4	9·5 6·6	15·5 10·7
36	10 24·0	10 25·7	9 55·6	3·6 2·5	9·6 6·6	15·6 10·8
37	10 24·3	10 26·0	9 55·8	3·7 2·6	9·7 6·7	15·7 10·8
38	10 24·5	10 26·2	9 56·1	3·8 2·6	9·8 6·8	15·8 10·9
39	10 24·8	10 26·5	9 56·3	3·9 2·7	9·9 6·9	15·9 11·0
40	10 25·0	10 26·7	9 56·5	4·0 2·8	10·0 6·9	16·0 11·1
41	10 25·3	10 26·9	9 56·8	4·1 2·8	10·1 7·0	16·1 11·1
42	10 25·5	10 27·2	9 57·0	4·2 2·9	10·2 7·1	16·2 11·2
43	10 25·8	10 27·4	9 57·2	4·3 3·0	10·3 7·1	16·3 11·3
44	10 26·0	10 27·7	9 57·5	4·4 3·0	10·4 7·2	16·4 11·3
45	10 26·3	10 28·0	9 57·7	4·5 3·1	10·5 7·3	16·5 11·4
46	10 26·5	10 28·2	9 58·0	4·6 3·2	10·6 7·3	16·6 11·5
47	10 26·8	10 28·5	9 58·2	4·7 3·3	10·7 7·4	16·7 11·6
48	10 27·0	10 28·7	9 58·4	4·8 3·3	10·8 7·5	16·8 11·6
49	10 27·3	10 29·0	9 58·7	4·9 3·4	10·9 7·5	16·9 11·7
50	10 27·5	10 29·2	9 58·9	5·0 3·5	11·0 7·6	17·0 11·8
51	10 27·8	10 29·5	9 59·2	5·1 3·5	11·1 7·7	17·1 11·8
52	10 28·0	10 29·7	9 59·4	5·2 3·6	11·2 7·7	17·2 11·9
53	10 28·3	10 30·0	9 59·6	5·3 3·7	11·3 7·8	17·3 12·0
54	10 28·5	10 30·2	9 59·9	5·4 3·7	11·4 7·9	17·4 12·0
55	10 28·8	10 30·5	10 00·1	5·5 3·8	11·5 8·0	17·5 12·1
56	10 29·0	10 30·7	10 00·3	5·6 3·9	11·6 8·0	17·6 12·2
57	10 29·3	10 31·0	10 00·6	5·7 3·9	11·7 8·1	17·7 12·2
58	10 29·5	10 31·2	10 00·8	5·8 4·0	11·8 8·2	17·8 12·3
59	10 29·8	10 31·5	10 01·1	5·9 4·1	11·9 8·2	17·9 12·4
60	10 30·0	10 31·7	10 01·3	6·0 4·2	12·0 8·3	18·0 12·5

of stars, in order of sidereal hour angle, for every month of the year. In this listing, the magnitude and declination of each star are given. Because you can work your star sights with the information on the daily pages, you generally don't need to use these lists. They are primarily a help in determining which stars will be visible at any time of the year.

Next are the Polaris tables. The Pole Star long has been used to establish latitude. The tables here provide all the information needed to convert a Polaris sight to latitude. Use of those tables is described in Chapter 11.

Finally, the *Almanac* provides a table for converting arc to time. It's another table that's rarely needed, although the relation of arc to time is covered fully in Chapter 15.

THE AIR ALMANAC

Some navigators prefer to use the *Air Almanac* rather than the *Nautical Almanac*. The *Air Almanac* also is published jointly by the United States Naval Observatory and Her Majesty's Stationery Office. It comes out twice a year, each edition covers a six-month period—January-June and July-December. Each edition costs more than the *Nautical Almanac,* so the yearly investment is considerably more than for the single-edition nautical version. Also, the values obtained by its use may not have the pinpoint precision possible with the *Nautical Almanac.* After all, if you're traveling hundreds of miles per hour, positioning to the nearest tenth of a mile is hardly necessary.

Daily Pages

The *Air Almanac* presents data for a single day on the two sides of a single page, with each side of the page covering a 12-hour span. The GHA is given at 10-minute intervals for the sun, moon, Aries, and three planets that are easiest to observe on that date. In some cases, only two planets will be listed, with the notation that others are too close to the sun for observation. Declination of the sun and moon are also given at 10-minute intervals; planet declinations are shown only for each whole hour.

The rate of change of GHA and declination for each body are shown at the bottom of the appropriate column. Also shown are the semidiameter (S.D.) form the sun and moon. Since S.D. is not included in the main corrections shown in the *Air Almanac,* these figures must be added to apparent altitude for lower limb sights or subtracted from Ha for upper limb sights.

The first page for each date runs from 0000 GMT or 1150 GMT. Times of moonrise are shown along with corrections for *Moon's Parallax in Altitude.* This figure also must be added to the apparent altitude in order to get observed altitude of the moon (see Table 3-9). The second page for each date runs from 1200 GMT to 2350 GMT, and times of moonset are shown.

Inside Front Cover

The 57 navigational stars are listed just once, on the inside front cover, in alphabetical order. Magnitude, SHA, and declination are given for each star. Also there's an *Interpolation of GHA* table. With this, times between tabulated 10-minute intervals on daily pages can be corrected to the nearest whole minute of arc of GHA. This is somewhat more convenient than the nautical version, as you go from the daily page to the inside front cover, rather than flip through 30 pages of listings to find the appropriate minute and second. In the *Air Almanac,* the same figures are used for the sun, planets, and Aries. Separate ones are listed for the moon (see Table 3-10).

If more precision is required, separate tables for Aries and the sun are

Table 3-9. Daily Page from the *Air Almanac*.

454   (DAY 227) GREENWICH P. M. 1980 AUGUST 14 (THURSDAY)

GMT	☉ SUN GHA / Dec.	ARIES GHA ♈	VENUS −4.1 GHA / Dec.	MARS 1.4 GHA / Dec.	JUPITER −1.2 GHA / Dec.	☽ MOON GHA / Dec.
h m	° ′ ° ′	° ′	° ′ ° ′	° ′ ° ′	° ′ ° ′	° ′ ° ′
12 00	358 51.4 N14 13.8	143 03.6	46 10 N19 27	304 05 S 8 12	337 03 N 7 06	318 41 N 2 01
10	1 21.4 · 13.7	145 34.0	48 40	306 35	339 34	321 07 2 00
20	3 51.4 · 13.5	148 04.4	51 10	309 06	342 04	323 33 1 58
30	6 21.4 · 13.3	150 34.8	53 40 ·	311 36 ·	344 34 · ·	325 59 · 57
40	8 51.4 · 13.3	153 05.2	56 10	314 06	347 05	328 24 55
50	11 21.5 · 13.2	155 35.7	58 40	316 36	349 35	330 50 53
13 00	13 51.5 N14 13.0	158 06.1	61 10 N19 27	319 06 S 8 12	352 05 N 7 06	333 16 N 1 52
10	16 21.5 · 12.9	160 36.5	63 40	321 36	354 36	335 42 50
20	18 51.5 · 12.8	163 06.9	66 10	324 07	357 06	338 07 48
30	21 21.5 · 12.6	165 37.3	68 40 ·	326 37 ·	359 36 · ·	340 33 · 47
40	23 51.6 · 12.5	168 07.7	71 10	329 07	2 07	342 59 45
50	26 21.6 · 12.4	170 38.1	73 40	331 37	4 37	345 25 43
14 00	28 51.6 N14 12.2	173 08.5	76 10 N19 27	334 07 S 8 13	7 07 N 7 06	347 50 N 1 42
10	31 21.6 · 12.1	175 38.9	78 40	336 37	9 38	350 16 40
20	33 51.6 · 12.0	178 09.4	81 10	339 08	12 08	352 42 38
30	36 21.7 · 11.9	180 39.8	83 40 ·	341 38 ·	14 38 · ·	355 07 · 37
40	38 51.7 · 11.7	183 10.2	86 10	344 08	17 08	357 33 35
50	41 21.7 · 11.6	185 40.6	88 40	346 38	19 39	359 59 33
15 00	43 51.7 N14 11.5	188 11.0	91 10 N19 27	349 08 S 8 13	22 09 N 7 06	2 25 N 1 32
10	46 21.7 · 11.3	190 41.4	93 40	351 39	24 39	4 50 30
20	48 51.8 · 11.2	193 11.8	96 10	354 09	27 10	7 16 28
30	51 21.8 · 11.1	195 42.2	98 40 ·	356 39 ·	29 40 · ·	9 42 · 27
40	53 51.8 · 11.0	198 12.6	101 10	359 09	32 10	12 08 25
50	56 21.8 · 10.8	200 43.1	103 40	1 39	34 41	14 33 23
16 00	58 51.8 N14 10.7	203 13.5	106 11 N19 28	4 09 S 8 14	37 11 N 7 06	16 59 N 1 22
10	61 21.9 · 10.6	205 43.9	108 41	6 40	39 41	19 25 20
20	63 51.9 · 10.4	208 14.3	111 11	9 10	42 12	21 51 18
30	66 21.9 · 10.3	210 44.7	113 41 ·	11 40 ·	44 42 · ·	24 16 · 17
40	68 51.9 · 10.2	213 15.1	116 11	14 10	47 12	26 42 15
50	71 21.9 · 10.0	215 45.5	118 41	16 40	49 43	29 08 13
17 00	73 52.0 N14 09.9	218 15.9	121 11 N19 28	19 10 S 8 15	52 13 N 7 05	31 34 N 1 12
10	76 22.0 · 09.8	220 46.3	123 41	21 41	54 43	33 59 10
20	78 52.0 · 09.7	223 16.7	126 11	24 11	57 14	36 25 08
30	81 22.0 · 09.5	225 47.2	128 41 ·	26 41 ·	59 44 · ·	38 51 · 07
40	83 52.0 · 09.4	228 17.6	131 11	29 11	62 14	41 17 05
50	86 22.1 · 09.3	230 48.0	133 41	31 41	64 45	43 42 03
18 00	88 52.1 N14 09.1	233 18.4	136 11 N19 28	34 11 S 8 15	67 15 N 7 05	46 08 N 1 02
10	91 22.1 · 09.0	235 48.8	138 41	36 42	69 45	48 34 1 00
20	93 52.1 · 08.9	238 19.2	141 11	39 12	72 16	51 00 0 58
30	96 22.1 · 08.8	240 49.6	143 41 ·	41 42 ·	74 46 · ·	53 25 · 57
40	98 52.2 · 08.6	243 20.0	146 11	44 12	77 16	55 51 55
50	101 22.2 · 08.5	245 50.4	148 41	46 42	79 47	58 17 53
19 00	103 52.2 N14 08.4	248 20.9	151 11 N19 28	49 12 S 8 16	82 17 N 7 05	60 43 N 0 52
10	106 22.2 · 08.2	250 51.3	153 41	51 43	84 47	63 08 50
20	108 52.2 · 08.1	253 21.7	156 11	54 13	87 18	65 34 48
30	111 22.3 · 08.0	255 52.1	158 41 ·	56 43 ·	89 48 · ·	68 00 · 47
40	113 52.3 · 07.8	258 22.5	161 11	59 13	92 18	70 26 45
50	116 22.3 · 07.7	260 52.9	163 41	61 43	94 49	72 51 43
20 00	118 52.3 N14 07.6	263 23.3	166 11 N19 28	64 13 S 8 17	97 19 N 7 05	75 17 N 0 42
10	121 22.3 · 07.5	265 53.7	168 41	66 44	99 49	77 43 40
20	123 52.4 · 07.3	268 24.1	171 11	69 14	102 20	80 09 38
30	126 22.4 · 07.2	270 54.5	173 41 ·	71 44 ·	104 50 · ·	82 34 · 37
40	128 52.4 · 07.1	273 25.0	176 11	74 14	107 20	85 00 35
50	131 22.4 · 06.9	275 55.4	178 41	76 44	109 51	87 26 33
21 00	133 52.4 N14 06.8	278 25.8	181 11 N19 28	79 14 S 8 17	112 21 N 7 05	89 52 N 0 32
10	136 22.5 · 06.7	280 56.2	183 41	81 45	114 51	92 18 30
20	138 52.5 · 06.5	283 26.6	186 11	84 15	117 22	94 43 28
30	141 22.5 · 06.4	285 57.0	188 41 ·	86 45 ·	119 52 · ·	97 09 · 27
40	143 52.5 · 06.3	288 27.4	191 11	89 15	122 22	99 35 25
50	146 22.5 · 06.2	290 57.8	193 41	91 45	124 53	102 01 24
22 00	148 52.6 N14 06.0	293 28.2	196 11 N19 28	94 15 S 8 18	127 23 N 7 04	104 26 N 0 22
10	151 22.6 · 05.9	295 58.7	198 41	96 46	129 53	106 52 20
20	153 52.6 · 05.8	298 29.1	201 11	99 16	132 24	109 18 19
30	156 22.6 · 05.6	300 59.5	203 41 ·	101 46 ·	134 54 · ·	111 44 · 17
40	158 52.6 · 05.5	303 29.9	206 11	104 16	137 24	114 09 15
50	161 22.7 · 05.3	306 00.3	208 41	106 46	139 55	116 35 14
23 00	163 52.7 N14 05.2	308 30.7	211 12 N19 28	109 17 S 8 18	142 25 N 7 04	119 01 N 0 12
10	166 22.7 · 05.1	311 01.1	213 42	111 47	144 55	121 27 10
20	168 52.7 · 04.9	313 31.5	216 12	114 17	147 26	123 52 09
30	171 22.7 · 04.9	316 01.9	218 42 ·	116 47 ·	149 56 · ·	126 18 · 07
40	173 52.8 · 04.7	318 32.4	221 12	119 17	152 26	128 44 05
50	176 22.8 · 04.6	321 02.8	223 42	121 47	154 57	131 10 04
Rate	15 00.1 S0 00.8		15 00.2 N0 00.1	15 01.0 S0 00.6	15 02.0 S0 00.2	14 34.5 S0 10.0

Lat. / Moon-set / Diff.

Lat.	Moon-set	Diff.
N	h m	m
72	21 06	−04
70	21 05	−02
68	21 05	+01
66	21 05	03
64	21 04	04
62	21 04	06
60	21 04	07
58	21 03	08
56	21 03	09
54	21 03	10
52	21 03	10
50	21 03	11
45	21 02	13
40	21 02	14
35	21 02	15
30	21 01	16
20	21 01	18
10	21 01	20
0	21 00	21
10	21 00	23
20	20 59	24
30	20 59	26
40	20 58	27
45	20 58	28
50	20 57	31
52	20 57	32
54	20 57	33
56	20 57	34
58	20 56	35
60	20 56	36
S		

Moon's P. in A.

Alt.	Corr.	Alt.	Corr.
°	′	°	′
0	54	57	29
9	53	58	28
14	52	59	27
18	51	60	26
21	50	61	25
24	49	63	24
26	48	64	23
28	47	65	22
30	46	66	21
32	45	67	20
34	44	68	19
36	43	70	18
38	42	71	17
40	41	72	16
41	40	73	15
43	39	74	14
44	38	75	13
46	37	76	12
49	36	77	11
50	35	78	11
50	34	79	10
51	33		
53	32		
54	31		
55	30		
58	29		

Sun SD 15.8
Moon SD 15′
Age 4d

Table 3-10. Stars and Interpolation of GHA from the *Air Almanac* Front Cover.

STARS, JULY—DEC., 1980

No.	Name		Mag.	S.H.A.	Dec.
				° ′	° ′
7*	Acamar		3·1	315 37	S. 40 23
5*	Achernar		0·6	335 45	S. 57 20
30*	Acrux		1·1	173 38	S. 62 59
19	Adhara	†	1·6	255 32	S. 28 57
10*	Aldebaran	†	1·1	291 18	N. 16 28
32*	Alioth		1·7	166 43	N. 56 04
34*	Alkaid		1·9	153 19	N. 49 25
55	Al Na'ir		2·2	28 15	S. 47 03
15	Alnilam	†	1·8	276 12	S. 1 13
25*	Alphard	†	2·2	218 21	S. 8 34
41*	Alphecca	†	2·3	126 32	N. 26 47
1*	Alpheratz	†	2·2	358 09	N. 28 59
51*	Altair	†	0·9	62 32	N. 8 49
2	Ankaa		2·4	353 40	S. 42 25
42*	Antares	†	1·2	112 57	S. 26 23
37*	Arcturus	†	0·2	146 19	N. 19 17
43	Atria		1·9	108 21	S. 69 00
22	Avior		1·7	234 28	S. 59 27
13	Bellatrix		1·7	278 59	N. 6 20
16*	Betelgeuse	†	0·1-1·2	271 28	N. 7 24
17*	Canopus		−0·9	264 07	S. 52 41
12*	Capella		0·2	281 11	N. 45 59
53*	Deneb		1·3	49 48	N. 45 13
28*	Denebola	†	2·2	182 59	N. 14 41
4*	Diphda	†	2·2	349 21	S. 18 06
27*	Dubhe		2·0	194 23	N. 61 51
14	Elnath	†	1·8	278 44	N. 28 35
47	Eltanin		2·4	90 58	N. 51 30
54*	Enif	†	2·5	34 11	N. 9 47
56*	Famalhaut	†	1·3	15 51	S. 29 43
31	Gacrux		1·6	172 29	S. 57 00
29*	Gienah	†	2·8	176 18	S. 17 26
35	Hadar		0·9	149 24	S. 60 17
6*	Hamal		2·2	328 29	N. 23 22
48	Kaus Aust.		2·0	84 17	S. 34 24
40*	Kochab		2·2	137 20	N. 74 14
57	Markab	†	2·6	14 03	N. 15 06
8*	Menkar	†	2·8	314 41	N. 4 01
36	Menkent		2·3	148 37	S. 36 16
24*	Miaplacidus		1·8	221 45	S. 69 38
9*	Mirfak		1·9	309 16	N. 49 47
50*	Nunki	†	2·1	76 29	S. 26 19
52*	Peacock		2·1	53 58	S. 56 48
21*	Pollux	†	1·2	243 58	N. 28 04
20*	Procyon	†	0·5	245 26	N. 5 17
46*	Rasalhague	†	2·1	96 30	N. 12 35
26*	Regulus	†	1·3	208 10	N. 12 04
11*	Rigel	†	0·3	281 36	S. 8 13
38*	Rigil Kent.		0·1	140 26	S. 60 45
44	Sabik	†	2·6	102 41	S. 15 42
3*	Schedar		2·5	350 09	N. 56 26
45*	Shaula		1·7	96 56	S. 37 05
18*	Sirius	†	−1·6	258 56	S. 16 41
33*	Spica	†	1·2	158 58	S. 11 03
23*	Suhail		2·2	223 11	S. 43 21
49*	Vega		0·1	80 56	N. 38 46
39	Zuben'ubi	†	2·9	137 33	S. 15 58

*Stars used in H.O. 249 (A.P. 3270) Vol. 1.
†Stars that may be used with Vols. 2 and 3.

INTERPOLATION OF G.H.A. F3

Increment to be added for intervals of G.M.T. to G.H.A. of: Sun, Aries (♈) and planets; Moon

SUN, etc.	MOON	SUN, etc.	MOON	SUN, etc.	MOON
m s	m s	m s	m s	m s	m s
00 00	00 00	03 17	03 25	06 37	06 52
01 00	0 00 / 00 02	21	0 50 / 03 29	41	1 40 / 06 56
05	0 01 / 00 06	25	0 51 / 03 33	45	1 41 / 07 00
09	0 02 / 00 10	29	0 52 / 03 37	49	1 42 / 07 04
13	0 03 / 00 14	33	0 53 / 03 41	53	1 43 / 07 08
17	0 04 / 00 18	37	0 54 / 03 45	06 57	1 44 / 07 13
21	0 05 / 00 22	41	0 55 / 03 49	07 01	1 45 / 07 17
25	0 06 / 00 26	45	0 56 / 03 54	05	1 46 / 07 21
29	0 07 / 00 31	49	0 57 / 03 58	09	1 47 / 07 25
33	0 08 / 00 35	53	0 58 / 04 02	13	1 48 / 07 29
37	0 09 / 00 39	03 57	0 59 / 04 06	17	1 49 / 07 33
41	0 10 / 00 43	04 01	1 00 / 04 10	21	1 50 / 07 37
45	0 11 / 00 47	05	1 01 / 04 14	25	1 51 / 07 42
49	0 12 / 00 51	09	1 02 / 04 19	29	1 52 / 07 46
53	0 13 / 00 55	13	1 03 / 04 23	33	1 53 / 07 50
00 57	0 14 / 01 00	17	1 04 / 04 27	37	1 54 / 07 54
01 01	0 15 / 01 04	21	1 05 / 04 31	41	1 55 / 07 58
05	0 16 / 01 08	25	1 06 / 04 35	45	1 56 / 08 02
09	0 17 / 01 12	29	1 07 / 04 39	49	1 57 / 08 06
13	0 18 / 01 16	33	1 08 / 04 43	53	1 58 / 08 11
17	0 19 / 01 20	37	1 09 / 04 48	07 57	1 59 / 08 15
21	0 20 / 01 24	41	1 10 / 04 52	08 01	2 00 / 08 19
25	0 21 / 01 29	45	1 11 / 04 56	05	2 01 / 08 23
29	0 22 / 01 33	49	1 12 / 05 00	09	2 02 / 08 27
33	0 23 / 01 37	53	1 13 / 05 04	13	2 03 / 08 31
37	0 24 / 01 41	04 57	1 14 / 05 08	17	2 04 / 08 35
41	0 25 / 01 45	05 01	1 15 / 05 12	21	2 05 / 08 40
45	0 26 / 01 49	05	1 16 / 05 17	25	2 06 / 08 44
49	0 27 / 01 53	09	1 17 / 05 21	29	2 07 / 08 48
53	0 28 / 01 58	13	1 18 / 05 25	33	2 08 / 08 52
01 57	0 29 / 02 02	17	1 19 / 05 29	37	2 09 / 08 56
02 01	0 30 / 02 06	21	1 20 / 05 33	41	2 10 / 09 00
05	0 31 / 02 10	25	1 21 / 05 37	45	2 11 / 09 04
09	0 32 / 02 14	29	1 22 / 05 41	49	2 12 / 09 09
13	0 33 / 02 18	33	1 23 / 05 46	53	2 13 / 09 13
17	0 34 / 02 22	37	1 24 / 05 50	08 57	2 14 / 09 17
21	0 35 / 02 27	41	1 25 / 05 54	09 01	2 15 / 09 21
25	0 36 / 02 31	45	1 26 / 05 58	05	2 16 / 09 25
29	0 37 / 02 35	49	1 27 / 06 02	09	2 17 / 09 29
33	0 38 / 02 39	53	1 28 / 06 06	13	2 18 / 09 33
37	0 39 / 02 43	05 57	1 29 / 06 10	17	2 19 / 09 38
41	0 40 / 02 47	06 01	1 30 / 06 15	21	2 20 / 09 42
45	0 41 / 02 51	05	1 31 / 06 19	25	2 21 / 09 46
49	0 42 / 02 56	09	1 32 / 06 23	29	2 22 / 09 50
53	0 43 / 03 00	13	1 33 / 06 27	33	2 23 / 09 54
02 57	0 44 / 03 04	17	1 34 / 06 31	37	2 24 / 09 58
03 01	0 45 / 03 08	21	1 35 / 06 35	41	2 25 / 10 00
05	0 46 / 03 12	25	1 36 / 06 39	45	2 26 /
09	0 47 / 03 16	29	1 37 / 06 44	49	2 27 /
13	0 48 / 03 20	33	1 38 / 06 48	53	2 28 /
17	0 49 / 03 25	37	1 39 / 06 52	09 57	2 29 /
03 21	0 50 / 03 29	06 41	1 40 / 06 56	10 00	2 30 /

39

provided at the back of the book. Corrections are given for every second through a 10-minute period.

Inside Back Cover

The inside back cover carries a table of contents, some corrections to be used with bubble sextants (ignore these), and dip corrections at whole-minute intervals for heights-of-eye ranging from sea level to 2,655 feet above sea level. You will be concerned only with the low end of this scale. As with the *Nautical Almanac,* you find the heights that bracket your actual eye height, and use the dip correction shown between these (see Table 3-11).

There's a table of refraction corrections to be applied to marine sextant readings on the page facing the inside back cover (Table 3-12). These corrections apply to all bodies. Semidiameter (S.D.) corrections for sun and moon are added separately. Since the *Almanac* is designed for use on aircraft, these corrections are based on heights above sea level, in units of 1,000 feet, ranging from sea level to 55,000 feet above sea level. The small-boat skipper, of course, will use only the figures in the first column because, for all practical purposes, your sights are presumed to be taken from sea level when dealing with refraction.

Note that the corrections are given in increments of whole minutes of arc and that they are to be *subtracted* from the sextant reading. On the same page, there's another table giving additional refraction corrections to be used in extreme conditions of temperature. A Coriolis correction table is also provided; ignore it unless you're navigating an airplane.

Back of the Book

Following the daily pages in the *Air Almanac,* there's a fold-out navigational star chart, showing SHA, declination, and magnitude of all the stars you're ever likely to use. The chart is backed, on one half, by the same information that's on the inside front cover. The other half has a table for interpolation of moonrise and moonset, plus a star index in order of sidereal hour angle.

Following this chart, there's information on the book's use, standard times at major cities and countries around the world, and data concerning sunrise, sunset, and duration of twilight.

Sky diagrams make up a major section of the back pages of the *Air Almanac.* With these it is possible to quickly preselect a pair of celestial bodies that will provide lines of position that will cross at close to a 90-degree angle for an accurate fix.

AN ALMANAC COMPARISON

Here are some hypothetical sextant sights that are corrected by both the *Nautical* and *Air Almanacs* to see what differences, if any, there might be:

Table 3-11. Dip Corrections Are Inside the Rear Cover of the *Air Almanac.*

STANDARD DOME REFRACTION

To be *subtracted* from sextant altitude when using sextant suspension in a perspex dome

Alt.	Refn.	Alt.	Refn.
°	′	°	′
10	8	50	4
20	7	60	4
30	6	70	3
40	5	80	3

This table must not be used if a calibration table is fitted to the dome, or if a flat glass plate is provided, or for non-standard domes.

BUBBLE SEXTANT ERROR

Sextant Number	Alt.	Corr.
	°	′

CORRECTIONS TO BE APPLIED TO MARINE SEXTANT ALTITUDES

CORRECTION FOR DIP OF THE HORIZON
To be subtracted from sextant altitude

Ht.	Dip	Ht.	Dip	Ht.	Dip	Ht.	Dip	Ht.	Dip
Ft.	′	Ft.	′	Ft.	′	Ft.	′	Ft.	′
0	1	114	11	437	21	968	31	1 707	41
2	2	137	12	481	22	1 033	32	1 792	42
6	3	162	13	527	23	1 099	33	1 880	43
12	4	189	14	575	24	1 168	34	1 970	44
21	5	218	15	625	25	1 239	35	2 061	45
31	6	250	16	677	26	1 311	36	2 155	46
43	7	283	17	731	27	1 386	37	2 251	47
58	8	318	18	787	28	1 463	38	2 349	48
75	9	356	19	845	29	1 543	39	2 449	49
93	10	395	20	906	30	1 624	40	2 551	50
114		437		968		1 707		2 655	

CORRECTIONS
In addition to sextant error and dip, corrections are to be applied for:
Refraction
Semi-diameter (for the Sun and Moon)
Parallax (for the Moon)
Dome refraction (if applicable)

MARINE SEXTANT ERROR

Sextant Number

Index Error

LIST OF CONTENTS

Table 3-12. Just the First Column at the Left of the Refraction Table Is Needed for Nautical Sights.

A104 CORRECTIONS TO BE APPLIED TO SEXTANT ALTITUDE

REFRACTION

To be subtracted from sextant altitude (referred to as observed altitude in A.P. 3270)

Height above sea level in units of 1 000 ft. — Sextant Altitude

R_0	0	5	10	15	20	25	30	35	40	45	50	55
0	90	90	90	90	90	90	90	90	90	90	90	90
1	63	59	55	51	46	41	36	31	26	20	17	13
2	33	29	26	22	19	16	14	11	9	7	6	4
3	21	19	16	14	12	10	8	7	5	4	2 40	1 40
4	16	14	12	10	8	7	6	5	3 10	2 20	1 30	0 40
5	12	11	9	8	7	5	4 00	3 10	2 10	1 30	0 39	+0 05
6	10	9	7	5 50	4 50	3 50	3 10	2 20	1 30	0 49	+0 11	-0 19
7	8 10	6 50	5 50	4 50	4 00	3 00	2 20	1 50	1 10	0 24	-0 11	-0 38
8	6 50	5 50	5 00	4 00	3 10	2 30	1 50	1 20	0 38	+0 04	-0 28	-0 54
9	6 00	5 10	4 10	3 20	2 40	2 00	1 30	1 00	0 19	-0 13	-0 42	-1 08
10	5 20	4 30	3 40	2 50	2 10	1 40	1 10	0 35	+0 03	-0 27	-0 53	-1 18
12	4 30	3 40	2 50	2 20	1 40	1 10	0 37	+0 11	-0 16	-0 43	-1 08	-1 31
14	3 30	2 50	2 10	1 40	1 10	0 34	+0 09	-0 14	-0 37	-1 00	-1 23	-1 44
16	2 50	2 10	1 40	1 10	0 37	+0 10	-0 13	-0 34	-0 53	-1 14	-1 35	-1 56
18	2 20	1 40	1 20	0 43	+0 15	-0 08	-0 31	-0 52	-1 08	-1 27	-1 46	-2 05
20	1 50	1 20	0 49	+0 23	-0 02	-0 26	-0 46	-1 06	-1 22	-1 39	-1 57	-2 14
25	1 12	0 44	+0 19	-0 06	-0 28	-0 48	-1 09	-1 27	-1 42	-1 58	-2 14	-2 30
30	0 34	+0 10	-0 13	-0 36	-0 55	-1 14	-1 32	-1 51	-2 06	-2 21	-2 34	-2 49
35	+0 06	-0 16	-0 37	-0 59	-1 17	-1 33	-1 51	-2 07	-2 23	-2 37	-2 51	-3 04
40	-0 18	-0 37	-0 58	-1 16	-1 34	-1 49	-2 06	-2 22	-2 35	-2 49	-3 03	-3 16
45			-0 53	-1 14	-1 31	-1 47	-2 03	-2 18	-2 33	-2 47	-2 59	-3 13
50			-1 10	-1 28	-1 44	-1 59	-2 15	-2 28	-2 43	-2 56	-3 08	-3 22
55			-1 40	-1 53	-2 09	-2 24	-2 38	-2 52	-3 04	-3 17	-3 29	-3 41
60				-2 03	-2 18	-2 33	-2 46	-3 01	-3 12	-3 25	-3 37	-3 48
							-2 53	-3 07			-3 42	-3 53

$R = R_0 \times f$

R_0	f = 0.9	1.0	1.1	1.2
			R	
0	0	0	0	0
1	1	1	1	1
2	2	2	2	2
3	3	3	3	4
4	4	4	4	5
5	5	5	5	6
6	5	6	7	7
7	6	7	8	8
8	7	8	9	10
9	8	9	10	11
10	9	10	11	12
12	11	12	13	14
14	13	14	15	17
16	14	16	18	19
18	16	18	20	22
20	18	20	22	24
25	22	25	28	30
30	27	30	33	36
35	31	35	38	42
40	36	40	44	48
45	40	45	50	54
50	45	50	55	60
55	49	55	60	66
60	54	60	66	72

Temperature in °C.

f	0	5	10	15	20	25	30	35	40	45	50	55
0.9	+47	+36	+27	+18	+10	+3	-5	-13				
1.0	+26	+16	+6	-4	-13	-22	-31	-40				
1.1	+5	-5	-15	-25	-36	-46	-57	-68				
1.2	-16	-25	-36	-46	-58	-71	-83	-95				
	-37	-45	-56	-67	-81	-95						

For these heights no temperature correction is necessary, so use $R = R_0$

Where R_0 is less than 10′ or the height greater than 35 000 ft. use $R = R_0$

Choose the column appropriate to height, in units of 1 000 ft., and find the range of altitude in which the sextant altitude lies; the corresponding value of R_0 is the refraction, to be subtracted from sextant altitude, unless conditions are extreme. In that case find f from the lower table, with critical argument temperature. Use the table on the right to form the refraction, $R = R_0 \times f$.

CORIOLIS (Z) CORRECTION

To be applied by moving the position line a distance Z to starboard (right) of the track in northern latitudes and to port (left) in southern latitudes.

G/S KNOTS	0° 10°	20° 30°	40° 50°	60° 70°	80° 90°	G/S KNOTS	0° 10°	20° 30°	40° 50°	60° 70°	80° 90°
150	0 1	1 2	3 3	3 4	4 4	550	0 3	5 7	9 11	12 14	14 14
200	0 1	2 3	3 4	5 5	5 5	600	0 3	5 8	10 12	14 15	16 16
250	0 1	2 3	4 5	6 6	6 7	650	0 3	6 9	11 13	15 16	17 17
300	0 1	3 4	5 6	7 7	8 8	700	0 3	6 9	12 14	16 17	18 18
350	0 2	3 5	6 7	8 9	9 9	750	0 3	7 10	13 15	17 18	19 20
400	0 2	4 5	7 8	9 10	10 10	800	0 4	7 10	13 16	18 20	21 21
450	0 2	4 6	8 9	10 11	12 12	850	0 4	8 11	14 17	19 21	22 22
500	0 2	4 7	8 10	11 12	13 13	900	0 4	8 12	15 18	20 22	23 24

42

Sun LL (lower limb), 21-36-28 GMT, 14 August 1980. Sextant altitude is 38°42.7′, index error is +1.7′, and eye height is 18 feet.

Nautical Almanac		Air Almanac	

Observed Altitude

Nautical Almanac		Air Almanac	
Hs	38°42.7′	Hs	38°42.7′
IC	− 1.7′	IC	− 1.7′
Dip	− 4.1′	Dip	− 4.0′
Ha	38°36.9′	Ha	38°37.0′
Main	+ 14.8′	Ref.	− 1.0′
		S.D.	+ 15.8′
Ho	38°51.7′	Ho	38°51.8′

GHA

Nautical Almanac		Air Almanac	
18h	88°52.1′	18h 30m	96°22.1′
36m 28s	9°07.0′	6m	1°37.0′
GHA	97°59.1′	GHA	97°59.1′

Declination

Nautical Almanac		Air Almanac	
18h	N 14°09.2′ (d−0.8′)	18h 30m	N 14°08.8′
d corr.	− 0.5′		
Dec.	N 14°08.7′	Dec.	N 14°08.8′

The differences here are certainly not significant. Try a moon shot, lower limb, on 19 September 1980, at GMT 10-43-42 that results in a sextant altitude of 47°21.3′. Index error is −2.6′ and eye height is 7.5 feet above sea level.

Nautical Almanac		Air Almanac	

Observed Altitude

Nautical Almanac		Air Almanac	
Hs	47°21.3′	Hs	47°21.3′
IC	+ 2.6′	IC	+ 2.6′
Dip	− 2.6′	Dip	− 3.0′
Ha	47°21.3′	Ha	47°20.9′
Main	+ 48.8′	Ref.	− 1.0′
HP	+ 4.7′	S.D.	+ 16.0′
		P in A	+ 38.0′
Ho	48°14.8′	Ho	48°13.9′

GHA

Nautical Almanac		Air Almanac	
10h	218°48.3′ (v+8.3′)	10h 40m	228°26.0′
43m 42s	10°25.6′	3m 42s	54.0′
v corr.	+ 6.0′		
GHA	229°19.9′	GHA	229°30.0′

Nautical Almanac			Air Almanac		
		Declination			
10h	S 19°29.7'	(d −2.0')	10h 40m	S 19°28.0'	
d corr.	−1.5'				
Dec.	S 19°28.2'		Dec.	S 19°28.0'	

There is a difference here in the observed altitude that will be a difference of about a mile in the plotted LOP. While the *Nautical Almanac* figure is considered to be more accurate, a position difference of less than a mile, at sea, is really fairly insignificant.

Now for a planet sight. On 28 July 1980, you take a sight on the extremely bright Venus (magnitude −4.2) at GMT 21-08-12. Sextant altitude is 63°14.9', with an index error of +4.6', and eye height at 22.5 feet above sea level.

Nautical Almanac   Air Almanac

Observed Altitude

Hs	63°14.9'		Hs	63°14.9'
IC	−4.6'		IC	−4.6'
Dip	−4.6'		Dip	−5.0'
Ha	63°05.7'		Ha	63°05.3'
Main	−0.5'		Ref.	0.0'
Ho	63°05.2'		Ho	63°05.3'

GHA

21h	178°09.8'	(v +0.9')	21h 00m	178°10.0'
8m 12s	2°03.0'		8m 12s	2°03.0'
v corr.	+0.1'			
GHA	180°12.9'		GHA	180°13.0'

Declination

21h	N 18°29.3'	(d +0.2')	21h	N 18°29.0'
d corr.	+0.0'			
Dec.	N 18°29.3'		Dec.	N 18°29.3'

Again, the results are extremely close. Finally, we will try a star sight. The constellation Orion is in the night sky in winter and *Sirius,* associated with this star group, is one of the brightest objects in the heavens (magnitude −1.6). The GMT is 01-38-53 and the sextant reads 21°08.4', with an error of −0.9'. Eye height is 9.0 feet.

Nautical Almanac		Air Almanac	

Observed Altitude

Hs	21°08.4′	Hs	21°08.4′
IC	+0.9′	IC	+0.9′
Dip	− 2.9′	Dip	− 3.0′
Ha	21°06.4′	Ha	21°06.3′
Main	− 2.5′	Ref.	− 4.0′
Ho	21°03.9′	Ho	21°02.3′

GHA

SHA Sirius	258°55.2′	SHA Sirius	258°56.0′
GHA Aries		GHA Aries	
01h	109°41.1′	01h 30m	117°12.3′
38m 53s	9°44.8′	8m 53s	2°13.0′
GHA	378°21.1′	GHA	378°21.3′
	− 360°		− 360°
GHA Sirius	18°21.1′	GHA Sirius	18°21.3′

Declination

Dec. S 16°41.5′   Dec. S 16°41.0′

Again, there is no significant difference in the results. Choose the *Almanac* that you believe will be easiest for you to use.

REED'S ALMANAC

There is one other almanac you might consider. It is the U.S. editions of *Reed's Nautical Almanac* published in Great Britain and sold in this country. While they provide piloting information (tide and current tables, for example) only for the east and west coasts of this country, plus the Caribbean, the celestial navigation section can be used anywhere on earth. An advantage of these books is that they contain both the almanac information and sight reduction tables. A separate reference book is not needed.

On daily pages, GHA of the sun and Aries are presented at two-hour intervals, along with sun declination. General sun, moon, and star information for each month precedes the daily pages for that month. Separate pages for planets and the moon follow the month's daily pages. There are also separate pages for altitude (refraction) corrections and GHA increments, and dip tables.

The organization of the material seems a little scattered in comparison to the *Nautical* or *Air Almanacs,* but once you locate the tables you need they are easy to use. An explanation of the material is given in the book (with examples). The results are accurate and they can be used with any sight reduction method.

Chapter 4

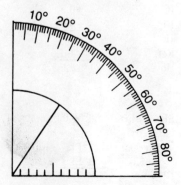

A Quick Look at the Time

We're almost ready to show how the information obtained from the *Almanac* is used to establish position, but first we need to discuss the subject of time and its relation to position.

All times listed in the *Almanacs* are Greenwich mean time (GMT)— now sometimes called *universal time*. This means that all data are referenced to the Greenwich meridian of longitude. Your observations are made according to time in your local time zone. This is called *zone time* (ZT). You have to convert zone time to GMT in order to look up the appropriate information in the *Almanac*.

TIME ZONES

In ages past, all time was *local apparent time*. If you had a sundial that was properly set up at Point A, it was exactly noon when the sun cast its shadow on the 12 o'clock mark on the dial. Noon at Point A was not the same time as noon at Point B, 20 miles to the east, or Point C, 20 miles to the west. As timepieces became more accurate, and people traveled more, time for a wide area—perhaps an entire European nation—became standardized. And after accurate chronometers were invented, it was possible to establish a vessel's longitude by comparing chronometer time (set to GMT) to the zone time for the area in which the ship was operating.

Since the earth is 360 degrees in equatorial circumference, and the sun takes 24 hours to make its apparent trip around the earth (remember, for purposes of navigation, the earth doesn't move; everything out there revolves around it), the sun is traveling 15 degrees of longitude each hour. Each time zone, then, covers a 15-degree arc of longitude, and the zones are numbered from zero through 12 consecutively, both east and west of the

Greenwich meridian. The central meridian of each zone is the zone meridian (ZM), and each zone extends 7.5 degrees (7°30.0') to each side of the zone meridian, as shown in Table 4-1 and in Fig. 4-1.

When standard time is in effect, zone time is equal to GMT plus or minus the hours of difference from GMT to the zone number (zone description, or ZD). Zones west of Greenwich are identified by a plus (+) sign, and GMT = ZT + ZD. Zones east of Greenwich are identified by a minus (−) sign, and GMT = ZT − ZD. When daylight saving time is in effect, the practice is to use the ZD of the zone immediately to the east of the one you're in.

Here are some time zone problems: You are at a longitude of 17°E, cruising in the Mediterranean,and the zone time is 1215. You need to determine GMT and you are in ZD − 1, which runs from 7.5° E to 22.5° E. The ZD is negative, so it is subtracted from ZT to get GMT:

$$
\begin{array}{ll}
\text{ZT} & 1215 \\
\text{ZD} & -\underline{1} \\
\text{GMT} & \overline{1115}
\end{array}
$$

You are out in the Pacific, at Lo 161°W, and the Zone Time is 1744 on 12 November 1980. What is the GMT and date at Greenwich? Your ZD is + 11 (Lo 157.5°W− Lo 172.5°W).

Table 4-1. Zone Numbers, Limits, and Meridians.

Zone	Zone Limits	Zone Meridian
0	7.5°E—7.5°W	0° Greenwich Meridian
1	7.5°E—22.5°E	15°E
	7.5°W—22.5°W	15°W
2	22.5°E—37.5°E	30°E
	22.5°W—37.5°W	30°W
3	37.5°E—52.5°E	45°E
	37.5°W—52.5°W	45°W
4	52.5°E—67.5°E	60°E
	52.5°W—67.5°W	60°W
5	67.5°E—82.5°E	75°E
	67.5°W—82.5°W	75°W
6	82.5°E—97.5°E	90°E
	82.5°W—97.5°W	90°W
7	97.5°E—112.5°E	105°E
	97.5°W—112.5°W	105°W
8	112.5°E—127.5°E	120°E
	112.5°W—127.5°W	120°W
9	127.5°E—142.5°E	135°E
	127.5°W—142.5°W	135°W
10	142.5°E—157.5°E	150°E
	142.5°W—157.5°W	150°W
11	157.5°E—172.5°E	165°E
	157.5°W—172.5°W	165°W
12	172.5°E—172.5°W	180°International Date Line

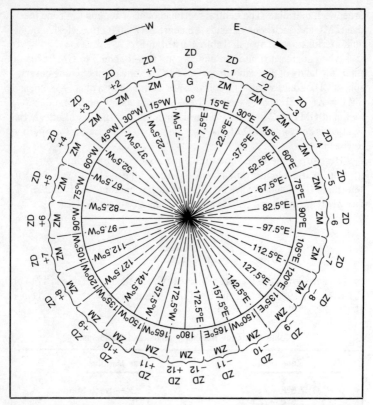

Fig. 4-1. Time zones: solid lines indicate zone meridians; dashed lines indicate zone boundaries.

$$
\begin{array}{lll}
\text{ZT} + & 1744 & 12 \text{ Nov} \\
\text{ZD} & \underline{11} & \\
\text{GMT} & 2844 & 12 \text{ Nov}
\end{array}
$$

Because there are only 24 hours in the day, subtract 24 hours from the time and add these—as a day—to the date:

$$
\begin{array}{ll}
\text{GMT } 2844 & 12 \text{ Nov} \\
\underline{-24} & \\
\text{GMT } 0444 & 13 \text{ Nov}
\end{array}
$$

Here are some problems you can work out to get the GMT and date. All zone times are standard time; answers are in the Appendix:

1. Your longitude is 87° W; date is 18 June; ZT is 0558.
2. Longitude 44° E; date is 12 April; time is 1702.

3. Longitude 130° E; 22 September; ZT 1159.
4. Longitude 68° W; 14 November; ZT 1022.
5. Longitude 147° W; 8 August; ZT 0604.

INTERNATIONAL DATE LINE

You can get into some interesting problems involving the international date line, which primarily runs down the center of ZD 12 (Lo 180°). It does zigzag slightly to keep all territory of some geographic units—such as Alaska's Aleutian Islands—in the same time zone. First we must show how to determine ZT based on GMT and also how to add elapsed time (ET) into the calculation.

Finding zone time on the basis of GMT is the reverse of the procedure used to find GMT on the basis of zone time. To avoid confusion, the correct ZD sign—plus or minus—is maintained, with the abbreviation (Rev) for "Reversed" added before ZD to show that you are going to reverse the action of the sign. Thus if GMT is 0841 and the ZD is + 3, ZT is:

$$
\begin{array}{ll}
\text{GMT} & 0841 \\
\text{(Rev) ZD} + & \underline{3} \\
\text{ZT} & 0541
\end{array}
$$

If, in the above example the ZD had been − 3, ZT would be:

$$
\begin{array}{ll}
\text{GMT} & 0841 \\
\text{(Rev) ZD} - & \underline{3} \\
\text{ZT} & 1141
\end{array}
$$

When elapsed time (ET) is introduced into a problem, it can be added in at any point, but as a matter of convenience and standardization, tack it onto GMT once this is established.

Now we can try a couple of problems based on the international date line. In each case, determine if New Year's Eve was celebrated once, twice, or not at all:

1. A ship is at Lo 179°08.7′W at ZT 1623 on 31 December 1980. Exactly six hours later it crosses the date line. What is the date and time one minute later?

2. A ship is at Lo 179°08.7′E at ZT 1623 on 1 January 1981. Six hours later it is at Lo 179°31.3′W. What is the ZT and date?

Here's how the problems are worked:

1.
$$
\begin{array}{ll}
\text{ZT} & 1623 \ 31 \ \text{December} \ 1980 \\
\text{ZD} + & \underline{12} \\
\text{GMT} & 2823 \ 13 \ \text{December} \ 1980 \\
\text{GMT} & 0423 \ 1 \ \text{Jan} \ 1981 \ \text{(This is GMT} \\
& \qquad \text{at 1623 ZT in ZD} + 12)
\end{array}
$$

```
          ET    601
          GMT  1024  1 Jan 1981 (GMT after
                           elapsed time of 6h 0lm)
    (Rev) ZD −  12      (Action of sign is
                           reversed)
          ZT    2224  1 Jan 1981
```

Since the vessel crossed the line before midnight on 31 December, and it was late in the evening of 1 January of the next year the instant the line was crossed, there was no New Year's Eve celebration.

```
2.              ZT    1623  1 Jan 1981
                ZD −  12
                GMT   0423  1 Jan 1981
                ET     600
                GMT   1023  1 Jan 1981
                GMT   3423  31 December 1980 (24 hours
                                 added to time so ZD can be
                                 subtracted)
        (Rev)   ZD +  12
                ZT    2223  31 December 1980
```

Obviously New Year's Eve would have been celebrated as the vessel was heading toward the Date Line, as it is already 1 January 1981 before the line is reached. Because it is again 31 December 1980 *after* the line has been crossed, New Year's Eve can be celebrated again—if the ship's skipper has a disposition that will permit this.

Just keep in mind that to find GMT when ZT and ZD are known, simply add or subtract the ZD as called for by its sign. To find ZT when GMT and ZD are known, show the proper ZD sign but reverse its action, with (Rev) in front of the ZD as a reminder—and as evidence—that you reversed the action.

Here are some problems you can work out for yourself. The answers are in the Appendix:

6. GMT is 1924 on 6 November; ZD is − 7; find ZT and date.

7. GMT 0022 18 October; ZD + 11; find ZT and date.

8. GMT 0022 18 October; ZD − 11; find ZT and date.

9. GMT 1313 18 August; ZD + 5; find ZT and date.

10. GMT 0924 1 January; ZD +8; find ZT and date.

11. ZT 1442 6 April; ZD + 6; ET 8h 15m; new ZD +7; find new ZT and date.

12. ZT 0631 21 June; ZD − 10; ET 27h 23m; new ZD − 9; find ZT and date.

13. ZT 1200 16 July; ZD + 9; ET 12h 00m; new ZD + 8; find ZT and date.

14. ZT 1924 24 December; ZD + 12; ET 7h 15m; new ZD − 12; find ZT and date.

15. ZT 0408 6 February; ZD − 4; ET 0h 18m; new ZD − 3; find ZT and date.

DAYLIGHT SAVING TIME

Where daylight saving time is in effect in longitudes west of Greenwich, simply use a ZD one hour less than the ZD for standard time. Where daylight saving time is in effect in longitude east of Greenwich, use the next larger ZD. In each case, you are using the standard ZD of the next zone to the east of the one in which you are located. Of course, at sea, you have the option of keeping standard time even during the months when daylight saving time may be in effect in an area.

WATCH ERROR

The time you record for a sight is not necessarily the correct zone time unless your watch keeps perfect time or it has just been set to correct time within an hour or so before the sights were taken. The recorded time is *watch time* (WT), and any difference between this and zone time is *watch error* (WE). The direction and extent of watch error must be known in order to correct watch time to zone time.

The best way to establish watch error is to pick up a radio time signal broadcast by the Bureau of Standards on short wave frequencies of 2.5, 5, 10, and 15 MHz; or the Canadian time signals broadcast on 3330, 7335, and 14670 kHz. Some watches can be set to match the time signals and thus eliminate watch error. In any case, check for watch error as soon as conveniently possible before taking sights.

If a chronometer is carried, WE is the difference between WT and corrected chronometer time. Here we will assume there is no chronometer error and you can establish WE by a simple comparison of watch time to chronometer time. Do keep in mind, however, that chronometers normally are set to GMT and that your watch error is the difference only between watch time and zone time.

APPLYING THE ERROR

If a watch is fast, subtract WE from watch time to get zone time. If a watch is slow, add WE to WT for ZT. Because watch error usually varies slightly from day to day, it is best to log WE—and identify it as such—along with sextant index error on the top of the sheet on which sight times and sextant altitudes are first recorded. It is recommended that this information be transferred to a permanent sight log that contains all the data pertaining to each sight. This makes it possible to re-check calculations later, if this should become necessary.

When dealing with hours, minutes, and seconds (not just hours and minutes in the 24-hour clock system as used in the above time examples), it's a convention to separate the units with a hyphen. This helps to identify the figures as units of time rather than units of arc.

Assume that you have determined a watch error of 2m 13s slow, and you record a run of sights on the sun at WT 11-56-22, 11-57-17, 11-57-57, 11-58-39, and 11-59-12. Since WE is slow, add it to WT to get ZT. Therefore:

WT	11-56-22	11-57-17	11-57-57	11-58-39	11-59-12
WE(s)	+ 2-13	+ 2-13	+ 2-13	+ 2-13	+ 2-13
ZT	11-58-35	11-59-30	12-00-10	12-00-52	12-01-25

In adding, remember that there are 60 seconds to the minute and 60 minutes to the hour.

In the above case, if the watch were fast by the same amount, what would be the ZT of the sights?

WT	11-56-22	11-57-17	11-57-57	11-58-39	11-59-12
WE(f)	− 2-13	− 2-13	− 2-13	− 2-13	− 2-13
ZT	11-54-09	11-55-04	11-55-44	11-56-26	11-56-59

Note that the letter (f) or (s) appears in parenthesis alongside the WE, and the plus or minus sign as required. This is a simple way of being doubly sure that you are applying the watch error in the correct direction.

Suppose the run of sights above, with the WE of 2m 13s fast, was used to establish meridian passage of the sun (to be explained in Chapter 5). Your recorded readings show that the highest altitude, indicating meridian passage, was at 11-57-17. You are in ZD + 7, and the date is 23 March 1980. Here's how to establish GMT on your sight reduction worksheet:

WT	11-57-17 23 Mar 1980
WE(f)	− 2-13
ZT	11-55-04
ZD	+ 7
GMT	18-55-04 23 Mar

In the *Nautical Almanac* or *Air Almanac*, you will look up the sun data for 18-55-04 on 23 March to establish the GP of the sun for your watch time of 11-57-17.

Find the GMT and date for the following watch times. See the Appendix for the answers:

16. WT 06-08-14; WE (f) 0-12; ZD + 5; 23 April.
17. WT 19-38-00 (DST); WE (s) 3-32; ZD + 5; 23 April.
18. WT 05-54-41 (standard time); WE (s) 1-12; ZD − 3; 19 July.
19. WT 07-23-45; WE (f) 1-51; ZD − 9; 12 October.
20. WT 16-45-13; WE (f) 0-28; ZD + 6; 28 September.

TIME DIAGRAMS

You now have all the basic information you need to establish GMT of each sight you take, but there is one more step you can and should take to make sure you're on the right track. This is to draw a *time diagram*. The time diagram graphically shows the relationships between your meridian, the Greenwich meridian, the *Greenwich hour angle* (GHA) of the celestial body used for the sight, the GHA of Aries if it is a star sight, and the GHA of the sun, regardless of the type of sight. The GHA of the sun in particular helps to verify the time, as is discussed in detail in Chapter 5. One new term—*local hour angle* (LHA) will be introduced.

Use a drawing compass or circular protractor to draw a circle (this is provided on some sight reduction forms, such as those used by the United States Power Squadrons). From the center of the circle, draw a solid line to the top. This represents your meridian. Your position is at the top of the line and it is marked "M." A dashed line from the center of the circle is marked "m", it is the meridian 180 degrees from your position (see Fig. 4-2).

Now establish the Greenwich meridian on the diagram. It can't be done with pinpoint accuracy—for one thing, you're supposed to be determining your position—but you should be within one degree, and that's accurate enough. Use the DR position longitude for the time of the sight. This is your angle east or west of Greenwich. Use the protractor and measure this distance clockwise from M if your longitude is west. Measure the distance

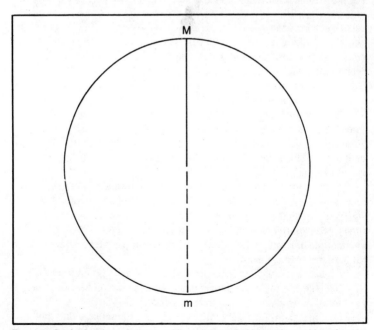

Fig. 4-2. Start of a time diagram, with meridian of observer's position shown.

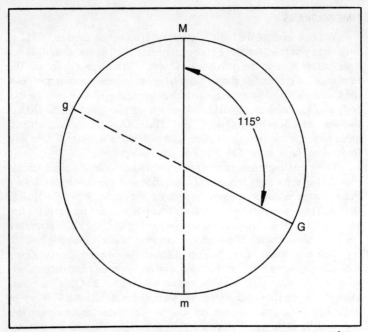

Fig. 4-3. Greenwich meridian added to time diagram, with longitude at 115° W.

counterclockwise if your longitude is east of Greenwich. In effect, it's as if you were looking up at the earth from the bottom, with the center of the circle representing the South Pole, and the circumference of the circle representing the equator. The various lines that you draw represent meridians of longitude, in effect, on the surface of the earth.

In Fig. 4-3, longitude is W 115°, so the Greenwich meridian is drawn 115 degrees clockwise from M as a solid line from the center to the edge. The line is continued in the opposite direction as a dashed line. The Greenwich point is marked "G" and a small "g" is used at the end of the dashed line. This point is assumed to be the international date line, ignoring its odd zigzags.

Next, establish the position of the sun at the time of the sight determining its GHA from the *Almanac*. If the sight is of a body other than the sun, you can use the GHA sun figure for the whole hour closest to the actual time *(Nautical Almanac)* or the nearest 10-minute interval *(Air Almanac)*. On the time diagram, measure this distance west (counterclockwise) from the Greenwich meridian G.

Assume a star sight was taken in evening twilight, giving a sun GHA of 215°; add the line to the diagram at this point as shown on Fig. 4-4. Outside the circle, draw a very small circle to indicate the sun.

It is assumed—and generally works out in practice—that anything within 90 degrees east or west of M is above your horizon (not taking refraction into account, or the earth's tilt in relation to the plane of its orbit about the sun). A body more than 90 degrees from M is below the horizon and out of sight. Figure 4-4 indicates the sun has indeed set, but as it's only 10 degrees below the horizon (100° west of M) there's still enough light in the sky to provide a good horizon for a sextant sight.

Since this is a star sight, and the GHA of a star is based on the GHA of Aries plus the sidereal hour angle (SHA) of the star, we need to show the GHA of Aries on the diagram. In this case, GHA of Aries at the time of the sight is determined to be 255°. Again, this angle is measured west (counterclockwise) from G, and the meridian is drawn from the center to the edge of the circle. Just outside the circle draw the Aries symbol (♈), which represents the horns of Aries, the ram in the Zodiac (see Fig. 4-5).

For this example, the star used is Sirius, which has a sidereal hour angle of almost 259°. Measure this distance west from the *Aries* meridian *not* from Greenwich. Again draw the meridian line, and put an asterisk (∗) outside the circle to indicate the star.

The GHA of Sirius equals the GHA of Aries plus the SHA of Sirius, or 255° + 259° = 514°. Of course 360 degrees is subtracted, giving a GHA Sirius of 154°. This can be confirmed on the diagram by measurement of the angle west from the Greenwich meridian to the meridian for Sirius (see Fig. 4-6).

This completes the time diagram for this sight. A sun, moon, or planet sight is diagrammed in the same manner, except that the GHA for Aries is

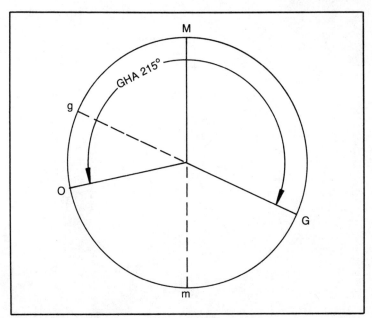

Fig. 4-4. Meridian of the sun added, with the GHA of 215°.

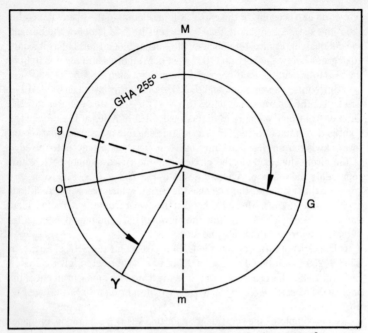

Fig. 4-5. Meridian of Aries is added to the diagram; its GHA is 255°.

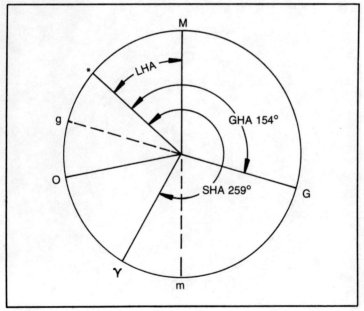

Fig. 4-6. Meridian of the star is added, so its LHA can be determined.

omitted, and the GHA of the body is measured directly west from Greenwich. Be sure to include the sun in each diagram; it is a good check on the accuracy of your observation and of the diagram itself.

LOCAL HOUR ANGLE

One of the most important bits of information needed in sight reduction is the angle in arc between your meridian and the GHA of the body. This is the local hour angle (LHA). It is always measured west from your meridian to the meridian of the body. If your longitude, for example, is W 105° and the GHA of a sighted body is 104°, the LHA of the body is 359°. For some reduction methods, the angle is measured east or west from your meridian. In such a case, the angle is identified as "t," not LHA. And the W or E is added as necessary.

$$
\begin{array}{r}
359.60 \\
33° 18.1 \\
\hline
336.41.9
\end{array}
$$

$$
\begin{array}{r}
66 \quad 74.2 \text{ w} \\
43 ° 56.1 \\
\hline
23 ° 18.1
\end{array}
$$

Chapter 5

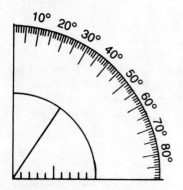

The Noon Sight

Before accurate timepieces were available to navigators, most captains relied extensively on a single daily observation to establish position. This was the noon sight. It provided accurate latitude information. The practice was to run generally north or south until you reached the latitude of your destination and then head directly east or west until you got there. Skippers may have gone farther north or south than required to pick up favorable trade winds or avoid shoals as necessary. On military and commercial vessels, the noon position is still considered the key position for each day's run, and every effort is made to ensure its accuracy.

Unfortunately, before chronometers were available, longitude could not be determined accurately. A captain relied on his dead reckoning, his knowledge of currents in the waters he was traversing, and a feeling for whatever leeway he might have made due to wind conditions.

A single observation can establish latitude quite simply. If you can establish the altitude of a body—the sun is most convenient—at the instant it crosses your meridian, you can add or subtract its declination to its zenith distance, as required, and the result is your latitude.

For example, on 21 June (the date on which the sun reaches its most northerly declination), you know from your *Almanac* that declination is N 23°26.4'. By taking a series of sights in the time spanning the anticipated time of meridian passage, you determine that the highest observed altitude was 64°39.7'. Subtract this from 90 degrees to get zenith distance:

$$\begin{array}{r} 90°00.0' \\ -64°39.7' \\ \hline 25°20.3' = \text{Zenith Distance} \end{array}$$

I assume that you are in the Northern Hemisphere so the GP of the sun, at its meridian passage, is 25°20.3′ due south of you. This means that its zenith distance, plus its declination, equals the total angle of arc to the equator, and this angle is your latitude:

$$
\begin{array}{ll}
25°20.3' & \text{Zenith Distance} \\
+N\ \underline{23°26.4'} & \text{Declination} \\
N\ 48°46.7' & \text{Latitude}
\end{array}
$$

Figure 5-1 illustrates this example. A simple diagram of this type, even with the angles of arc drawn approximately in position, gives a quick indication of the location of the equator and shows graphically whether you need to add or subtract declination from zenith distance—or in some cases subtract zenith distance from declination—to get your latitude.

Suppose you are south of the sun when it crosses your meridian and you record the same observed altitude on the same date as for the above

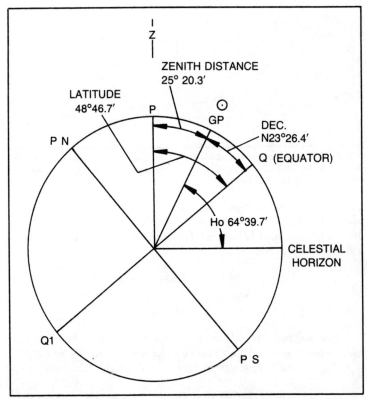

Fig. 5-1. This type of diagram is based on the plane of the celestial horizon, and here it is used to establish latitude. In this case latitude is zenith distance plus declination.

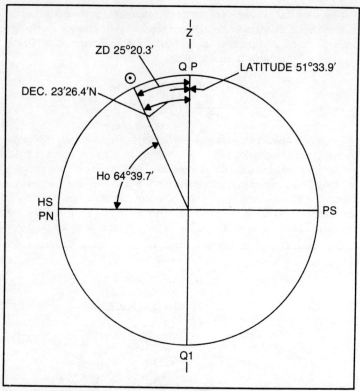

ZD 25°20.3'

DEC. 23'26.4'N

LATITUDE 51°33.9'

Q P

Z

Ho 64°39.7'

HS
PN

PS

Q1

Fig. 5-2. When a vessel is in southern latitudes, it's a convention to show the equator to the left of Z, and the South Pole elevated above the celestial horizon on the right side of the diagram.

problem. You get the same zenith distance (25°20.3') and your diagram shows that the equator is between you and the CP of the sun. This means the declination must be *subtracted* from zenith distance to obtain your latitude (see Fig. 5-2).

$$
\begin{array}{ll}
25°20.3' & \text{Zenith Distance} \\
-\text{N } 23°26.4' & \text{Declination} \\
\hline
\text{S } 1°33.9' & \text{Latitude}
\end{array}
$$

Now suppose it is 21 December, time of the winter solstice, and the sun's declination is S 23°26.4'. You're in the Northern Hemisphere and Ho is 48°37.2'. What is your latitude? First, establish zenith distance:

$$
\begin{array}{ll}
90°00.0' & \\
-48°37.2' & \text{Ho} \\
\hline
41°22.8' & \text{Zenith Distance}
\end{array}
$$

60

The diagram (Fig. 5-3) again shows the equator is between you and the GP of the sun. So subtract declination from zenith distance:

$$
\begin{array}{ll}
41°22.8' & \text{Zenith Distance} \\
-\text{S } 23°26.4' & \text{Declination} \\
\hline
\text{N } 17°56.4' & \text{Latitude}
\end{array}
$$

Here's one more example: Observed altitude is 79°47.7' and declination is N 18°36.8'. You are in the Northern Hemisphere and the sun crosses the meridian to the north of your position. What is your latitude?

$$
\begin{array}{ll}
90°00.0' & \\
-79°47.7' & \text{Ho} \\
\hline
10°12.3' & \text{Zenith Distance}
\end{array}
$$

Figure 5-4 shows that you are between the GP of the body and the equator. In this case zenith distance must be subtracted from the declination:

$$
\begin{array}{ll}
\text{N } 18°36.8' & \text{Declination} \\
-\underline{10°12.3'} & \text{Zenith Distance} \\
\text{N } 8°24.5' & \text{Latitude}
\end{array}
$$

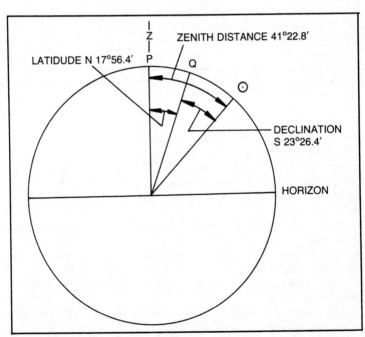

Fig. 5-3. The diagram can be simplified by omitting the polar axis line, as the needed angles are measured between Z and the horizon.

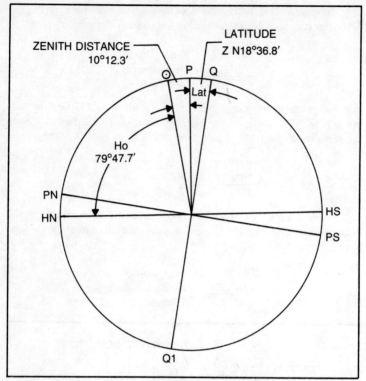

Fig. 5-4. Here's an example of a meridian passage diagram with the boat's position between the equator and the GP of the sun.

Here are some latitude problems you can work out yourself. For each, draw a diagram to show the relation of observed altitude to zenith distance and declination. It will show the way the problem is to be worked and leave you with just the simple arithmetic. To find zenith distance, it's simplest to express the full 90°00.0′ as 89°60.0′. Remember that there are only 60′ of arc to each degree. Now, find the latitude for the following:

21. Northern Hemisphere; sun crosses to the south of your position; Ho 55°12.6′; declination N 4°54.2′.

22. Northern Hemisphere; sun crosses to the south; Ho 68°17.0′; declination S 11°21.4′.

23. Northern Hemisphere; sun crosses to the north; Ho 78°17.0′; declination N 19°00.6′.

24. Southern Hemisphere; sun crosses to the north; Ho 60°09.3′; declination N 20°16.7′.

25. Southern Hemisphere; sun crosses to the north; Ho 75°23.4′; declination N 8°20.0′.

ESTABLISHING MERIDIAN PASSAGE

When you're at sea, it's not possible to predetermine the exact instant at which your boat and the sun will be on the same meridian. However, you can work from the time of meridian passage shown at the lower right hand corner of the right hand daily pages of the *Nautical Almanac*. This is the time (in hours and whole minutes) at which the sun crosses the zone meridian of each time zone. By determining your DR position for this time, you can determine approximately how far east or west of the zone meridian you are at that moment. On that basis, you can get the approximate time of meridian passage at your DR position.

Round your DR position to the nearest whole degree of longitude. Determine the difference between this point and the zone meridian in whole degrees of arc. Since the sun covers 15 degrees of arc each hour, it moves 1 degree of arc every 4 minutes. By multiplying the degrees between your DR position and the zone meridian by four, and adding or subtracting the result from the time of meridian passage at the zone meridian, you have the approximate time of meridian passage at your DR position.

If you are east of the zone meridian, subtract the time difference, as the sun will pass your position before it reaches the zone meridian. If you are to the west, add the time difference, as the sun will cross the zone meridian before it reaches you. You should wind up with a time that's within 10 minutes or so of the actual meridian passage at your position. Then start taking sights, starting about 10 minutes or so before this calculated time of meridian passage, and continue until you are sure the sun has passed your meridian. By recording these sights and times, you get the actual time of meridian passage. First we'll show how to take some practice sights that will accustom you to the phenomena of meridian passage.

Obviously, altitude of the sun increases as it rises toward your meridian, and decreases after it has passed the meridian. There is a brief period of just a few seconds at the peak where there is no apparent change in altitude. You can observe this by taking a number of practice sights from a known location.

Establish the exact longitude of this position from a nautical chart. From the *Nautical Almanac* or *Air Almanac* you can determine the time at which the sun will pass this longitude on the date you will take the sights. Here's how it works with the *Nautical Almanac* (the procedure with the *Air Almanac* is similar and should present no problems).

Assuming your longitude is 73°16.7'W, you want to find, in the *Almanac*, the time at which the sun's GHA is 73°16.7'. You will take the sights on 28 July 1980. You know that this longitude puts you in Zone +5, but since daylight saving time is in effect, your time calculation will be based on ZD +4.

In the *Almanac*, you see that the sun has a GHA of 58°23.8' at 1600 on 28 July, and that at 1700 it is 73°23.9'—just west of your position (Table 5-1). You need to know how long it will take to get from the GHA at 1600 to your meridian. The difference, in angle of arc, is:

Table 5-1. Sun Data for 1600 on 28 July 1980.

G.M.T.	SUN G.H.A.	SUN Dec.	MOON G.H.A.	v	Dec.	d	H.P.
26 00	178 23.3	N19 27.1	23 02.3	7.5	S19 40.1	0.2	57.8
01	193 23.3	26.6	37 28.8	7.5	19 39.9	0.3	57.8
02	208 23.3	26.0	51 55.3	7.5	19 39.6	0.5	57.8
03	223 23.3	·· 25.5	66 21.8	7.4	19 39.1	0.5	57.8
04	238 23.3	24.9	80 48.2	7.4	19 38.6	0.7	57.9
05	253 23.3	24.4	95 14.6	7.3	19 37.9	0.8	57.9
06	268 23.3	N19 23.8	109 40.9	7.3	S19 37.1	1.0	58.0
S 07	283 23.3	23.3	124 07.2	7.3	19 36.1	1.0	58.0
A 08	298 23.3	22.7	138 33.5	7.3	19 35.1	1.2	58.0
T 09	313 23.3	·· 22.2	152 59.8	7.2	19 33.9	1.3	58.0
U 10	328 23.3	21.6	167 26.0	7.2	19 32.6	1.4	58.1
R 11	343 23.3	21.1	181 52.2	7.1	19 31.2	1.6	58.1
D 12	358 23.3	N19 20.5	196 18.3	7.1	S19 29.6	1.6	58.1
A 13	13 23.3	19.9	210 44.4	7.1	19 28.0	1.8	58.2
Y 14	28 23.3	19.4	225 10.5	7.1	19 26.2	1.9	58.2
15	43 23.3	·· 18.8	239 36.6	7.1	19 24.3	2.1	58.2
16	58 23.3	18.3	254 02.7	7.0	19 22.2	2.1	58.3
17	73 23.3	17.7	268 28.7	7.0	19 20.1	2.3	58.3
18	88 23.3	N19 17.1	282 54.7	7.0	S19 17.8	2.4	58.3
19	103 23.4	16.6	297 20.7	7.0	19 15.4	2.5	58.3
20	118 23.4	16.0	311 46.7	6.9	19 12.9	2.7	58.4
21	133 23.4	·· 15.5	326 12.6	6.9	19 10.2	2.8	58.4
22	148 23.4	14.9	340 38.5	7.0	19 07.4	2.9	58.4
23	163 23.4	14.3	355 04.5	6.9	19 04.5	3.0	58.5
27 00	178 23.4	N19 13.8	9 30.4	6.9	S19 01.5	3.1	58.5
01	193 23.4	13.2	23 56.3	6.8	18 58.4	3.3	58.5
02	208 23.4	12.6	38 22.1	6.9	18 55.1	3.4	58.5
03	223 23.4	·· 12.1	52 48.0	6.8	18 51.7	3.5	58.6
04	238 23.4	11.5	67 13.8	6.9	18 48.2	3.6	58.6
05	253 23.4	10.9	81 39.7	6.8	18 44.6	3.7	58.6
06	268 23.4	N19 10.4	96 05.5	6.8	S18 40.9	3.9	58.7
07	283 23.4	09.8	110 31.3	6.9	18 37.0	4.0	58.7
08	298 23.4	09.2	124 57.2	6.8	18 33.0	4.1	58.7
S 09	313 23.5	·· 08.7	139 23.0	6.8	18 28.9	4.3	58.7
U 10	328 23.5	08.1	153 48.8	6.8	18 24.6	4.3	58.8
N 11	343 23.5	07.5	168 14.6	6.8	18 20.3	4.5	58.8
D 12	358 23.5	N19 07.0	182 40.4	6.8	S18 15.8	4.6	58.8
A 13	13 23.5	06.4	197 06.2	6.8	18 11.2	4.7	58.8
Y 14	28 23.5	05.8	211 32.0	6.8	18 06.5	4.9	58.9
15	43 23.5	·· 05.2	225 57.8	6.8	18 01.6	4.9	58.9
16	58 23.5	04.7	240 23.6	6.8	17 56.7	5.1	58.9
17	73 23.5	04.1	254 49.4	6.9	17 51.6	5.2	58.9
18	88 23.5	N19 03.5	269 15.3	6.8	S17 46.4	5.3	59.0
19	103 23.6	02.9	283 41.1	6.8	17 41.1	5.4	59.0
20	118 23.6	02.4	298 06.9	6.8	17 35.7	5.6	59.0
21	133 23.6	·· 01.8	312 32.7	6.9	17 30.1	5.7	59.0
22	148 23.6	01.2	326 58.6	6.8	17 24.4	5.7	59.1
23	163 23.6	00.6	341 24.4	6.9	17 18.7	5.9	59.1
28 00	178 23.6	N19 00.0	355 50.3	6.9	S17 12.8	6.0	59.1
01	193 23.6	18 59.5	10 16.2	6.9	17 06.8	6.2	59.1
02	208 23.6	58.9	24 42.1	6.9	17 00.6	6.2	59.2
03	223 23.7	·· 58.3	39 08.0	6.9	16 54.4	6.3	59.2
04	238 23.7	57.7	53 33.9	6.9	16 48.1	6.5	59.2
05	253 23.7	57.2	67 59.8	7.0	16 41.6	6.6	59.2
06	268 23.7	N18 56.6	82 25.8	6.9	S16 35.0	6.7	59.3
07	283 23.7	56.0	96 51.7	7.0	16 28.3	6.8	59.3
M 08	298 23.7	55.4	111 17.7	7.0	16 21.5	6.9	59.3
O 09	313 23.7	·· 54.8	125 43.7	7.0	16 14.6	7.0	59.3
N 10	328 23.8	54.3	140 09.7	7.0	16 07.6	7.1	59.3
N 11	343 23.8	53.7	154 35.7	7.1	16 00.5	7.2	59.5
D 12	358 23.8	N18 53.1	169 01.8	7.1	S15 53.3	7.3	59.4
A 13	13 23.8	52.5	183 27.9	7.1	15 46.0	7.5	59.4
Y 14	28 23.8	51.9	197 54.0	7.1	15 38.5	7.5	59.4
15	43 23.8	·· 51.3	212 20.1	7.1	15 31.0	7.7	59.4
16	58 23.8	50.7	226 46.2	7.2	15 23.3	7.7	59.4
17	73 23.9	50.2	241 12.4	7.2	15 15.6	7.9	59.5
18	88 23.9	N18 49.6	255 38.6	7.2	S15 07.7	7.9	59.5
19	103 23.9	49.0	270 04.8	7.2	14 59.8	8.1	59.5
20	118 23.9	48.4	284 31.0	7.3	14 51.7	8.1	59.5
21	133 23.9	·· 47.8	298 57.3	7.3	14 43.6	8.3	59.5
22	148 24.0	47.2	313 23.6	7.4	14 35.3	8.3	59.5
23	163 24.0	46.6	327 49.9	7.3	14 27.0	8.4	59.6
	S.D. 15.8	d 0.6	S.D. 15.8		16.0		16.2

Lat.	Naut.	Civil	Sunrise	Moonrise 26	27	28	29
N 72	□	□	□	23 16	22 42	22 27	
N 70	□	□	□	22 13	22 14	22 12	22 09
68	////	////	01 48	21 16	21 38	21 49	21 55
66	////	////	02 26	20 42	21 13	21 31	21 43
64	////	01 05	02 52	20 18	20 53	21 17	21 34
62	////	01 54	03 13	19 58	20 36	21 04	21 25
60	////	02 24	03 29	19 42	20 23	20 54	21 18
N 58	00 59	02 46	03 43	19 29	20 11	20 45	21 12
56	01 44	03 04	03 55	19 17	20 01	20 37	21 06
54	02 12	03 19	04 05	19 07	19 52	20 29	21 01
52	02 33	03 32	04 14	18 58	19 44	20 23	20 56
50	02 49	03 43	04 22	18 50	19 37	20 17	20 52
45	03 21	04 05	04 40	18 32	19 21	20 05	20 43
N 40	03 45	04 23	04 54	18 18	19 08	19 54	20 35
35	04 03	04 37	05 06	18 06	18 58	19 45	20 29
30	04 18	04 50	05 16	17 56	18 48	19 37	20 23
20	04 42	05 10	05 34	17 37	18 32	19 24	20 13
N 10	05 00	05 26	05 49	17 22	18 17	19 12	20 04
0	05 16	05 41	06 03	17 07	18 04	19 00	19 54
S 10	05 29	05 55	06 17	16 52	17 50	18 49	19 48
20	05 42	06 08	06 32	16 36	17 36	18 37	19 39
30	05 54	06 23	06 49	16 18	17 19	18 23	19 29
35	06 01	06 31	06 58	16 08	17 10	18 15	19 23
40	06 07	06 40	07 09	15 56	16 59	18 06	19 16
45	06 15	06 50	07 22	15 42	16 46	17 55	19 08
S 50	06 23	07 02	07 38	15 24	16 30	17 42	18 59
52	06 26	07 08	07 46	15 16	16 23	17 36	18 54
54	06 30	07 14	07 54	15 07	16 14	17 30	18 50
56	06 34	07 20	08 03	14 57	16 05	17 22	18 44
58	06 39	07 28	08 14	14 45	15 55	17 14	18 38
S 60	06 44	07 37	08 26	14 31	15 43	17 04	18 31

Lat.	Sunset	Civil	Naut.	Moonset 26	27	28	29
N 72	□	////	////			01 30	04 02
N 70	23 36	////	////	24 33	00 33	02 31	04 31
68	22 20	////	////	00 16	01 30	03 06	04 53
66	21 44	////	////	00 54	02 03	03 31	05 09
64	21 18	23 00	////	01 20	02 27	03 50	05 23
62	20 58	22 15	////	01 41	02 47	04 06	05 35
60	20 42	21 46	////	01 58	03 02	04 19	05 44
N 58	20 29	21 24	23 07	02 12	03 16	04 30	05 53
56	20 17	21 07	22 25	02 24	03 27	04 40	06 00
54	20 07	20 53	21 59	02 35	03 37	04 49	06 07
52	19 58	20 40	21 38	02 44	03 46	04 56	06 13
50	19 50	20 29	21 22	02 53	03 54	05 03	06 18
45	19 33	20 07	20 50	03 10	04 11	05 18	06 30
N 40	19 19	19 49	20 27	03 25	04 25	05 30	06 39
35	19 07	19 35	20 09	03 37	04 37	05 41	06 47
30	18 57	19 23	19 54	03 48	04 47	05 50	06 55
20	18 39	19 03	19 31	04 07	05 05	06 05	07 06
N 10	18 24	18 46	19 12	04 23	05 20	06 18	07 17
0	18 10	18 32	18 57	04 38	05 34	06 31	07 27
S 10	17 56	18 18	18 44	04 53	05 48	06 43	07 37
20	17 41	18 05	18 31	05 09	06 03	06 56	07 47
30	17 25	17 50	18 19	05 27	06 21	07 12	07 59
35	17 15	17 42	18 13	05 38	06 31	07 20	08 06
40	17 04	17 33	18 06	05 50	06 42	07 30	08 13
45	16 51	17 23	17 59	06 04	06 56	07 42	08 22
S 50	16 35	17 11	17 50	06 22	07 12	07 56	08 33
52	16 28	17 05	17 47	06 30	07 20	08 02	08 38
54	16 20	17 00	17 43	06 39	07 28	08 09	08 43
56	16 10	16 53	17 39	06 49	07 38	08 17	08 49
58	16 00	16 46	17 35	07 01	07 48	08 26	08 56
S 60	15 48	16 38	17 30	07 15	08 01	08 36	09 04

Day	SUN Eqn. of Time 00h	12h	Mer. Pass.	MOON Mer. Pass. Upper	Lower	Age	Phase
26	06 27	06 27	12 06	23 20	10 52	14	
27	06 26	06 26	12 06	24 17	11 49	15	
28	06 26	06 25	12 06	00 17	12 46	16	○

Lo 73°16.7'

GHA   58°23.8'   16h   28 Jul

Diff.   14°52.9'

At the back of the *Nautical Almanac* under Increments and Correction, look up this difference under the Sun-Planets column. The closest figure is 14°52.8' at 59m 32s (Table 5-2) so your worksheet will show:

Lo 73°16.7'

GHA	58°23.8'	16h	28 Jul
Diff.	14°52.8'	59m 31s	
GHA	73°16.6'	16h 59m 31s	

Since you are working to the nearest second of time, the difference of one-tenth minute of arc is immaterial. At GMT 16-59-32 on 28 July the sun will cross your meridian. Correct this to zone time:

Table 5-2. GHA Increment for 59m 31s.

58ᵐ	SUN PLANETS	ARIES	MOON	v or Corrⁿ d	v or Corrⁿ d	v or Corrⁿ d
00	14 30·0	14 32·4	13 50·4	0·0 0·0	6·0 5·9	12·0 11·7
01	14 30·3	14 32·6	13 50·6	0·1 0·1	6·1 5·9	12·1 11·8
02	14 30·5	14 32·9	13 50·8	0·2 0·2	6·2 6·0	12·2 11·9
03	14 30·8	14 33·1	13 51·1	0·3 0·3	6·3 6·1	12·3 12·0
04	14 31·0	14 33·4	13 51·3	0·4 0·4	6·4 6·2	12·4 12·1
05	14 31·3	14 33·6	13 51·6	0·5 0·5	6·5 6·3	12·5 12·2
06	14 31·5	14 33·9	13 51·8	0·6 0·6	6·6 6·4	12·6 12·3
07	14 31·8	14 34·1	13 52·0	0·7 0·7	6·7 6·5	12·7 12·4
08	14 32·0	14 34·4	13 52·3	0·8 0·8	6·8 6·6	12·8 12·5
09	14 32·3	14 34·6	13 52·5	0·9 0·9	6·9 6·7	12·9 12·6
10	14 32·5	14 34·9	13 52·8	1·0 1·0	7·0 6·8	13·0 12·7
11	14 32·8	14 35·1	13 53·0	1·1 1·1	7·1 6·9	13·1 12·8
12	14 33·0	14 35·4	13 53·2	1·2 1·2	7·2 7·0	13·2 12·9
13	14 33·3	14 35·6	13 53·5	1·3 1·3	7·3 7·1	13·3 13·0
14	14 33·5	14 35·9	13 53·7	1·4 1·4	7·4 7·2	13·4 13·1
15	14 33·8	14 36·1	13 53·9	1·5 1·5	7·5 7·3	13·5 13·2
16	14 34·0	14 36·4	13 54·2	1·6 1·6	7·6 7·4	13·6 13·3
17	14 34·3	14 36·6	13 54·4	1·7 1·7	7·7 7·5	13·7 13·4
18	14 34·5	14 36·9	13 54·7	1·8 1·8	7·8 7·6	13·8 13·5
19	14 34·8	14 37·1	13 54·9	1·9 1·9	7·9 7·7	13·9 13·6
20	14 35·0	14 37·4	13 55·1	2·0 2·0	8·0 7·8	14·0 13·7
21	14 35·3	14 37·6	13 55·4	2·1 2·0	8·1 7·9	14·1 13·7
22	14 35·5	14 37·9	13 55·6	2·2 2·1	8·2 8·0	14·2 13·8
23	14 35·8	14 38·1	13 55·9	2·3 2·2	8·3 8·1	14·3 13·9
24	14 36·0	14 38·4	13 56·1	2·4 2·3	8·4 8·2	14·4 14·0
25	14 36·3	14 38·6	13 56·3	2·5 2·4	8·5 8·3	14·5 14·1
26	14 36·5	14 38·9	13 56·6	2·6 2·5	8·6 8·4	14·6 14·2
27	14 36·8	14 39·2	13 56·8	2·7 2·6	8·7 8·5	14·7 14·3
28	14 37·0	14 39·4	13 57·0	2·8 2·7	8·8 8·6	14·8 14·4
29	14 37·3	14 39·7	13 57·3	2·9 2·8	8·9 8·7	14·9 14·5
30	14 37·5	14 39·9	13 57·5	3·0 2·9	9·0 8·8	15·0 14·6
31	14 37·8	14 40·2	13 57·8	3·1 3·0	9·1 8·9	15·1 14·7
32	14 38·0	14 40·4	13 58·0	3·2 3·1	9·2 9·0	15·2 14·8
33	14 38·3	14 40·7	13 58·2	3·3 3·2	9·3 9·1	15·3 14·9
34	14 38·5	14 40·9	13 58·5	3·4 3·3	9·4 9·2	15·4 15·0
35	14 38·8	14 41·2	13 58·7	3·5 3·4	9·5 9·3	15·5 15·1
36	14 39·0	14 41·4	13 59·0	3·6 3·5	9·6 9·4	15·6 15·2
37	14 39·3	14 41·7	13 59·2	3·7 3·6	9·7 9·5	15·7 15·3
38	14 39·5	14 41·9	13 59·4	3·8 3·7	9·8 9·6	15·8 15·4
39	14 39·8	14 42·2	13 59·7	3·9 3·8	9·9 9·7	15·9 15·5
40	14 40·0	14 42·4	13 59·9	4·0 3·9	10·0 9·8	16·0 15·6
41	14 40·3	14 42·7	14 00·1	4·1 4·0	10·1 9·8	16·1 15·7
42	14 40·5	14 42·9	14 00·4	4·2 4·1	10·2 9·9	16·2 15·8
43	14 40·8	14 43·2	14 00·6	4·3 4·2	10·3 10·0	16·3 15·9
44	14 41·0	14 43·4	14 00·9	4·4 4·3	10·4 10·1	16·4 16·0
45	14 41·3	14 43·7	14 01·1	4·5 4·4	10·5 10·2	16·5 16·1
46	14 41·5	14 43·9	14 01·3	4·6 4·5	10·6 10·3	16·6 16·2
47	14 41·8	14 44·2	14 01·6	4·7 4·6	10·7 10·4	16·7 16·3
48	14 42·0	14 44·4	14 01·8	4·8 4·7	10·8 10·5	16·8 16·4
49	14 42·3	14 44·7	14 02·1	4·9 4·8	10·9 10·6	16·9 16·5
50	14 42·5	14 44·9	14 02·3	5·0 4·9	11·0 10·7	17·0 16·6
51	14 42·8	14 45·2	14 02·5	5·1 5·0	11·1 10·8	17·1 16·7
52	14 43·0	14 45·4	14 02·8	5·2 5·1	11·2 10·9	17·2 16·8
53	14 43·3	14 45·7	14 03·0	5·3 5·2	11·3 11·0	17·3 16·9
54	14 43·5	14 45·9	14 03·3	5·4 5·3	11·4 11·1	17·4 17·0
55	14 43·8	14 46·2	14 03·5	5·5 5·4	11·5 11·2	17·5 17·1
56	14 44·0	14 46·4	14 03·7	5·6 5·5	11·6 11·3	17·6 17·2
57	14 44·3	14 46·7	14 04·0	5·7 5·6	11·7 11·4	17·7 17·3
58	14 44·5	14 46·9	14 04·2	5·8 5·7	11·8 11·5	17·8 17·4
59	14 44·8	14 47·2	14 04·4	5·9 5·8	11·9 11·6	17·9 17·5
60	14 45·0	14 47·4	14 04·7	6·0 5·9	12·0 11·7	18·0 17·6

59ᵐ	SUN PLANETS	ARIES	MOON	v or Corrⁿ d	v or Corrⁿ d	v or Corrⁿ d
00	14 45·0	14 47·4	14 04·7	0·0 0·0	6·0 6·0	12·0 11·9
01	14 45·3	14 47·7	14 04·9	0·1 0·1	6·1 6·0	12·1 12·0
02	14 45·5	14 47·9	14 05·2	0·2 0·2	6·2 6·1	12·2 12·1
03	14 45·8	14 48·2	14 05·4	0·3 0·3	6·3 6·2	12·3 12·2
04	14 46·0	14 48·4	14 05·6	0·4 0·4	6·4 6·3	12·4 12·3
05	14 46·3	14 48·7	14 05·9	0·5 0·5	6·5 6·4	12·5 12·4
06	14 46·5	14 48·9	14 06·1	0·6 0·6	6·6 6·5	12·6 12·5
07	14 46·8	14 49·2	14 06·4	0·7 0·7	6·7 6·6	12·7 12·6
08	14 47·0	14 49·4	14 06·6	0·8 0·8	6·8 6·7	12·8 12·7
09	14 47·3	14 49·7	14 06·8	0·9 0·9	6·9 6·8	12·9 12·8
10	14 47·5	14 49·9	14 07·1	1·0 1·0	7·0 6·9	13·0 12·9
11	14 47·8	14 50·2	14 07·3	1·1 1·1	7·1 7·0	13·1 13·0
12	14 48·0	14 50·4	14 07·5	1·2 1·2	7·2 7·1	13·2 13·1
13	14 48·3	14 50·7	14 07·8	1·3 1·3	7·3 7·2	13·3 13·2
14	14 48·5	14 50·9	14 08·0	1·4 1·4	7·4 7·3	13·4 13·3
15	14 48·8	14 51·2	14 08·3	1·5 1·5	7·5 7·4	13·5 13·4
16	14 49·0	14 51·4	14 08·5	1·6 1·6	7·6 7·5	13·6 13·5
17	14 49·3	14 51·7	14 08·7	1·7 1·7	7·7 7·6	13·7 13·6
18	14 49·5	14 51·9	14 09·0	1·8 1·8	7·8 7·7	13·8 13·7
19	14 49·8	14 52·2	14 09·2	1·9 1·9	7·9 7·8	13·9 13·8
20	14 50·0	14 52·4	14 09·5	2·0 2·0	8·0 7·9	14·0 13·9
21	14 50·3	14 52·7	14 09·7	2·1 2·1	8·1 8·0	14·1 14·0
22	14 50·5	14 52·9	14 09·9	2·2 2·2	8·2 8·1	14·2 14·1
23	14 50·8	14 53·2	14 10·2	2·3 2·3	8·3 8·2	14·3 14·2
24	14 51·0	14 53·4	14 10·4	2·4 2·4	8·4 8·3	14·4 14·3
25	14 51·3	14 53·7	14 10·6	2·5 2·5	8·5 8·4	14·5 14·4
26	14 51·5	14 53·9	14 10·9	2·6 2·6	8·6 8·5	14·6 14·5
27	14 51·8	14 54·2	14 11·1	2·7 2·7	8·7 8·6	14·7 14·6
28	14 52·0	14 54·4	14 11·4	2·8 2·8	8·8 8·7	14·8 14·7
29	14 52·3	14 54·7	14 11·6	2·9 2·9	8·9 8·8	14·9 14·8
30	14 52·5	14 54·9	14 11·8	3·0 3·0	9·0 8·9	15·0 14·9
31	14 52·8	14 55·2	14 12·1	3·1 3·1	9·1 9·0	15·1 15·0
32	14 53·0	14 55·4	14 12·3	3·2 3·2	9·2 9·1	15·2 15·1
33	14 53·3	14 55·7	14 12·6	3·3 3·3	9·3 9·2	15·3 15·2
34	14 53·5	14 55·9	14 12·8	3·4 3·4	9·4 9·3	15·4 15·3
35	14 53·8	14 56·2	14 13·0	3·5 3·5	9·5 9·4	15·5 15·4
36	14 54·0	14 56·4	14 13·3	3·6 3·6	9·6 9·5	15·6 15·5
37	14 54·3	14 56·7	14 13·5	3·7 3·7	9·7 9·6	15·7 15·6
38	14 54·5	14 56·9	14 13·8	3·8 3·8	9·8 9·7	15·8 15·7
39	14 54·8	14 57·2	14 14·0	3·9 3·9	9·9 9·8	15·9 15·8
40	14 55·0	14 57·5	14 14·2	4·0 4·0	10·0 9·9	16·0 15·9
41	14 55·3	14 57·7	14 14·5	4·1 4·1	10·1 10·0	16·1 16·0
42	14 55·5	14 58·0	14 14·7	4·2 4·2	10·2 10·1	16·2 16·1
43	14 55·8	14 58·2	14 14·9	4·3 4·3	10·3 10·2	16·3 16·2
44	14 56·0	14 58·5	14 15·2	4·4 4·4	10·4 10·3	16·4 16·3
45	14 56·3	14 58·7	14 15·4	4·5 4·5	10·5 10·4	16·5 16·4
46	14 56·5	14 59·0	14 15·7	4·6 4·6	10·6 10·5	16·6 16·5
47	14 56·8	14 59·2	14 15·9	4·7 4·7	10·7 10·6	16·7 16·6
48	14 57·0	14 59·5	14 16·1	4·8 4·8	10·8 10·7	16·8 16·7
49	14 57·3	14 59·7	14 16·4	4·9 4·9	10·9 10·8	16·9 16·8
50	14 57·5	15 00·0	14 16·6	5·0 5·0	11·0 10·9	17·0 16·9
51	14 57·8	15 00·2	14 16·9	5·1 5·1	11·1 11·0	17·1 17·0
52	14 58·0	15 00·5	14 17·1	5·2 5·2	11·2 11·1	17·2 17·1
53	14 58·3	15 00·7	14 17·3	5·3 5·3	11·3 11·2	17·3 17·2
54	14 58·5	15 01·0	14 17·6	5·4 5·4	11·4 11·3	17·4 17·3
55	14 58·8	15 01·2	14 17·8	5·5 5·5	11·5 11·4	17·5 17·4
56	14 59·0	15 01·5	14 18·0	5·6 5·6	11·6 11·5	17·6 17·5
57	14 59·3	15 01·7	14 18·3	5·7 5·7	11·7 11·6	17·7 17·6
58	14 59·5	15 02·0	14 18·5	5·8 5·8	11·8 11·7	17·8 17·7
59	14 59·8	15 02·2	14 18·8	5·9 5·9	11·9 11·8	17·9 17·8
60	15 00·0	15 02·5	14 19·0	6·0 6·0	12·0 11·9	18·0 17·9

$$\text{(Rev)} \quad \begin{array}{ll} \text{GMT} & 16\text{-}59\text{-}31 \\ \text{ZD} & +\ 4 \\ \hline \text{ZT} & 12\text{-}59\text{-}31 \end{array}$$

Add or subtract watch error as necessary to get the exact watch time of the sight.

With an assistant to call the time (the reverse of the usual procedure), sight on the sun about a minute prior to meridian passage. Keep adjusting the micrometer drum to hold the sun on the horizon as altitude increases. As the time of meridian passage approaches, your assistant can start counting down the time, first at 5-second intervals, and the last 10 seconds at 1-second intervals. At "zero," take the sextant reading, and then immediately take one or two more sights to make sure that altitude starts to decrease after the apparent momentary pause at the peak. If you can do this on two or three occasions, you can get a good feel for the meridian passage.

RUN OF SIGHTS GRAPH

Now try taking a run of sights. Start about five minutes before the calculated time of meridian passage. Record each sight as it is taken. Have your assistant record altitudes and times in the normal manner. Take the sights in quick succession, but do not sacrifice speed for accuracy. Do this until it is about five minutes past the time of meridian passage. You should end up with about 8 to 12 sights, spaced about a minute apart, with the sextant altitudes gradually increasing, then decreasing.

Next plot these sights, on graph paper, against the time for each sight. Connect the plotted sights with a fair curve. Allow for sights that obviously do not fall on the curve and represent erroneous readings. The high point of the curve is meridian passage and its time can be taken from the graph (see Fig. 5-5). This should match your precalculated time within a second or two, because you are working from a known location. At sea, the method might not be quite so accurate, but it still should come within about five seconds of actual meridian passage. For practical purposes that is close enough.

Suppose you are at sea on 13 May 1980. You know you're near longitude 65° W, which puts you in ZD +4 with a zone meridian of 60° W. The *Nautical Almanac* shows that on 13 May meridian passage will be at ZT 1156 at the zone meridian. Based on your last known position and your course and speed through the water, you determine that your DR position at ZT 1156 will be L 52°38.2' N, Lo 65°27.4' W. Your eye height is 9.5 feet above sea level and sextant error is −2.3'. Your watch is 1m 08s slow.

Since the difference between your DR Lo and the zone meridian is about five full degrees, your estimate time of meridian passage at the DR position will be 20 minutes (5 degrees × 4 minutes) later than at the zone meridian—since you are west of it. This would put it at ZT 1216.

About 10 minutes before this—ZT 1216—you start taking sights. It is not until just after WT 1220 that you note that altitude is decreasing. Your next few sights bear this out. In plotting the sights on the graph, you use

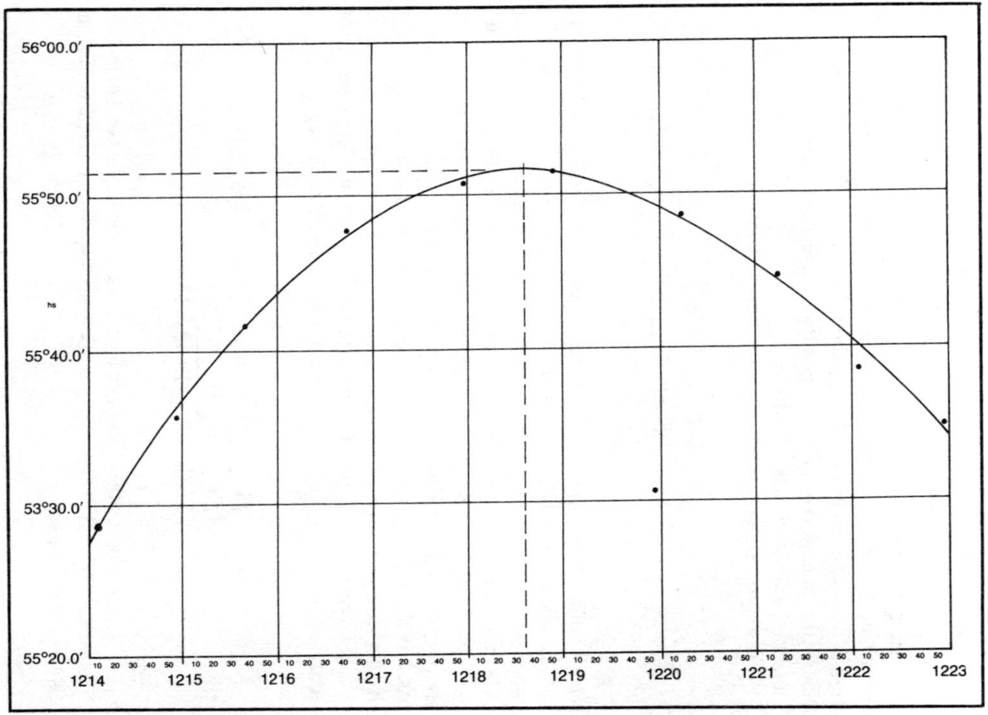

Fig. 5-5. Run of sights plotted on a graph to determine sextant altitude and time of meridian passage.

enough sights before and after the peak to establish your curve (Fig. 5-5). Here are the plotted sights for this example:

Time	Sextant Altitude
12-14-06	55°28.6'
12-14-55	55°35.7'
12-15-39	55°41.7'
12-16-43	55°47.6'
12-17-52	55°50.8'
12-18-55	55°51.3'
12-20-07	55°48.9'
12-21-15	55°45.2'
12-22-03	55°38.0'
12-22-56	55°34.8'

After drawing a curve that reflects the position of these points, you determine that the peak of the curve—meridian passage—is at WT 12-18-34. The curve shows graphically the relatively flat line in this section, where the sun, when under observation, appears to "hang" for a few moments without change of altitude.

Watch time corrected to zone time gives you:

WT	12-18-34
WE(s)	+ 1-08
ZT	12-19-42

This is almost four minutes later than your estimated time of meridian passage, based on your DR position, but when you are able to establish a good curve, as in this example, the result is very accurate. It is possible to determine the exact instant of meridian passage for the DR longitude, just as for the practice sights from a known location, but the method outlined above is a practical shortcut. Even when exact time is known for meridian passage at the DR position, the chances are your boat won't actually be at that location at the calculated time.

By converting ZT to GMT, and with a little help from the *Almanac,* you establish your exact longitude at the time of meridian passage:

ZT	12-19-42	13 May 1980
ZD	+ 4	
GMT	16-19-42	13 May

The *Almanac,* on the daily page for this date, shows the GHA of the sun at 1600 to be 60°55.8'. The buff pages at the back show an increment for 19m 42s to be 4°55.5', so:

GHA	16h	60°55.8′
	19m 42s	4°55.5′
GHA		65°51.3′

Since your boat is on the meridian that matches the GHA of the sun at the instant of meridian passage, your exact longitude at that time is 65°51.3′ W.

Next, determine your latitude at this point. From the peak of the curve on your graph pick off the altitude of the sun at its meridian passage. Depending on the accuracy of the curve, this may be off a few tenths of minute of arc. This will make a difference of the equivalent tenths of mile of distance in the calculated latitude. That's close enough for all practical purposes. Work the latitude out as shown earlier:

Hs	53°51.7′
IC	+ 2.3′
Dip	− 3.2′
Ha	55°50.8′
Main	+ 15.3′
Ho	56°06.1′

	90°00.0′	
−	56°06.1′	Ho
	33°53.9′	Zenith Distance

For 13 May, the *Almanac* shows declination of the sun to be N 18°31.8′ at GMT 1600, with a d correction of +0.6′. The correction for this, on the same page as the increment in GHA for 19m 42s, is +0.2°, so declination of the sun, at the instant of meridian passage, is N 18°32.0′. Since you are in the Northern Hemisphere and you are north of the GP of the sun, add declination to zenith distance to get latitude:

ZD	33°53.9′
Dec.	+N 18°32.0′
Lat.	N 52°25.9′

This is your exact latitude at ZT 12-19-42 (see Fig. 5-6). In plotting the position on your chart, round the time to the nearest whole minute—1220. Your chart, and your boat's log, will show a noon fix on 13 May of L 52°25.9′ N; Lo 65°51.3′ W. Compare this to your DR position for 1216 of L 52°38.2′ N; Lo 65°27.4 W. Not bad. Since your last fix, you've been set to the south and west of your DR course. The new fix lets you adjust course, if necessary, to maintain the desired track to your destination.

After first establishing the approximate time of meridian passage, some navigators prefer a different sextant method in determining the exact

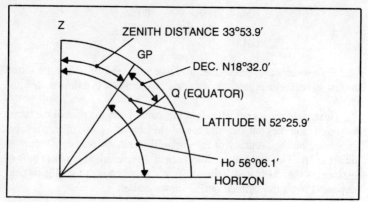

Fig. 5-6. A much simplified diagram based on the plane of the celestial horizon.

instant of meridian passage. About 5 or 10 minutes before this point, a sight is taken and recorded with the lower limb of the sun just touching the horizon. Then they continue to observe the sun without changing the sextant reading. As the sun continues to gain altitude, it appears to sink below the horizon as seen in the horizon mirror. Then, as altitude starts to decrease, the sun appears to rise in the mirror. When the sun's lower limb again touches the horizon, the time is again recorded.

Time of meridian passage is the midpoint between the first and second times taken for the sight. Unless a run of sights is taken before and after these points, so a curve of altitude can be drawn on a graph, the method does not provide an accurate indication of altitude at the time of meridian passage. A lot depends on the navigator's ability to sense the flattening of the curve of altitude before "locking" the sextant on the desired sight.

Chapter 6

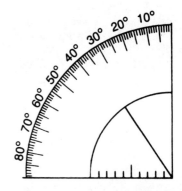

Sight Reduction

With the availability of chronometers, quartz timepieces (including some highly accurate wristwatches, and radio time signals, it is possible to establish the exact geographical position of a celestial body at any time with the aid of the *Nautical* or *Air Almanac*. Your sextant reading gives you your distance, in arc, from this GP. By "reducing" the sight, you determine the azimuth (bearing) to the GP and the difference in nautical miles between your actual position and a nearby "known location" (your DR position or an assumed position (AP) used in working the sight).

SIGHT REDUCTION TABLES

Use your *Almanac* to determine the observed altitude and the GP of the body, as explained in Chapter 3. Then, using one of the sight reduction methods to be discussed, determine the *calculated altitude* (Hc). This is the distance, in arc, to the GP of the body from an assumed position or your DR position at the time of the sight. A knowledge of spherical trigonometry was once essential for this process, but now all you need are the sight reduction tables available from the government or private publication such as *Reed's Nautical Almanac*.

Sight reduction tables in various formats have been published by the government. The one discussed here is the latest version: *Sight Reduction Tables for Marine Navigation*. It is *Pub. 229* from the Defense Mapping Agency Hydrographic/Topographic Center (DMAHTC), and is available from major suppliers of government charts and publications throughout the country. It is actually six separate volumes. Each covers 15 degrees of latitude. Most cruising skippers need just one or two of the books (three at most). Examples given here are based on Vol. 3, 30° to 45° inclusive.

Because *Pub. 229* (as the book is commonly called) tables are based on whole degrees of latitude and whole degrees of local hour angle (LHA), you need to establish an assumed position that will provide these figures. The AP is located at the whole degree of latitude nearest your DR position for the time of the sight, and at the nearest longitude that will provide whole degrees of LHA when it is added to (if east of Greenwich) or subtracted from (if west of Greenwich) the GHA of the observed body.

ESTABLISHING LOCAL HOUR ANGLE

The first step in finding AP longitude is to establish the GHA of the body at the time of the sight, as described in Chapter 3 (*Nautical Almanac*). If you are in west longitude, set your AP at a position equal, in minutes and tenths of minutes of arc, to the GHA of the body and within 30' of longitudinal arc from your DR position. For example, you're at DR Lo 47°18.3 W, and you establish the GHA of the sun as 84°06.6'. Set your AP at Lo 47°06.6' W; this gives you:

GHA	84°06.6'
AP Lo	47°06.6'
LHA	37°00.0'

This is diagrammed in Fig. 6-1. For the sight reduction tables, you will use LHA 37°.

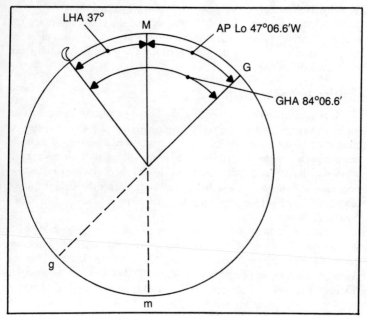

Fig. 6-1. Time diagram shows how AP longitude is set so that whole degrees of LHA result when AP Lo is subtracted from GHA of the body.

Always be sure to use the AP longitude that is closest to your DR position. Generally it will result in the shortest possible *intercept*. The intercept is difference in miles between the observed altitude and the calculated altitude based on the sight reduction tables.

In the above example, if you had a DR longitude of 47°48.3′ W, the difference between this and an AP Lo of 47°06.6′ is 41.7′ of arc. By making your AP Lo 48°06.6′ W, the difference is only 18.3′ of arc. This changes LHA, of course:

$$
\begin{array}{ll}
\text{GHA} & 84°06.6' \\
\text{AP Lo} & -\,\underline{48°06.6'\text{W}} \\
\text{LHA} & 36°00.0'
\end{array}
$$

In some cases the figure for GHA will be less than that for the AP Lo. Simply add 360 degrees to the GHA figure before subtracting your AP longitude. For example, GHA of the sun is 62°38.4′ and your AP Lo is 68°38.4′ W.

$$
\begin{array}{ll}
\text{GHA} & 62°38.4' \\
& +\,\underline{360°} \\
& 422°38.4' \\
\text{AP Lo} & -\,68°38.4'\,\text{W} \\
\text{LHA} & \overline{354°00.0'}
\end{array}
$$

This is shown in Fig. 6-2. You should recognize these drawings as the time diagrams discussed in Chapter 4. In practice, they are not drawn on the basis of your AP longitude, but that of your DR position. Because you can't accurately pinpoint minutes and fractions of minutes on such a diagram in any case, the plotted angles of GHA, longitude, LHA—and SHA if required—are close approximations at best. They provide all the proper angle relationships and will quickly indicate if any of your calculations are wrong.

If you are east of Greenwich, you can't set AP longitude at the same minutes and tenths of minutes of arc as the GHA of the body. AP Lo must be added to GHA to get your local hour angle. Suppose you're at DR Lo 56°37.2′ E and the GHA of the sun is 265°16.7′. If you set your AP Lo at 56°16.7′ E, you get:

$$
\begin{array}{ll}
\text{GHA} & 265°16.7' \\
\text{AP Lo} & +\,\underline{56°16.7'\,\text{E}} \\
\text{LHA} & 321°33.4'
\end{array}
$$

Since this isn't whole degrees of LHA, you can't use it for the *Pub. 229* tables. The trick is to first subtract the minutes and tenths of minutes of GHA from 60′ of arc:

$$
\begin{array}{r}
60.0' \\
-\,\underline{16.7'} \\
43.3'
\end{array}
$$

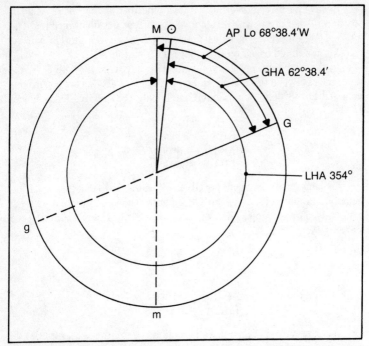

Fig. 6-2. LHA is always measured west from the meridian of the observer, even though the body may be just a few degrees east of this meridian.

This gives you the minutes and tenths for your AP Lo. In this case, it would be 56°43.3′ E, and:

$$
\begin{array}{ll}
\text{GHA} & 265°16.7′ \\
\text{AP Lo} \; + & \underline{\;56°43.3′} \\
\text{LHA} & 322°00.0′
\end{array}
$$

Enter the tables with the LHA 322° figure (see Fig. 6-3). In some cases LHA will exceed 360 degrees; simply subtract this figure from the result and you have LHA for the tables.

Here are some LHA problems you can work out for yourself. Draw a time diagram for each. The answers are in the Appendix.

26. DR Lo is 84°22.8′ W; GHA of Aldebaran is 13°17.6′. Find the LHA of the star.

27. DR Lo is 28°47.3′ W; GHA of the moon is 331°12.1′. Find the LHA of the moon.

28. DR Lo 42°08.7 E; GHA sun is 19°28.4′. Find LHA.

29. DR Lo 136°53.2′ E; GHA sun 142°17.7′. Find LHA.

30. DR Lo 151°16.8′ W; GHA Spica 128°29.4′. Find LHA.

PUB. 229 MAIN PAGES

Each volume of *Pub. 229* is divided into two main sections, each covering eight degrees of latitude. In Vol. 3, the first section runs from 30° to 37°, and the second section from 38° to 45°. Each section covers LHA from 0° to 360°. Left hand pages range from LHA 0° to 90°, with each page also used for its reciprocal figure, ranging from LHA 360° to 270°. The page for LHA 16°, for example, is also the page for LHA 344° (see Table 6-1). Note that the left hand pages are used only when AP latitude is the same as the declination of the observed body (both are north or both are south).

The top part of each right-hand page (Table 6-2), above the separation lines, carries tables for the same LHA figures as the facing left-hand page. These tables are used when the AP latitude is contrary to the declination of the sighted body (one is north and the other is south). In sight reduction, you will be using either a left-hand page or the top section of a right-hand page. This means the LHA you establish must be in the 0° to 90° range, or the 270° to 360° range. If your LHA falls between 90° and 270°, you've miscalculated. Such an LHA would indicate the body was below your horizon.

The bottom section of the right-hand pages carry LHA figures ranging from 180° to 270° and 180° to 90°. These tables can be used to determine distances and courses to destinations on the other side of the world from you.

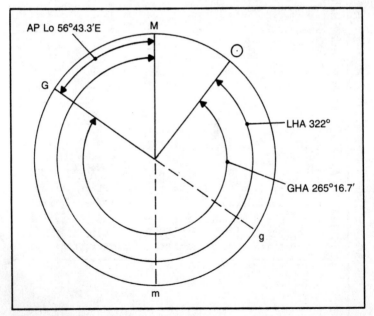

Fig. 6-3. Note that in eastern longitudes, AP longitude must be added to GHA of the body to establish whole degrees of LHA.

Table 6-1. Typical Left-Hand Page from Pub. 229; This Is Used When Latitude and Declination Are Both North or Both South.

16°, 344° L.H.A. LATITUDE SAME NAME AS DECLINATION

N. Lat. { L.H.A. greater than 180° .... Zn=Z ; L.H.A. less than 180° .... Zn=360°−Z }

Dec.	30° Hc	30° d	30° Z	31° Hc	31° d	31° Z	32° Hc	32° d	32° Z	33° Hc	33° d	33° Z	34° Hc	34° d	34° Z	35° Hc	35° d	35° Z	36° Hc	36° d	36° Z	37° Hc	37° d	37° Z	Dec.
0	56 21.2	+54.0	150.9	55 29.0	+54.4	150.9	54 36.4	+54.8	151.9	53 43.5	+55.1	152.2	52 50.2	+55.5	152.9	51 56.7	+55.7	153.4	51 02.9	+56.0	154.0	50 08.9	+56.2	154.5	0
1	57 15.2	53.7	149.9	56 23.4	54.1	150.0	55 31.2	54.5	150.9	54 38.6	54.9	151.5	53 45.7	55.2	152.2	52 52.4	55.6	152.8	51 58.9	55.9	153.4	51 05.1	56.1	154.0	1
2	58 08.9	53.3	148.5	57 17.5	53.8	149.4	56 25.7	54.2	150.1	55 33.5	54.4	150.9	54 40.9	55.0	151.5	53 48.0	55.1	152.1	52 54.8	55.6	152.8	52 01.2	56.0	153.4	2
3	59 02.2	53.0	148.5	58 11.3	53.1	148.1	57 19.9	53.6	149.3	56 28.1	54.1	150.1	55 35.9	54.5	150.8	54 43.3	55.1	151.5	53 50.4	55.5	152.2	52 57.2	55.7	152.8	3
4	59 55.2	52.6	146.7	59 04.8	53.1	147.6	58 13.8	53.6	148.5	57 22.5	54.0	149.3	56 30.6	54.5	150.1	55 38.4	54.9	150.8	54 45.9	55.2	151.5	53 52.9	55.6	152.2	4
5	60 47.8	+52.1	145.8	57 57.9	+52.7	146.7	59 07.4	+53.3	147.7	58 16.5	+53.7	148.5	57 25.1	+54.2	149.3	56 33.3	+54.6	150.1	55 41.1	+55.0	150.9	54 48.5	+55.3	151.5	5
6	61 39.9	51.6	144.7	60 50.6	52.2	145.8	60 00.7	52.8	146.8	59 10.2	53.4	147.7	58 19.3	53.8	148.5	57 27.9	54.3	149.4	56 36.1	54.7	150.1	55 43.8	55.1	150.9	6
7	62 31.5	50.9	143.5	61 42.8	51.5	144.5	60 53.5	52.5	145.8	60 03.6	52.8	146.8	59 13.1	53.4	147.7	58 22.2	53.9	148.6	57 30.8	54.4	149.4	56 38.9	54.7	150.1	7
8	63 22.5	50.6	142.5	62 34.5	51.1	143.7	61 45.9	51.8	144.8	60 56.6	52.5	145.8	60 06.5	52.9	146.8	59 16.1	53.5	147.8	58 25.2	53.9	148.6	57 33.6	54.5	149.4	8
9	64 13.0	49.8	141.3	63 25.8	50.6	142.5	62 37.8	51.4	143.7	61 49.1	52.1	144.8	60 59.8	52.7	145.8	60 09.8	53.3	146.8	59 19.3	53.8	147.8	58 28.4	54.2	148.6	9
10	65 02.8	+49.1	140.0	64 16.4	+49.3	141.3	63 29.2	+50.8	142.5	62 41.2	+51.6	143.7	61 52.5	+52.1	144.8	61 03.1	+52.9	145.9	60 13.1	+53.5	146.9	59 22.6	+54.0	147.8	10
11	65 51.9	48.3	138.6	65 06.4	49.3	140.0	64 20.0	50.2	141.3	63 32.8	51.0	142.6	62 44.7	51.8	143.8	61 56.0	52.4	144.9	61 06.6	53.0	145.9	60 16.6	53.5	146.9	11
12	66 40.2	47.4	137.1	65 55.7	48.5	138.6	65 10.2	49.5	140.1	64 23.8	50.3	141.4	63 36.5	51.1	142.7	62 48.4	51.9	143.8	61 59.6	52.6	145.0	61 10.1	53.2	146.0	12
13	67 27.6	46.5	135.5	66 44.2	47.7	137.2	65 59.7	48.7	138.7	65 14.1	49.7	140.1	64 27.6	50.6	141.5	63 40.3	51.3	142.7	62 52.2	52.0	143.9	62 03.3	52.8	145.0	13
14	68 14.1	45.3	133.8	67 31.9	46.6	135.6	66 48.4	47.8	137.2	66 03.8	48.9	138.8	65 18.2	49.9	140.2	64 31.6	50.8	141.5	63 44.2	51.6	142.8	62 56.1	52.2	144.0	14
15	68 59.4	+44.2	132.0	68 18.5	+45.6	133.9	67 36.2	+46.9	135.7	66 52.7	+48.1	137.3	66 08.1	+49.1	138.8	65 22.4	+50.1	140.3	64 35.8	+50.9	141.6	63 48.3	+51.8	142.9	15
16	69 43.6	42.8	130.2	69 04.1	44.6	132.0	68 23.1	45.8	133.9	67 40.8	47.1	135.8	66 57.2	48.3	137.4	66 12.5	49.3	138.9	65 26.7	50.3	140.4	64 40.1	51.1	141.7	16
17	70 26.4	41.8	128.1	69 48.5	43.4	130.2	69 08.9	44.5	132.0	68 27.9	45.8	133.8	67 45.5	47.0	135.8	67 01.8	48.5	137.5	66 17.0	49.6	139.1	65 31.2	50.5	140.5	17
18	71 07.7	39.6	125.9	70 31.5	41.6	128.2	69 53.6	43.3	130.3	69 13.7	44.8	132.2	68 32.5	46.0	134.0	67 50.3	47.3	135.9	67 06.6	48.5	137.6	66 21.7	49.5	139.2	18
19	71 47.3	37.7	123.5	71 13.1	40.6	126.0	70 36.9	41.8	128.3	69 58.5	43.6	130.4	69 18.5	45.2	132.4	68 37.9	46.6	134.3	67 55.3	47.8	136.1	67 11.5	48.9	137.8	19
20	72 25.0	+35.7	121.0	71 53.0	+38.0	123.6	71 18.7	+40.2	126.1	70 42.4	+42.1	128.4	70 04.3	+43.8	130.5	69 24.5	+45.1	132.6	68 43.1	+46.8	134.5	68 00.4	+48.1	136.2	20
21	73 00.7	33.5	118.3	72 31.0	36.0	121.1	71 58.9	38.3	123.7	71 24.5	40.5	126.2	70 48.1	42.3	128.5	70 09.9	44.1	130.7	69 30.0	45.6	132.7	68 48.5	47.1	134.6	21
22	73 34.2	30.8	115.4	73 07.0	33.7	118.4	72 37.2	36.2	121.2	72 05.0	38.6	123.8	71 30.5	40.8	126.3	70 54.0	42.7	128.6	70 15.6	44.3	130.8	69 35.6	45.9	132.9	22
23	74 05.0	28.1	112.3	73 40.7	31.4	115.5	73 13.5	34.0	118.5	72 43.6	36.6	121.3	72 11.3	38.7	123.9	71 36.7	41.0	126.5	71 00.0	43.0	128.8	70 21.5	44.6	131.0	23
24	74 33.1	25.1	109.0	74 12.1	28.4	112.4	73 47.5	31.5	115.6	73 20.2	34.3	118.6	72 50.2	36.8	121.4	72 17.7	39.3	124.1	71 43.0	41.4	126.6	71 06.2	43.3	129.0	24
25	74 58.2	+21.7	105.4	74 40.5	+24.2	109.1	74 19.2	+26.7	112.5	73 54.5	+28.7	115.7	73 27.1	+31.0	118.7	72 57.0	+32.7	121.6	72 24.4	+34.5	124.3	71 49.5	+36.2	126.8	25
26	75 19.9	18.3	101.6	75 04.7	20.7	105.5	74 45.9	23.3	109.2	74 23.2	25.8	112.6	73 58.1	28.0	115.9	73 29.7	30.3	118.8	72 58.9	32.1	121.7	72 25.7	33.9	124.4	26
27	75 38.1	14.5	98.1	75 25.4	17.4	102.0	75 09.2	20.3	105.7	74 49.0	22.5	109.4	74 26.1	25.0	112.8	74 00.0	27.3	116.0	73 31.0	29.4	119.0	72 59.6	31.3	121.8	27
28	75 52.6	10.4	94.2	75 42.8	14.7	98.1	75 29.5	16.7	102.0	75 11.5	19.2	105.8	74 51.1	22.0	109.5	74 27.3	24.3	112.9	74 00.4	26.4	116.2	73 30.9	28.6	119.1	28
29	76 03.0	6.3	90.1	75 55.9	10.4	94.1	75 46.2	13.4	98.1	75 30.7	15.7	102.1	75 13.1	18.7	106.0	74 51.6	21.3	109.6	74 26.8	23.5	113.1	73 59.5	25.8	116.3	29
30	76 09.3	+2.1	86.0	76 11.4	+6.4	89.9	76 02.8	+15.1	94.1	75 46.4	+12.0	98.1	75 31.8	+15.3	102.1	75 12.9	+18.1	106.0	74 44.0	+20.9	109.6	74 24.8	+23.3	113.1	30
31	76 11.4	-2.2	81.8	76 17.8	+2.2	85.9	76 17.9	+11.0	90.0	75 58.4	+8.0	94.1	75 47.1	+11.2	98.1	75 31.0	+14.9	102.1	74 58.5	+17.8	106.1	74 39.0	+20.6	109.8	31
32	76 09.2	-6.4	77.6	76 20.0	-2.2	81.8	76 28.9	+6.6	85.7	76 06.4	+3.8	89.9	75 58.3	+7.5	94.0	75 45.9	+11.2	98.2	75 16.3	+14.8	102.3	74 59.6	+17.6	106.3	32
33	76 02.8	-10.5	73.5	76 17.9	-6.4	77.6	76 35.5	+2.1	81.5	76 10.2	-0.6	85.7	76 05.8	+2.9	89.8	75 57.1	+7.5	94.0	75 31.1	+11.2	98.2	75 17.2	+14.6	102.4	33
34	75 52.3	-14.5	69.4	76 11.7	-10.5	73.5	76 37.6	-2.1	77.2	76 09.6	-4.7	81.3	76 08.7	-1.4	85.5	76 04.9	+2.9	89.9	75 42.3	+7.8	94.0	75 31.8	+11.4	98.2	34
35	75 37.8	-18.3	65.5	76 01.5	-14.7	69.1	76 35.5	-6.3	72.9	76 04.9	-9.0	77.0	76 07.3	-5.7	81.1	76 07.8	-1.9	85.4	75 50.1	+3.2	89.7	75 43.2	+6.9	94.0	35
36	75 19.5	-22.1	61.7	75 46.7	-18.0	65.3	76 28.9	-10.6	68.7	75 55.9	-13.5	72.8	76 01.6	-10.0	76.8	76 05.9	-6.4	81.0	75 53.3	-1.9	85.3	75 50.1	+2.7	89.6	36
37	74 57.7	-25.5	58.1	75 28.0	-22.0	61.3	76 18.3	-14.5	64.8	75 42.4	-17.3	68.5	75 51.6	-13.9	72.5	76 00.5	-10.4	76.3	75 51.4	-6.5	80.6	75 52.8	-1.8	85.1	37
38	74 32.6	-28.2	54.6	75 05.2	-25.5	57.6	76 03.8	-18.7	61.0	75 25.1	-21.0	64.4	75 37.7	-17.8	68.2	75 50.1	-14.0	72.1	75 44.9	-10.5	76.3	75 51.0	-6.5	80.6	38
39	74 04.4	-30.9	51.3	74 39.0	-28.1	54.0	75 45.1	-22.0	57.3	75 04.1	-24.6	60.5	75 19.2	-21.1	64.1	75 36.1	-18.1	67.9	75 34.4	-14.1	71.9	75 44.5	-10.8	76.1	39
40	73 33.5	-33.5	48.2	74 08.6	-31.1	51.0	75 23.2	-25.5	53.7	74 39.5	-27.7	56.8	74 54.7	-25.5	60.2	75 22.9	-19.0	63.7	75 37.0	-11.8	94.0	75 55.1	-11.8	94.0	40
41	73 00.2	-35.7	45.4	73 34.1	-33.7	47.8	75 06.3	-28.7	50.4	74 11.8	-30.4	53.3	74 32.2	-25.9	56.4	75 22.9	-22.6	59.8	75 32.5	-19.2	63.4	75 56.9	-7.3	89.6	41
42	72 24.3	-37.8	42.7	73 07.0	-35.9	44.9	74 39.2	-33.8	47.3	73 41.4	-33.0	49.8	74 05.3	-28.9	52.8	75 13.1	-22.9	56.0	75 13.4	-22.9	59.3	75 42.4	-19.4	63.0	42
43	71 46.5	-39.8	40.1	72 31.7	-38.0	42.2	73 54.3	-34.2	44.4	73 08.4	-35.3	46.8	73 37.4	-31.8	49.5	75 15.2	-29.2	52.4	75 05.0	-26.3	55.5	75 23.0	-23.1	58.9	43
44	71 06.8	-41.3	37.8	71 53.7	-39.9	39.6	73 09.2	-38.3	41.7	72 33.1	-37.5	43.9	73 05.8	-34.4	46.3	74 40.0	-32.1	49.0	75 24.2	-29.5	51.9	75 59.9	-26.5	55.0	44

Dec.	30°	31°	32°	33°	34°	35°	36°	37°	Dec.
45	70 25.5 - 42.8 35.6	71 13.8 - 41.6 37.3	72 00.9 - 40.1 39.1	72 46.8 - 38.5 41.2	73 31.2 - 36.7* 43.4	74 13.9 - 34.6* 45.8	74 54.7 - 32.3* 48.5	75 33.4 - 29.8* 51.4	**45**
46	69 42.7 44.2 33.5	70 32.2 43.0 35.1	71 20.8 41.7 36.8	72 08.3 40.3 38.6	72 54.5 38.7 40.7	73 39.3 36.9* 42.9	74 22.4 34.9* 45.3	75 03.6 32.6* 48.0	**46**
47	68 58.5 45.4 31.6	69 49.2 44.3 33.0	70 39.1 43.2 34.6	71 28.0 42.0 36.3	72 15.8 40.6 38.1	73 02.4 39.0 40.1	73 47.5 37.2* 42.3	74 31.0 35.2* 44.8	**47**
48	68 13.1 46.4 29.8	69 04.9 45.6 31.1	69 55.9 44.6 32.5	70 46.0 43.4 34.0	71 35.2 42.2 35.7	72 23.4 40.8 37.6	73 10.3 39.2 39.6	73 55.8 37.5* 41.8	**48**
49	67 26.7 47.5 28.1	68 19.3 46.6 29.3	69 11.3 45.8 30.6	70 02.6 44.8 32.0	70 53.0 43.7 33.5	71 42.6 42.5 35.2	72 31.1 41.1 37.0	73 18.3 39.5 39.0	**49**
50	66 39.2 - 48.3 26.6	67 32.7 - 47.7 27.6	68 25.5 - 46.8 28.8	69 17.8 - 46.0 30.1	70 09.3 - 45.0 31.5	71 00.1 - 43.9 33.0	71 50.0 - 42.8 34.6	72 38.8 - 41.4 36.4	**50**
51	65 50.9 49.1 25.1	66 45.0 48.4 26.1	67 38.7 47.8 27.1	68 31.8 47.0 28.3	69 24.3 46.1 29.5	70 16.2 45.3 30.9	71 07.2 44.1 32.4	71 57.4 43.0 34.1	**51**
52	65 01.8 49.8 23.7	65 56.6 49.3 24.6	66 50.9 48.7 25.6	67 44.8 48.0 26.6	68 38.2 47.3 27.8	69 30.9 46.4 29.0	70 23.1 45.5 30.4	71 14.4 44.4 31.8	**52**
53	64 12.0 50.5 22.4	65 07.3 50.0 23.2	66 02.2 49.4 24.1	66 56.8 48.8 25.1	67 50.9 48.2 26.1	68 44.5 47.4 27.2	69 37.6 46.6 28.5	70 30.0 45.7 29.8	**53**
54	63 21.5 51.1 21.2	64 17.3 50.6 21.9	65 12.8 50.2 22.7	66 08.0 49.7 23.6	67 02.7 49.0 24.5	67 57.1 48.4 25.6	68 51.0 47.7 26.7	69 44.3 46.9 27.9	**54**
55	62 30.4 - 51.6 20.0	63 26.7 - 51.3 20.7	64 22.6 - 50.7 21.4	65 18.3 - 50.3 22.2	66 13.7 - 49.8 23.1	67 08.7 - 49.2 24.0	68 03.3 - 48.6 25.0	68 57.4 - 47.9 26.1	**55**
56	61 38.8 52.1 18.9	62 35.4 51.7 19.6	63 31.9 51.4 20.2	64 28.0 50.9 21.0	65 23.9 50.5 21.7	66 19.5 50.0 22.6	67 14.7 49.4 23.5	68 09.5 48.8 24.5	**56**
57	60 46.7 52.6 17.9	61 43.7 52.3 18.5	62 40.5 51.9 19.1	63 37.1 51.5 19.7	64 33.4 51.1 20.5	65 29.5 50.7 21.2	66 25.3 50.2 22.0	67 20.7 49.6 22.9	**57**
58	59 54.1 52.9 16.9	60 51.4 52.6 17.5	61 48.6 52.4 18.0	62 45.6 52.1 18.6	63 42.3 51.7 19.3	64 38.8 51.2 19.9	65 35.1 50.8 20.7	66 31.1 50.4 21.5	**58**
59	59 01.2 53.4 16.0	59 58.8 53.1 16.5	60 56.2 52.8 17.0	61 53.5 52.5 17.5	62 50.6 52.1 18.1	63 47.6 51.9 18.8	64 44.3 51.5 19.4	65 40.7 51.0 20.2	**59**
60	58 07.8 - 53.7 15.1	59 05.7 - 53.5 15.6	60 03.4 - 53.2 16.0	61 01.0 - 52.9 16.5	61 58.5 - 52.7 17.1	62 55.7 - 52.3 17.6	63 52.8 - 52.0 18.2	64 49.7 - 51.6 18.9	**60**
61	57 14.1 54.0 14.3	58 12.2 53.8 14.7	59 10.2 53.6 15.1	60 08.1 53.4 15.6	61 05.8 53.1 16.1	62 03.4 52.8 16.6	63 00.8 52.5 17.1	63 58.1 52.2 17.7	**61**
62	56 20.1 54.3 13.5	57 18.4 54.1 13.9	58 16.6 53.9 14.2	59 14.7 53.7 14.7	60 12.7 53.4 15.1	61 10.6 53.2 15.6	62 08.3 52.9 16.1	63 05.9 52.7 16.6	**62**
63	55 25.8 54.5 12.7	56 24.3 54.4 13.1	57 22.7 54.2 13.4	58 21.0 54.0 13.8	59 19.3 53.9 14.2	60 17.4 53.6 14.6	61 15.4 53.4 15.1	62 13.2 53.1 15.6	**63**
64	54 31.3 54.8 12.0	55 29.9 54.6 12.3	56 28.5 54.5 12.6	57 27.0 54.3 13.0	58 25.4 54.1 13.3	59 23.8 54.0 13.7	60 22.0 53.7 14.1	61 20.1 53.5 14.6	**64**
65	53 36.5 - 55.1 11.3	54 35.3 - 54.9 11.6	55 34.0 - 54.7 11.9	56 32.7 - 54.6 12.2	57 31.3 - 54.4 12.5	58 29.8 - 54.2 12.9	59 28.3 - 54.1 13.3	60 26.6 - 53.8 13.7	**65**
66	52 41.4 55.2 10.7	53 40.4 55.2 10.9	54 39.3 55.0 11.2	55 38.1 54.9 11.5	56 36.9 54.8 11.8	57 35.6 54.6 12.1	58 34.2 54.4 12.4	59 32.8 54.2 12.8	**66**
67	51 46.2 55.5 10.0	52 45.2 55.3 10.3	53 44.3 55.3 10.5	54 43.2 55.1 10.7	55 42.1 54.9 11.0	56 41.0 54.8 11.3	57 39.8 54.7 11.6	58 38.6 54.6 11.9	**67**
68	50 50.7 55.6 9.4	51 49.9 55.5 9.6	52 49.0 55.4 9.8	53 48.1 55.3 10.1	54 47.2 55.2 10.3	55 46.2 55.1 10.6	56 45.1 54.9 10.9	57 44.0 54.8 11.2	**68**
69	49 55.1 55.8 8.8	50 54.4 55.8 9.0	51 53.6 55.6 9.2	52 52.8 55.5 9.4	53 52.0 55.4 9.6	54 51.1 55.3 9.9	55 50.2 55.2 10.1	56 49.2 55.0 10.4	**69**
70	48 59.3 - 56.0 8.3	49 58.6 - 55.9 8.4	50 58.0 - 55.8 8.6	51 57.3 - 55.7 8.8	52 56.6 - 55.7 9.0	53 55.8 - 55.5 9.2	54 55.0 - 55.4 9.4	55 54.2 - 55.3 9.7	**70**
71	48 03.3 56.1 7.7	49 02.7 56.0 7.9	50 02.2 56.0 8.0	5: 01.6 55.9 8.2	52 00.9 55.8 8.4	53 00.3 55.7 8.6	53 59.6 55.6 8.8	54 58.9 55.5 9.0	**71**
72	47 07.2 56.3 7.2	48 06.7 56.2 7.3	49 06.2 56.1 7.5	50 05.7 56.1 7.6	51 05.1 55.9 7.8	52 04.6 55.9 8.0	53 04.0 55.8 8.1	54 03.4 55.8 8.3	**72**
73	46 10.9 56.4 6.7	47 10.5 56.3 6.8	48 10.1 56.3 6.9	49 09.6 56.2 7.1	50 09.2 56.2 7.2	51 08.7 56.1 7.4	52 08.2 56.0 7.5	53 07.6 55.9 7.7	**73**
74	45 14.5 56.5 6.2	46 14.2 56.5 6.3	47 13.8 56.4 6.4	48 13.4 56.3 6.5	49 13.0 56.2 6.7	50 12.6 56.2 6.8	51 12.2 56.2 7.0	52 11.7 56.0 7.1	**74**
75	44 18.0 - 56.6 5.7	45 17.7 - 56.5 5.8	46 17.4 - 56.5 5.9	47 17.1 - 56.5 6.0	48 16.8 - 56.5 6.2	49 16.4 - 56.4 6.3	50 16.0 - 56.3 6.4	51 15.7 - 56.3 6.5	**75**
76	43 21.4 56.7 5.3	44 21.2 56.7 5.4	45 20.9 56.6 5.4	46 20.6 56.6 5.5	47 20.3 56.5 5.6	48 20.0 56.5 5.8	49 19.7 56.4 5.9	50 19.4 56.4 6.0	**76**
77	42 24.7 56.8 4.8	43 24.5 56.8 4.9	44 24.3 56.8 5.0	45 24.0 56.7 5.1	46 23.8 56.7 5.2	47 23.5 56.6 5.3	48 23.3 56.6 5.4	49 23.0 56.5 5.5	**77**
78	41 27.9 57.0 4.4	42 27.7 56.9 4.5	43 27.5 56.8 4.5	44 27.3 56.8 4.6	45 27.1 56.8 4.7	46 26.9 56.7 4.8	47 26.7 56.7 4.9	48 26.5 56.7 5.0	**78**
79	40 30.9 57.0 4.0	41 30.8 57.0 4.0	42 30.7 57.0 4.1	43 30.5 56.9 4.2	44 30.3 56.9 4.2	45 30.2 56.9 4.3	46 30.0 56.8 4.4	47 29.8 56.8 4.5	**79**
80	39 33.9 - 57.1 3.6	40 33.8 - 57.0 3.6	41 33.7 - 57.0 3.7	42 33.6 - 57.0 3.7	43 33.4 - 56.9 3.8	44 33.3 - 56.9 3.9	45 33.2 - 57.0 3.9	46 33.0 - 56.9 4.0	**80**
81	38 36.8 57.1 3.2	39 36.8 57.2 3.2	40 36.7 57.2 3.3	41 36.6 57.1 3.3	42 36.5 57.1 3.4	43 36.4 57.1 3.4	44 36.2 57.0 3.5	45 36.1 57.0 3.5	**81**
82	37 39.7 57.3 2.8	38 39.6 57.2 2.8	39 39.5 57.2 2.9	40 39.5 57.2 2.9	41 39.4 57.2 2.9	42 39.3 57.1 3.0	43 39.2 57.1 3.0	44 39.1 57.1 3.1	**82**
83	36 42.4 57.3 2.4	37 42.4 57.3 2.4	38 42.3 57.3 2.5	39 42.3 57.3 2.5	40 42.2 57.2 2.5	41 42.2 57.3 2.6	42 42.1 57.2 2.6	43 42.0 57.2 2.7	**83**
84	35 45.1 57.4 2.0	36 45.1 57.4 2.1	37 45.0 57.3 2.1	38 45.0 57.3 2.1	39 45.0 57.4 2.1	40 44.9 57.3 2.2	41 44.9 57.3 2.2	42 44.8 57.3 2.2	**84**
85	34 47.7 - 57.4 1.7	35 47.7 - 57.4 1.7	36 47.7 - 57.4 1.7	37 47.7 - 57.4 1.7	38 47.6 - 57.4 1.8	39 47.6 - 57.4 1.8	40 47.6 - 57.4 1.8	41 47.5 - 57.3 1.8	**85**
86	33 50.3 57.5 1.3	34 50.3 57.5 1.3	35 50.3 57.5 1.4	36 50.3 57.5 1.4	37 50.2 57.4 1.4	38 50.2 57.4 1.4	39 50.2 57.4 1.4	40 50.2 57.5 1.5	**86**
87	32 52.8 57.5 1.0	33 52.8 57.5 1.0	34 52.8 57.6 1.0	35 52.8 57.6 1.0	36 52.8 57.6 1.0	37 52.8 57.6 1.0	38 52.8 57.6 1.1	39 52.7 57.5 1.1	**87**
88	31 55.3 57.6 0.6	32 55.3 57.6 0.7	33 55.2 57.5 0.7	34 55.2 57.6 0.7	35 55.2 57.6 0.7	36 55.2 57.6 0.7	37 55.2 57.6 0.7	38 55.2 57.6 0.7	**88**
89	30 57.7 57.7 0.3	31 57.7 57.7 0.3	32 57.7 57.7 0.3	33 57.6 57.6 0.3	34 57.6 57.6 0.3	35 57.6 57.6 0.3	36 57.6 57.6 0.3	37 57.6 57.6 0.3	**89**
90	30 00.0 - 57.7 0.0	31 00.0 - 57.7 0.0	32 00.0 - 57.7 0.0	33 00.0 - 57.7 0.0	34 00.0 - 57.7 0.0	35 00.0 - 57.7 0.0	36 00.0 - 57.7 0.0	37 00.0 - 57.7 0.0	**90**
	30°	31°	32°	33°	34°	35°	36°	37°	

16°, 344° L.H.A. LATITUDE **SAME** NAME AS DECLINATION

Declination, in one-degree intervals, is given down the left- and right-hand columns of each page, in groups of five degrees. Across the page are columns giving Hc (calculated altitude) and Z (azimuth) for each degree of AP latitude covered by the page. A d figure (altitude difference) is shown between each Hc and Z.

TABULATED Hc

Suppose your AP latitude is N 37°, declination of the sun at the time of a sight is S 16°37.5′, and you have determined LHA to be 344°. Latitude is contrary to declination so the right-hand page is used. See Table 6-2. At the 16° declination line, in the 37° latitude column, you find:

Hc	d	Z
34°53.7′	−57.8	161.2

All three figures will be placed on your worksheet, but we will deal with azimuth later. The Hc figure, however, is the *tabulated calculated altitude*. It is the actual distance, in arc, from your AP to the GP of the body if the body is exactly on a whole degree of declination.

DECLINATION INCREMENTS

Actual declination of an observed body will seldom fall on a whole degree. Invariably there are increments of minutes and tenths of minutes of arc to account for. Once you have established tabulated calculated altitude (Tab Hc), the d figure is used to find what must be added or subtracted to get Hc for the actual declination. The d factor shows the extent that Hc changes with each additional degree of declination.

While it is possible to work out the correction mathematically, it is not necessary. It has been done for you in interpolation tables that appear on the inside of the front and rear covers of *Pub. 229*. The front pages cover declination increments from 0.0′ to 31.9′; those at the rear cover an increase from 28.0′ to 59.9′. You can see that there is some overlap; there's no significance to this. It just makes the book a little easier to work with in some cases. Table 6-3 shows the table on the page facing the inside rear cover.

Here's how the tables are used. Using the example above, the d factor of −57.8 indicates that Hc angle is declining as declination increases from S 16° to S 17°. Note in Table 6-3 that altitude differences are expressed in units of 10, in decimals, and then single units. This means the d figure must be broken down into two components: 50 (units of 10) and 7.8, the remainder (the single units and decimal place).

In Table 6-3, the line for a declination increment of 38.7′ shows the figure 32.3 under the "50" column. Put this down on your worksheet as −32.3′ since d is negative. The decimal and units figures to the right apply to the entire declination range at the left, in this case 38.0′ to 38.9′. Go down

Table 6-2. Right-Hand Page from *Pub. 229*. Data Above the C-S (Contrary-Same) Lines Is Used When Declination and Latitude Are Contrary Names. Data Below the Lines Is Seldom Needed for Sight Reduction.

LATITUDE CONTRARY NAME TO DECLINATION — L.H.A. 16°, 344°

Dec.	30° Hc	30° d	30° Z	31° Hc	31° d	31° Z	32° Hc	32° d	32° Z	33° Hc	33° d	33° Z	34° Hc	34° d	34° Z	35° Hc	35° d	35° Z	36° Hc	36° d	36° Z	37° Hc	37° d	37° Z	Dec.
0	56 21.2	54.2	150.2	55 29.0	54.7	150.2	54 36.4	55.0	151.6	53 43.5	55.4	152.2	52 50.2	55.6	152.9	51 56.7	55.9	153.4	51 02.9	56.2	154.0	50 08.9	56.5	154.5	0
1	55 27.0	54.6	150.9	54 34.3	54.9	150.9	53 41.4	55.2	152.3	52 48.1	55.5	152.8	51 54.6	55.8	153.5	51 00.8	56.1	154.0	50 06.7	56.3	154.5	49 12.4	56.5	155.0	1
2	54 32.4	54.8	151.7	53 39.4	55.1	152.0	52 46.2	55.5	152.9	51 52.6	55.7	153.5	50 58.8	56.0	154.1	50 04.7	56.1	154.6	49 10.4	56.5	155.1	48 15.9	56.5	155.6	2
3	53 37.6	55.0	152.3	52 44.3	55.1	152.7	51 50.7	55.6	153.5	50 56.9	55.8	154.1	50 02.8	56.2	154.6	49 08.5	56.3	155.1	48 13.9	56.6	155.6	47 19.2	56.8	156.0	3
4	52 42.6	55.3	153.0	51 48.9	55.5	153.3	50 55.1	55.8	154.1	50 01.1	56.1	154.7	49 06.6	56.3	155.2	48 12.1	56.4	155.6	47 17.3	56.7	156.1	46 22.4	56.9	156.5	4
5	51 47.3	55.4	153.6	50 53.4	55.7	154.0	49 59.3	56.0	154.7	49 04.9	56.2	155.2	48 10.3	56.4	155.7	47 15.6	56.7	156.1	46 20.6	56.8	156.6	45 25.5	57.1	157.0	5
6	50 51.9	55.7	154.3	49 57.7	55.9	154.8	49 03.3	56.1	155.3	48 08.7	56.4	155.7	47 13.9	56.6	156.2	46 18.9	56.6	156.6	45 23.8	57.0	157.0	44 28.4	57.1	157.4	6
7	49 56.2	55.8	154.8	49 01.8	56.0	155.3	48 07.2	56.3	155.8	47 12.3	56.4	156.3	46 17.3	56.8	156.7	45 22.2	56.8	157.1	44 26.8	57.0	157.5	43 31.3	57.2	157.8	7
8	49 00.4	55.9	155.4	48 05.8	56.0	155.8	47 10.9	56.4	156.3	46 15.9	56.6	156.7	45 20.7	56.8	157.1	44 25.3	56.9	157.5	43 29.8	57.2	157.9	42 34.1	57.3	158.2	8
9	48 04.5	56.2	156.0	47 09.6	56.1	156.4	46 14.5	56.5	156.8	45 19.3	56.8	157.2	44 23.9	56.9	157.6	43 28.3	57.1	158.0	42 32.6	57.2	158.3	41 36.8	57.4	158.6	9
10	47 08.3	56.1	156.5	46 13.2	56.4	156.9	45 18.0	56.7	157.3	44 22.5	56.9	157.7	43 27.0	57.0	158.0	42 31.2	57.1	158.4	41 35.4	57.3	158.7	40 39.4	57.4	159.0	10
11	46 12.2	56.5	157.0	45 16.8	56.6	157.4	44 21.3	56.7	157.8	43 25.9	56.9	158.1	42 30.0	57.1	158.5	41 34.1	57.3	158.8	40 38.1	57.4	159.1	39 42.0	57.6	159.4	11
12	45 15.7	56.5	157.5	44 20.2	56.8	157.9	43 24.6	56.9	158.2	42 28.8	57.0	158.6	41 32.9	57.2	158.9	40 36.8	57.3	159.2	39 40.7	57.5	159.5	38 44.4	57.5	159.8	12
13	44 19.2	56.6	158.0	43 23.5	56.8	158.3	42 27.7	57.0	158.7	41 31.8	57.2	159.0	40 35.7	57.3	159.3	39 39.5	57.5	159.6	38 43.2	57.5	159.9	37 46.9	57.7	160.1	13
14	43 22.6	56.7	158.4	42 26.7	56.9	158.8	41 30.7	57.0	159.1	40 34.6	57.1	159.4	39 38.4	57.3	159.7	38 42.1	57.5	160.0	37 45.7	57.6	160.2	36 49.2	57.7	160.5	14
15	42 25.9	56.9	158.9	41 29.8	56.9	159.2	40 33.7	57.1	159.5	39 37.5	57.3	159.8	38 41.1	57.4	160.1	37 44.6	57.5	160.3	36 48.1	57.7	160.6	35 51.5	57.8	160.8	15
16	41 29.0	56.9	159.3	40 32.9	57.1	159.6	39 36.6	57.2	159.9	38 40.2	57.4	160.2	37 43.7	57.5	160.5	36 47.1	57.6	160.7	35 50.5	57.8	160.9	34 53.7	57.9	161.2	16
17	40 32.1	57.1	159.7	39 35.8	57.1	160.0	38 39.4	57.3	160.3	37 42.8	57.4	160.5	36 46.2	57.5	160.8	35 49.5	57.6	161.0	34 52.7	57.7	161.3	33 55.9	57.9	161.5	17
18	39 35.0	57.1	160.1	38 38.6	57.2	160.4	37 42.1	57.4	160.7	36 45.4	57.5	160.9	35 48.7	57.5	161.1	34 51.9	57.7	161.4	33 55.0	57.9	161.6	32 58.0	57.9	161.8	18
19	38 38.0	57.1	160.5	37 41.4	57.3	160.8	36 44.7	57.4	161.0	35 47.9	57.5	161.3	34 51.1	57.7	161.5	33 54.2	57.8	161.7	32 57.2	57.9	161.9	32 00.1	58.0	162.1	19
20	37 40.9	57.3	160.9	36 44.1	57.4	161.1	35 47.3	57.5	161.4	34 50.4	57.6	161.6	33 53.4	57.7	161.8	32 56.4	57.8	162.0	31 59.3	57.9	162.2	31 02.1	58.0	162.4	20
21	36 43.6	57.3	161.3	35 46.7	57.4	161.5	34 49.8	57.5	161.7	33 52.8	57.7	162.0	32 55.7	57.7	162.2	31 58.6	57.9	162.3	31 01.4	58.0	162.5	30 04.1	58.0	162.7	21
22	35 46.3	57.4	161.6	34 49.3	57.5	161.9	33 52.3	57.6	162.1	32 55.1	57.7	162.3	31 58.0	57.8	162.5	31 00.7	57.9	162.7	30 03.4	58.0	162.8	29 06.1	58.1	163.0	22
23	34 48.9	57.5	162.0	33 51.8	57.5	162.2	32 54.7	57.7	162.4	31 57.4	57.7	162.6	31 00.2	57.9	162.8	30 02.8	57.9	163.0	29 05.4	58.0	163.1	28 08.0	58.1	163.3	23
24	33 51.5	57.5	162.3	32 54.3	57.6	162.5	31 57.0	57.7	162.7	30 59.7	57.8	162.9	30 02.3	57.9	163.1	29 04.9	58.0	163.3	28 07.4	58.1	163.4	27 09.9	58.2	163.6	24
25	32 54.0	57.6	162.7	31 56.7	57.7	162.9	30 59.3	57.7	163.1	30 01.9	57.9	163.2	29 04.4	58.0	163.4	28 06.9	58.0	163.5	27 09.3	58.1	163.7	26 11.7	58.2	163.8	25
26	31 56.4	57.6	163.0	30 59.0	57.7	163.2	30 01.6	57.8	163.4	29 04.0	57.8	163.5	28 06.5	58.0	163.7	27 08.9	58.1	163.8	26 11.2	58.1	164.0	25 13.5	58.2	164.1	26
27	30 58.8	57.7	163.4	30 01.3	57.7	163.5	29 03.8	57.8	163.7	28 06.2	57.9	163.8	27 08.5	58.0	164.0	26 10.8	58.1	164.1	25 13.1	58.2	164.2	24 15.3	58.2	164.4	27
28	30 01.2	57.7	163.7	30 01.3	57.7	163.8	29 05.9	57.9	164.0	28 08.2	57.9	164.1	26 10.5	58.0	164.3	25 12.7	58.1	164.4	24 14.9	58.2	164.5	23 17.1	58.3	164.4	28
29	29 03.5	57.8	164.0	28 05.8	57.8	164.1	27 08.1	58.0	164.3	26 10.3	58.0	164.4	25 12.5	58.1	164.5	24 14.6	58.1	164.7	23 16.7	58.2	164.8	22 18.8	58.3	164.9	29

Table 6-3. Information Inside the Front and Rear Covers of *Pub. 229* Corrects for the Increment of Declination Beyond the Nearest Whole Degree.

INTERPOLATION TABLE

Left half — Altitude Difference (d)

Tens columns:

Dec. Inc.	10'	20'	30'	40'	50'
28.0	4.6	9.3	14.0	18.6	23.3
28.1	4.7	9.3	14.0	18.7	23.4
28.2	4.7	9.4	14.1	18.8	23.5
28.3	4.7	9.4	14.1	18.9	23.6
28.4	4.7	9.5	14.2	18.9	23.7
28.5	4.8	9.5	14.3	19.0	23.8
28.6	4.8	9.5	14.3	19.1	23.8
28.7	4.8	9.6	14.4	19.2	23.9
28.8	4.8	9.6	14.4	19.2	24.0
28.9	4.9	9.7	14.5	19.3	24.1
29.0	4.8	9.6	14.5	19.3	24.1
29.1	4.8	9.7	14.5	19.4	24.2
29.2	4.8	9.7	14.6	19.4	24.3
29.3	4.9	9.8	14.6	19.5	24.4
29.4	4.9	9.8	14.7	19.6	24.5
29.5	4.9	9.8	14.8	19.7	24.6
29.6	4.9	9.9	14.8	19.7	24.7
29.7	5.0	9.9	14.9	19.8	24.8
29.8	5.0	10.0	14.9	19.9	24.9
29.9	5.0	10.0	15.0	20.0	25.0
30.0	5.0	10.0	15.0	20.0	25.0
30.1	5.0	10.0	15.0	20.1	25.1
30.2	5.0	10.1	15.1	20.2	25.2
30.3	5.0	10.1	15.1	20.2	25.3
30.4	5.1	10.2	15.2	20.3	25.4
30.5	5.1	10.2	15.3	20.3	25.4
30.6	5.1	10.2	15.3	20.4	25.5
30.7	5.1	10.3	15.4	20.5	25.6
30.8	5.2	10.3	15.4	20.6	25.7
30.9	5.2	10.3	15.5	20.6	25.8

Decimals / Units (0'–9'):

Dec.	0'	1'	2'	3'	4'	5'	6'	7'	8'	9'
.0	0.0	0.5	0.9	1.4	1.9	2.4	2.8	3.3	3.8	4.3
.1	0.0	0.5	1.0	1.5	1.9	2.4	2.9	3.4	3.8	4.3
.2	0.1	0.6	1.0	1.5	2.0	2.5	2.9	3.4	3.9	4.4
.3	0.1	0.6	1.1	1.6	2.0	2.5	3.0	3.5	3.9	4.4
.4	0.2	0.7	1.1	1.6	2.1	2.6	3.0	3.5	4.0	4.5
.5	0.2	0.7	1.2	1.7	2.2	2.6	3.1	3.6	4.0	4.5
.6	0.3	0.8	1.2	1.7	2.2	2.7	3.1	3.6	4.1	4.6
.7	0.3	0.8	1.3	1.8	2.3	2.7	3.2	3.7	4.1	4.6
.8	0.4	0.9	1.3	1.8	2.3	2.8	3.2	3.7	4.2	4.7
.9	0.4	0.9	1.4	1.9	2.4	2.8	3.3	3.8	4.2	4.7

Double Second Diff. and Corr.:

Diff.	Corr.
0.8	0.1
2.4	0.2
4.0	0.3
5.6	0.4
7.2	0.5
8.8	0.6
10.4	0.7
12.0	0.8
13.6	0.9
15.2	1.0
16.8	1.1
18.4	1.2
20.0	1.3
21.6	1.4
23.2	1.5
24.8	1.6
26.4	1.7
28.0	1.8
29.6	1.8
31.2	1.9
32.8	2.0
34.4	

Right half — Altitude Difference (d)

Tens columns:

Dec. Inc.	10'	20'	30'	40'	50'
36.0	6.0	12.0	18.0	24.0	30.0
36.1	6.0	12.0	18.0	24.0	30.1
36.2	6.0	12.1	18.1	24.1	30.1
36.3	6.0	12.1	18.1	24.2	30.2
36.4	6.1	12.1	18.2	24.3	30.3
36.5	6.1	12.2	18.3	24.3	30.4
36.6	6.1	12.2	18.3	24.4	30.5
36.7	6.1	12.3	18.4	24.5	30.6
36.8	6.2	12.3	18.4	24.6	30.7
36.9	6.2	12.3	18.5	24.6	30.8
37.0	6.1	12.3	18.5	24.6	30.8
37.1	6.2	12.3	18.5	24.7	30.9
37.2	6.2	12.4	18.6	24.8	31.0
37.3	6.2	12.4	18.6	24.9	31.1
37.4	6.2	12.5	18.7	24.9	31.2
37.5	6.3	12.5	18.8	25.0	31.3
37.6	6.3	12.5	18.8	25.1	31.3
37.7	6.3	12.6	18.9	25.2	31.4
37.8	6.3	12.6	18.9	25.2	31.5
37.9	6.4	12.7	19.0	25.3	31.6
38.0	6.3	12.6	19.0	25.3	31.6
38.1	6.3	12.7	19.0	25.4	31.7
38.2	6.3	12.7	19.1	25.4	31.8
38.3	6.4	12.8	19.1	25.5	31.9
38.4	6.4	12.8	19.2	25.6	32.0
38.5	6.4	12.8	19.3	25.7	32.1
38.6	6.4	12.9	19.3	25.7	32.2
38.7	6.5	12.9	19.4	25.8	32.3
38.8	6.5	13.0	19.4	25.9	32.4
38.9	6.5	13.0	19.5	26.0	32.5

Decimals / Units (0'–9'):

Dec.	0'	1'	2'	3'	4'	5'	6'	7'	8'	9'
.0	0.0	0.6	1.2	1.8	2.4	3.1	3.6	4.3	4.9	5.5
.1	0.1	0.7	1.3	1.9	2.5	3.1	3.7	4.3	4.9	5.5
.2	0.1	0.7	1.3	1.9	2.6	3.2	3.8	4.4	5.0	5.6
.3	0.2	0.8	1.4	2.0	2.6	3.2	3.8	4.5	5.0	5.7
.4	0.2	0.9	1.5	2.1	2.7	3.3	3.9	4.6	5.1	5.7
.5	0.3	0.9	1.5	2.1	2.7	3.3	4.0	4.6	5.2	5.8
.6	0.4	1.0	1.6	2.2	2.8	3.4	4.0	4.6	5.2	5.8
.7	0.4	1.1	1.7	2.3	2.9	3.5	4.1	4.7	5.3	5.9
.8	0.5	1.1	1.7	2.4	2.9	3.6	4.2	4.8	5.4	6.0
.9	0.6	1.2	1.8	2.5	3.1	3.7	4.3	5.0	5.6	6.2

Double Second Diff. and Corr.:

Diff.	Corr.
0.8	0.1
2.5	0.2
4.2	0.3
5.9	0.4
7.6	0.5
9.3	0.6
11.0	0.7
12.7	0.8
14.4	0.9
16.1	1.0
17.8	
19.5	1.1
21.2	1.2
22.8	1.3
24.5	1.4
26.2	1.5
27.9	1.7
29.6	1.8
31.3	1.9
33.0	2.0
34.7	
0.9	0.1
2.6	0.2
4.4	0.3
6.2	0.4
7.9	0.5
9.7	0.6
11.4	0.7
13.2	0.8
14.9	0.9
16.7	

the decimal column to .8 and then across to the 7' column under units. The figure here is 5.0. This is *added algebraically*—as −5.0'—to the −32.3' figure, so:

$$
\begin{array}{rr}
 & -32.5' \\
+ & -\;\underline{5.0'} \\
\text{d corr.} & -37.5'
\end{array}
$$

You will subtract 37.5' from the Tab Hc for 16° declination:

$$
\begin{array}{ll}
\text{Tab Hc} & 34°53.7' \\
\text{d corr.} & \underline{-37.5'} \\
\text{Hc} & 34°16.2'
\end{array}
$$

If you were exactly at your assumed position, an accurate sight would result in an observed altitude (Ho) of 34°16.2'. The chances are you will be within about 30 miles of it if your DR position is anywhere near accurate.

Table 6-4. Data for Latitude of N 42°, and a Declination of N 54°. Note That the Sign for the d Factor is Negative.

	38°			39°			40°			41°			42°		
50	60 32.9	-10.7	53.6	61 07.9	- 9.0	55.1	61 41.6	- 7.2	56.6	62 14.0	- 5.4*	58.2	62 44.9	- 3.4*	59.8
51	60 22.2	12.4	51.6	60 58.9	10.8	53.0	61 34.4	9.1	54.5	62 08.6	7.3	56.0	62 41.5	5.5*	57.6
52	60 09.8	14.1	49.6	60 48.1	12.6	51.0	61 25.3	10.9	52.4	62 01.3	9.2	53.9	62 36.0	7.4*	55.5
53	59 55.7	15.9	47.7	60 35.5	14.3	49.0	61 14.4	12.8	50.4	61 52.1	11.2	51.8	62 28.6	9.5	53.3
54	59 39.8	17.5	45.8	60 21.2	16.1	47.0	61 01.6	14.6	48.3	61 40.9	13.0	49.7	62 19.1	11.3	51.2
55	59 22.3	-19.1	43.9	60 05.1	-17.8	45.1	60 47.0	-16.3	46.3	61 27.9	-14.8	47.7	62 07.8	-13.3	49.1
56	59 03.2	20.7	42.0	59 47.3	19.3	43.2	60 30.7	18.1	44.4	61 13.1	16.6	45.6	61 54.5	15.1	47.0
57	58 42.5	22.2	40.2	59 28.0	21.0	41.3	60 12.6	19.6	42.4	60 56.5	18.3	43.7	61 39.4	16.9	44.9
58	58 20.3	23.6	38.4	59 07.0	22.5	39.5	59 53.0	21.3	40.6	60 38.2	20.0	41.7	61 22.5	18.6	42.9
59	57 56.7	25.0	36.7	58 44.5	23.9	37.7	59 31.7	22.8	38.7	60 18.2	21.6	39.8	61 03.9	20.4	40.9
60	57 31.7	-26.3	35.0	58 20.6	-25.4	35.9	59 08.9	-24.3	36.9	59 56.6	-23.2	37.9	60 43.5	-22.0	39.0
61	57 05.4	27.7	33.3	57 55.2	26.7	34.2	58 44.6	25.7	35.1	59 33.4	24.7	36.1	60 21.5	23.5	37.1
62	56 37.7	28.9	31.7	57 28.5	28.0	32.5	58 18.9	27.1	33.4	59 08.7	26.1	34.3	59 58.0	25.1	35.3
63	56 08.8	30.1	30.1	57 00.5	29.2	30.9	57 51.8	28.4	31.7	58 42.6	27.5	32.6	59 32.9	26.5	33.5
64	55 38.7	31.2	28.6	56 31.3	30.5	29.3	57 23.4	29.7	30.1	58 15.1	28.8	30.9	59 06.4	27.9	31.7
65	55 07.5	-32.2	27.1	56 00.8	-31.6	27.7	56 53.7	-30.8	28.5	57 46.3	-30.1	29.2	58 38.5	-29.3	30.0
66	54 35.3	33.4	25.6	55 29.2	32.7	26.2	56 22.9	32.0	26.9	57 16.2	31.2	27.6	58 09.2	30.5	28.3
67	54 01.9	34.3	24.2	54 56.5	33.7	24.8	55 50.9	33.1	25.4	56 45.0	32.5	26.0	57 38.7	31.7	26.7
68	53 27.6	35.2	22.8	54 22.8	34.7	23.3	55 17.8	34.1	23.9	56 12.5	33.5	24.5	57 07.0	32.9	25.1
69	52 52.4	36.1	21.4	53 48.1	35.6	21.9	54 43.7	35.1	22.5	55 39.0	34.5	23.0	56 34.1	33.9	23.6
70	52 16.3	-37.0	20.1	53 12.5	-36.5	20.6	54 08.6	-36.1	21.1	55 04.5	-35.6	21.6	56 00.2	-35.0	22.1
71	51 39.3	37.8	18.8	52 36.0	37.4	19.3	53 32.5	36.9	19.7	54 28.9	36.4	20.2	55 25.2	36.0	20.7
72	51 01.5	38.6	17.6	51 58.6	38.2	18.0	52 55.6	37.8	18.4	53 52.5	37.4	18.8	54 49.2	36.9	19.3
73	50 22.9	39.3	16.4	51 20.4	38.9	16.7	52 17.8	38.6	17.1	53 15.1	38.2	17.5	54 12.3	37.9	17.9
74	49 43.6	40.0	15.2	50 41.5	39.7	15.5	51 39.2	39.3	15.9	52 36.9	39.0	16.2	53 34.4	38.6	16.6
75	49 03.6	-40.6	14.1	50 01.8	-40.4	14.4	50 59.9	-40.1	14.7	51 57.9	-39.8	15.0	52 55.8	-39.5	15.3
76	48 23.0	41.3	13.0	49 21.4	41.0	13.2	50 19.8	40.8	13.5	51 18.1	40.5	13.8	52 16.3	40.2	14.1
77	47 41.7	41.9	11.9	48 40.4	41.7	12.1	49 39.0	41.4	12.4	50 37.6	41.2	12.6	51 36.1	40.9	12.9
78	46 59.8	42.5	10.8	47 58.7	42.2	11.0	48 57.6	42.1	11.2	49 56.4	41.8	11.5	50 55.2	41.6	11.7
79	46 17.3	42.9	9.8	47 16.5	42.9	10.0	48 15.5	42.6	10.2	49 14.6	42.5	10.4	50 13.6	42.3	10.6
80	45 34.4	-43.6	8.8	46 33.6	-43.3	8.9	47 32.9	-43.2	9.1	48 32.1	-43.0	9.3	49 31.3	-42.8	9.5
81	44 50.8	43.9	7.8	45 50.3	43.9	7.9	46 49.7	43.7	8.1	47 49.1	43.6	8.2	48 48.5	43.5	8.4
82	44 06.9	44.5	6.9	45 06.4	44.3	7.0	46 06.0	44.3	7.1	47 05.5	44.1	7.2	48 05.0	44.0	7.4
83	43 22.4	44.9	5.9	44 22.1	44.8	6.0	45 21.7	44.7	6.1	46 21.4	44.6	6.2	47 21.0	44.5	6.4
84	42 37.5	45.3	5.0	43 37.3	45.3	5.1	44 37.0	45.1	5.2	45 36.8	45.1	5.3	46 36.5	45.0	5.4
85	41 52.2	-45.7	4.1	42 52.0	-45.6	4.2	43 51.9	-45.6	4.3	44 51.7	-45.5	4.3	45 51.5	-45.4	4.4
86	41 06.5	46.1	3.3	42 06.4	46.1	3.3	43 06.3	46.0	3.4	44 06.2	46.0	3.4	45 06.1	46.0	3.5
87	40 20.4	46.5	2.4	41 20.3	46.4	2.5	42 20.3	46.4	2.5	43 20.2	46.4	2.5	44 20.1	46.3	2.6
88	39 33.9	46.8	1.6	40 33.9	46.8	1.6	41 33.9	46.8	1.6	42 33.8	46.7	1.7	43 33.8	46.7	1.7
89	38 47.1	47.1	0.8	39 47.1	47.1	0.8	40 47.1	47.1	0.8	41 47.1	47.1	0.8	42 47.1	47.1	0.8
90	38 00.0	- 47.4	0.0	39 00.0	- 47.4	0.0	40 00.0	- 47.4	0.0	41 00.0	- 47.5	0.0	42 00.0	- 47.5	0.0
	38°			**39°**			**40°**			**41°**			**42°**		

38°, 322° L.H.A.   LATITUDE SAME NAME AS DECl

It's a lot easier to plot a 30-mile intercept line from an AP than one that may be a thousand miles or more from the GP of the observed body.

Here's another problem in finding Hc. You have established your AP latitude as N 42°, LHA is calculated to be 38°, and declination of the sighted body is N 54°22.7′.

You'll need the rear section of Vol. 3, *Pub. 229*, which has the AP latitude range 38° to 45°. Use the left-hand page for LHA 38°, 322°, as latitude and declination both are north (Table 6-4). For the whole degree of declination at 54°, you have:

Hc	d	Z
62°19.1′	−11.3	51.2

The declination increment is 22.7′ of arc, so the interpolation table on the inside front cover is needed (Table 6-5). The d factor of 11.3 breaks down to 10 plus 1.3. In the 10′ column, at the 22.7′ line, the correction is 3.8 and the correction for an increment of 1.3 is 0.5. The correction factor is negative, so:

	−3.8′
+	−0.5′
d cor.	−4.3′

Tab Hc	62°19.1′
d corr.	− 4.3′
Hc	62°14.8′

It just takes a couple of minutes to work this out (using the tables). Be careful not to let your eyes jump to the wrong column or to confuse the d factor with the declination increment when you go to the interpolation tables.

AZIMUTH

As noted earlier, the difference between Hc and Ho is the length of the intercept, in nautical miles, that you will draw on your plotting sheet. Now you need to know the direction in which to draw it. The intercept tells you that you are closer to the body or farther from it than you would be if you were right at the assumed position. The line you draw will head toward the body or away from it. The Z (azimuth) figure given in the tables is for whole degrees of declination. When adjusted for the declination increment, it indicates the exact azimuth to the body from the AP.

Using the problem given above for Hc, note the Z figure of 51.2. This is degrees so on our worksheet it will be noted as 51.2°. Next, note the Z figure for the next higher degree of declination. Here it is 49.1 so you know that azimuth is decreasing 1.1° of arc for the one degree increase in declination.

82

Table 6-5. The Correction for a d Factor of 11.3 (See Table 6-4) Is 3.8 from the 10' Column, and 0.5 from the 1' Column in Line with .3. The 1.1° "z" Correction Is 0.4.

INTERPOLATION TABLE

Left half — Altitude Difference (d): Tens

Dec. Inc.	10'	20'	30'	40'	50'
16.0	2.6	5.3	8.0	10.6	13.3
16.1	2.7	5.3	8.0	10.7	13.4
16.2	2.7	5.4	8.1	10.8	13.5
16.3	2.7	5.4	8.1	10.9	13.6
16.4	2.7	5.5	8.2	10.9	13.7
16.5	2.8	5.5	8.3	11.0	13.8
16.6	2.8	5.5	8.3	11.1	13.8
16.7	2.8	5.6	8.4	11.1	13.9
16.8	2.8	5.6	8.4	11.2	14.0
16.9	2.9	5.7	8.5	11.3	14.1
17.0	2.8	5.6	8.5	11.3	14.1
17.1	2.8	5.7	8.5	11.4	14.2
17.2	2.8	5.7	8.6	11.4	14.3
17.3	2.9	5.8	8.6	11.5	14.4
17.4	2.9	5.8	8.7	11.6	14.5
22.5	3.8	7.5	11.3	15.0	18.8
22.6	3.8	7.5	11.3	15.1	18.8
22.7	3.8	7.6	11.4	15.2	18.9
22.8	3.8	7.6	11.4	15.2	19.0
22.9	3.9	7.7	11.5	15.3	19.1
23.0	3.8	7.6	11.5	15.3	19.1
23.1	3.8	7.7	11.5	15.4	19.2
23.2	3.9	7.7	11.6	15.4	19.3
23.3	3.9	7.8	11.6	15.5	19.4
23.4	3.9	7.8	11.7	15.6	19.5
23.5	3.9	7.8	11.8	15.7	19.6
23.6	3.9	7.9	11.8	15.7	19.7
23.7	4.0	7.9	11.8	15.8	19.8
23.8	4.0	8.0	11.9	15.9	19.9
23.9	4.0	8.0	12.0	16.0	20.0

Left half — Decimals / Units

Decimals	0'	1'	2'	3'	4'	5'	6'	7'	8'	9'
.0	0.0	0.3	0.5	0.8	1.1	1.4	1.6	1.9	2.2	2.5
.1	0.0	0.3	0.6	0.9	1.2	1.4	1.7	2.0	2.2	2.5
.2	0.1	0.3	0.6	0.9	1.2	1.4	1.7	2.0	2.3	2.6
.3	0.1	0.4	0.6	0.9	1.2	1.5	1.8	2.0	2.3	2.6
.4	0.1	0.4	0.7	0.9	1.2	1.5	1.8	2.0	2.3	2.6
.5	0.1	0.4	0.7	1.0	1.3	1.5	1.8	2.1	2.3	2.6
.6	0.2	0.4	0.7	1.0	1.3	1.5	1.8	2.1	2.4	2.6
.7	0.2	0.5	0.7	1.0	1.3	1.6	1.8	2.1	2.4	2.7
.8	0.2	0.5	0.8	1.1	1.3	1.6	1.9	2.1	2.4	2.7
.9	0.2	0.5	0.8	1.1	1.3	1.6	1.9	2.2	2.4	2.7
.0	0.0	0.3	0.6	0.9	1.2	1.5	1.7	2.0	2.3	2.6
.1	0.1	0.3	0.6	0.9	1.2	1.5	1.8	2.1	2.4	2.7
.2	0.1	0.3	0.6	0.9	1.2	1.5	1.8	2.1	2.4	2.7
.3	0.1	0.4	0.7	1.0	1.3	1.6	1.9	2.1	2.4	2.7
.4	0.1	0.4	0.7	1.0	1.3	1.6	1.9	2.2	2.4	2.7
.5	0.1	0.6	0.9	1.3	1.7	2.1	2.4	2.8	3.2	3.6
.6	0.2	0.6	1.0	1.3	1.7	2.1	2.5	2.8	3.2	3.6
.7	0.2	0.6	1.0	1.4	1.8	2.1	2.5	2.9	3.3	3.6
.8	0.3	0.7	1.0	1.4	1.8	2.2	2.5	2.9	3.3	3.7
.9	0.3	0.7	1.1	1.5	1.8	2.2	2.6	3.0	3.3	3.7
.0	0.0	0.4	0.8	1.2	1.6	2.0	2.3	2.7	3.1	3.5
.1	0.0	0.4	0.8	1.2	1.6	2.0	2.4	2.8	3.2	3.6
.2	0.1	0.5	0.9	1.3	1.6	2.0	2.4	2.8	3.2	3.6
.3	0.1	0.5	0.9	1.3	1.7	2.1	2.5	2.9	3.3	3.7
.4	0.2	0.5	0.9	1.3	1.7	2.1	2.5	2.9	3.3	3.7
.5	0.2	0.6	1.0	1.4	1.8	2.2	2.5	2.9	3.3	3.7
.6	0.2	0.6	1.0	1.4	1.8	2.2	2.6	3.0	3.4	3.8
.7	0.3	0.7	1.1	1.4	1.8	2.2	2.6	3.0	3.4	3.8
.8	0.3	0.7	1.1	1.5	1.9	2.3	2.7	3.1	3.5	3.9
.9	0.4	0.7	1.1	1.5	1.9	2.3	2.7	3.1	3.5	3.9

Left half — Double Second Diff. and Corr.

1.0	0.1
3.0	0.2
4.9	0.3
6.9	0.4
8.9	0.5
10.8	0.6
12.8	0.7
14.8	0.8
16.7	0.8
18.7	0.9
20.7	1.0
22.7	1.1
24.6	1.2
26.6	1.3
9.3	0.6
11.0	0.7
12.7	0.8
14.4	0.9
16.1	1.0
17.8	1.1
19.5	1.2
21.2	1.3
22.8	1.4
24.5	1.5
26.2	1.6
27.9	1.7
29.6	1.8
31.3	1.9
33.0	2.0
34.7	2.1

Right half — Altitude Difference (d): Tens

Dec. Inc.	10'	20'	30'	40'	50'
24.0	4.0	8.0	12.0	16.0	20.0
24.1	4.0	8.0	12.0	16.0	20.1
24.2	4.0	8.1	12.1	16.1	20.1
24.3	4.0	8.1	12.1	16.2	20.2
24.4	4.1	8.1	12.2	16.3	20.3
24.5	4.1	8.2	12.3	16.3	20.4
24.6	4.1	8.2	12.3	16.4	20.5
24.7	4.1	8.3	12.4	16.5	20.6
24.8	4.2	8.3	12.4	16.6	20.7
24.9	4.2	8.3	12.5	16.6	20.8
25.0	4.1	8.3	12.5	16.6	20.8
25.1	4.2	8.4	12.5	16.7	20.9
25.2	4.2	8.4	12.6	16.8	21.0
25.3	4.2	8.4	12.6	16.9	21.1
25.4	4.2	8.5	12.7	16.9	21.2
30.5	5.1	10.2	15.3	20.4	25.4
30.6	5.1	10.2	15.3	20.4	25.5
30.7	5.1	10.3	15.4	20.5	25.6
30.8	5.2	10.3	15.4	20.6	25.7
30.9	5.2	10.3	15.5	20.6	25.8
31.0	5.1	10.3	15.5	20.6	25.8
31.1	5.1	10.3	15.5	20.7	25.9
31.2	5.2	10.4	15.6	20.8	26.0
31.3	5.2	10.4	15.6	20.9	26.1
31.4	5.2	10.5	15.7	20.9	26.2
31.5	5.3	10.5	15.8	21.0	26.3
31.6	5.3	10.5	15.8	21.1	26.3
31.7	5.3	10.6	15.9	21.2	26.4
31.8	5.3	10.6	15.9	21.2	26.5
31.9	5.4	10.7	16.0	21.3	26.6

Right half — Decimals / Units

Decimals	0'	1'	2'	3'	4'	5'	6'	7'	8'	9'
.0	0.0	0.4	0.8	1.2	1.6	2.0	2.4	2.9	3.3	3.7
.1	0.0	0.4	0.9	1.3	1.7	2.1	2.5	2.9	3.3	3.7
.2	0.1	0.5	0.9	1.3	1.7	2.1	2.6	3.0	3.3	3.8
.3	0.1	0.6	0.9	1.3	1.8	2.2	2.6	3.0	3.4	3.8
.4	0.2	0.6	1.0	1.4	1.8	2.2	2.6	3.0	3.4	3.8
.5	0.2	0.6	1.0	1.4	1.8	2.2	2.7	3.1	3.5	3.9
.6	0.2	0.7	1.1	1.5	1.9	2.3	2.7	3.1	3.5	3.9
.7	0.2	0.7	1.1	1.5	1.9	2.3	2.7	3.1	3.5	4.0
.8	0.3	0.7	1.1	1.6	2.0	2.4	2.8	3.2	3.6	4.0
.9	0.3	0.8	1.2	1.6	2.0	2.4	2.8	3.2	3.6	4.0
.5	0.2	0.6	1.0	1.4	1.8	2.2	2.7	3.1	3.5	3.9
.6	0.2	0.7	1.1	1.5	1.9	2.3	2.7	3.1	3.5	3.9
.7	0.3	0.7	1.1	1.5	1.9	2.3	2.7	3.2	3.6	4.0
.8	0.3	0.7	1.2	1.6	2.0	2.4	2.8	3.2	3.6	4.0
.9	0.3	0.8	1.2	1.6	2.0	2.4	2.8	3.2	3.6	4.0
.0	0.3	0.8	1.3	1.8	2.3	2.8	3.3	3.8	4.3	4.8
.1	0.4	0.9	1.4	1.9	2.4	2.9	3.3	3.8	4.3	4.8
.2	0.4	0.9	1.4	1.9	2.4	2.9	3.4	3.9	4.4	4.9
.3	0.4	1.0	1.4	1.9	2.4	2.9	3.4	3.9	4.4	4.9
.4	0.5	1.0	1.5	2.0	2.5	3.0	3.5	4.0	4.5	5.0
.5	0.3	0.8	1.3	1.8	2.3	2.8	3.3	3.8	4.3	4.8
.6	0.3	0.8	1.3	1.8	2.3	2.8	3.3	3.8	4.3	4.8
.7	0.4	0.9	1.4	1.9	2.4	2.9	3.4	3.9	4.4	4.9
.8	0.4	0.9	1.4	1.9	2.4	2.9	3.4	3.9	4.4	4.9
.9	0.5	1.0	1.5	2.0	2.5	3.0	3.5	4.0	4.5	5.0
.5	0.3	0.8	1.3	1.8	2.2	2.7	3.1	3.6	4.1	4.5
.6	0.3	0.8	1.3	1.8	2.3	2.7	3.2	3.6	4.1	4.5
.7	0.4	0.9	1.4	1.8	2.3	2.7	3.2	3.6	4.1	4.6
.8	0.4	0.9	1.4	1.9	2.3	2.8	3.3	3.8	4.2	4.6
.9	0.5	1.0	1.5	2.0	2.6	3.1	3.6	4.1	4.6	5.2

Right half — Double Second Diff. and Corr.

0.8	0.1
2.5	0.2
4.1	0.2
5.8	0.3
7.4	0.4
9.1	0.5
10.7	0.6
12.3	0.7
14.0	0.8
15.6	0.9
17.3	1.0
18.9	1.1
20.6	1.2
22.2	1.3
23.9	1.4
25.5	1.5
8.8	0.6
10.4	0.7
12.0	0.8
13.6	0.9
15.2	0.9
16.8	1.0
18.4	1.1
20.0	1.2
21.6	1.3
23.2	1.4
24.8	1.5
26.4	1.7
28.0	1.7
29.6	1.8
31.2	1.9
32.8	2.0
34.4	2.1

The Double-Second-Difference correction (Corr.) is always to be added to the tabulated altitude.

Back at the interpolation table used for the declination increment for Hc, look up the correction for a factor of 1.1. Here it is 0.4. Since azimuth is decreasing, subtract this from the tabulated Z figure:

$$
\begin{array}{ll}
\text{Tab Z} & 51.2° \\
\text{Z corr.} & -\ \underline{0.4°} \\
\text{Z} & 50.8°
\end{array}
$$

This is the azimuth (bearing) of the body from your assumed position. However, it is an azimuth *in quadrant*, and it is not necessarily the azimuth from true north you need in order to draw the intercept on your plotting sheet. Fortunately, the rules for conversion of azimuth to azimuth north (Zn), appear on each spread in the main sections of *Pub. 229*. At the top of the left hand pages, you'll see:

LHA greater than 180°......Zn = Z

N Lat

LHA less than 180°......Zn = 360°−Z

At the bottom of the right hand pages is:

LHA greater than 180°......Zn = 180°−Z

S Lat

LHA less than 180°......Zn = 180°+Z

So there's never any question as to what you'll do. In the example I have just given, LHA is 38° and latitude is north so Zn = 360° −Z, or:

$$
\begin{array}{ll}
& 360.0° \\
\text{Z} - & \underline{50.8°} \\
\text{Zn} & 309.2°
\end{array}
$$

As a convention, Zn is always written as a three-digit number, such as 004°, for an azimuth north of 4 degrees.

PLOTTING THE INTERCEPT

If the calculated altitude is greater than observed altitude, you are closer to the sighted body than you would be if you were at the AP. This is because the distance calculated (using the tables) is greater than the distance to the GP of the body actually measured by the sextant. The intercept is *toward* the body. If Ho is greater than Hc, this situation is obviously the reverse, and intercept is *away* from the body. As a memory device, you can equate Coast Guard Academy with Calculated Greater Away.

When intercept is *toward* the body, it is drawn from the AP in the direction of azimuth north. Draw it as a dashed line for a distance equal to the difference between Ho and Hc. When the intercept is *away*, it is drawn from the AP in a direction 180 degrees from Zn. Actually, you line up your rules, triangle, or course protractor with the Zn angle and just draw the line in the opposite direction. Some navigators like to put a small arrowhead on the intercept to indicate direction toward the body, but this is not necessary. It is not required in the plotting procedures developed jointly in recent years by the Navy, Coast Guard, Merchant Marine, and organizations such as the Coast Guard Auxiliary and U.S. Power Squadrons.

In Fig. 6-4. AP is at Latitude N 42°, and Lo is 64°28.2′W. Intercept distance is 14.7 miles toward the body at a Zn of 309. 2°. *Note*: You can generally eyeball the decimal fraction of Zn within about half a degree, and this is certainly accurate enough for your purposes. In Fig. 6-5, AP is L N 33°, Lo 48°06.6W; intercept distance is 7.4 miles away from the body at a Zn of 086.4°.

Line of Position

Your actual position may be several miles from the point at which the intercept ends, but on an arc that has its center at the GP of the sighted body. At the scale of your plotting chart, this arc can be drawn as a straight line perpendicular to the intercept and extending several miles to each side of it. Although this line is considered to be a short segment of an arc, it is called a line of position (LOP). It is identified by the name of the sighted body and the time, in hours and minutes, of the sight.

Estimated Position

Your plotting sheet will carry your DR position for the time of the sight. Drop a perpendicular dashed line from the DR to the LOP. The intersection of the perpendicular with the LOP is presumed to be your estimated position (EP), and is indicated as such by a small square. Since the square indicates an EP, no further identification is needed. Note in Figs. 6-4 and 6-5 that the perpendiculars are always parallel to the intercepts.

AN AFTERNOON SUN SIGHT

Now let's take a typical afternoon sun sight and follow all the steps from the sextant observation to the establishment of an estimated position. The sky and horizon are clear, eye height is 12.5 feet above sea level, you have found sextant error to be + 2.8′, and you know that watch error is (f) − 1-17. You get the sun in the horizon mirror of your sextant and bring it down until the lower limb just touches the horizon. You call "mark." Your assistant notes the time: 15-29-52 and your sextant reading is 47°19.7′.

It is 18 August 1980 and you are in ZD +4. Because your watch is on daylight saving time, you will use ZD + 3 for your time calculation. Based on your boat speed through the water, the heading you have maintained, and the time since your last position fix, you determine your DR position to be L 37°58.7′N, Lo 57°29.9′W. All this information is entered in your sight log:

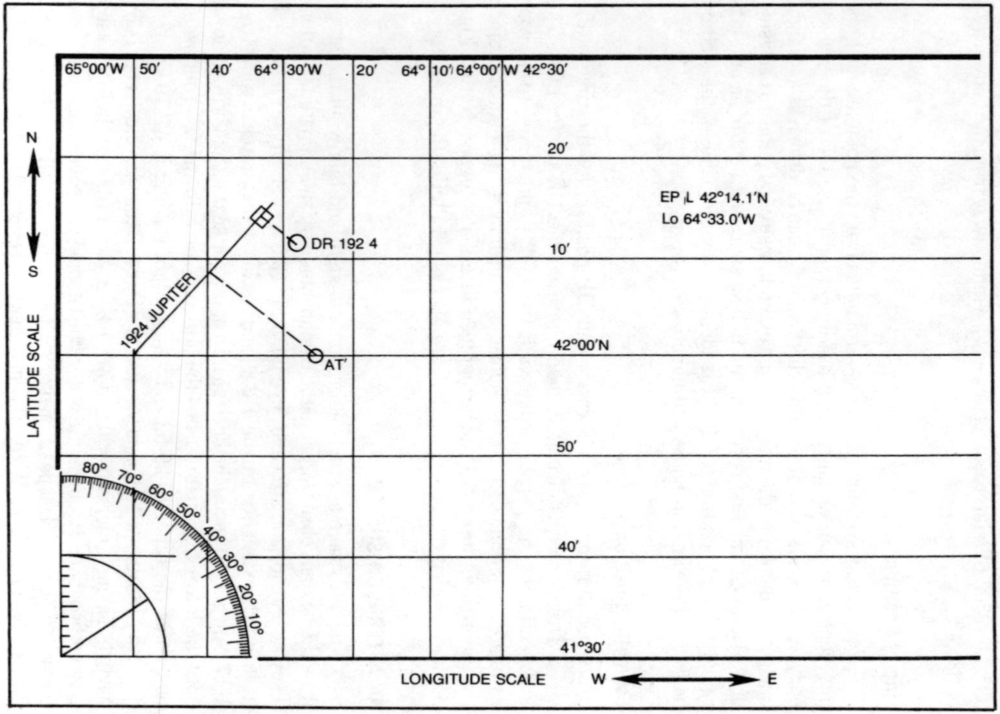

Fig. 6-4. Plot of the Jupiter sight.

EP L 42°14.1'N
Lo 64°33.0'W

DR 192 4

AT'

1924 JUPITER

LONGITUDE SCALE   W ⟷ E

LATITUDE SCALE   N ⟷ S

80° 70° 60° 50° 40° 30° 20° 10°

65°00'W   50'   64° 40'   30'W   20'   64° 10'64°00'W 42°30'
20'
10'
42°00'N
50'
40'
41°30'

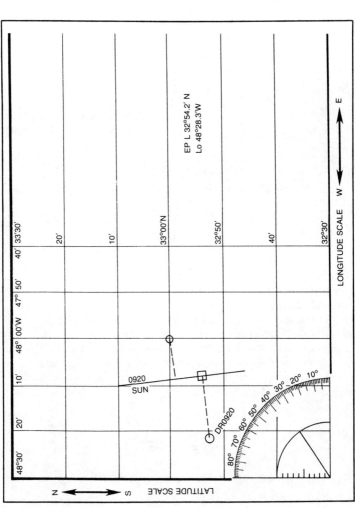

Fig. 6-5. Plot of morning sun sight.

Date	WT	WE	ZD	Body
18 Aug	15-29-52	(f)−1 17	+3	Sun LL

		Eye	
Hs	**IE**	**Height**	**DR**
47°19.7′	+2.8′	12′6″	L 37°58.7′N
			Lo 57°29.9′W

If you have been taking a run of sights—a good policy—the information for each will be recorded in sequence under its heading.

Without recourse to your references, you can quickly establish the GMT of the sight and enter this on your worksheet:

WT	15-29-52	18 Aug
WE	(f)−1-17	
ZT	15-28-35	
ZD	+ 3	
GMT	18-28-35	

From the inside front cover of the *Nautical Almanac*, you get the information needed to transform sextant altitude to observed altitude (Table 6-6):

Hs	47°19.7′
IC	− 2.8′
Dip	− 3.4′
Ha	47°13.5′
Main	+ 15.1′
Ho	47°28.6′

This all goes on the worksheet. Note that the index correction is the reverse of the index error and that the dip and main correction figures are from the *Almanac*.

Turn to the daily page for 18 August (Table 6-7). At 1800, GHA of the sun is 89°04.7′ and declination is N 12°52.8′. The d factor of 0.8′ is negative as an inspection of the table shows that declination is decreasing at this time. It will show on your worksheet as −0.8′. Enter the GHA and declination figures on your worksheet and turn to the buff page at the back for an increment of 38m 35s (Table 6-8). This is 7°08.8′. GHA of the sun is:

18h	89°04.7′
28m 35s	7°08.8′
GHA Sun	96°13.5′

Also correct the declination of the sun, using the same buff page, where the d factor for −0.8′ is −0.4′:

Table 6-6. Dip and Refraction Corrections for a Sun Sight.

A2 ALTITUDE CORRECTION TABLES 10°-90°—SUN, STARS, PLANETS

SUN

OCT.–MAR. App. Alt.	Lower Limb	Upper Limb	APR.–SEPT. App. Alt.	Lower Limb	Upper Limb
9 34	+10·8	−21·5	9 39	+10·6	−21·2
9 45	+10·9	−21·4	9 51	+10·7	−21·1
9 56	+11·0	−21·3	10 03	+10·8	−21·0
10 08	+11·1	−21·2	10 15	+10·9	−20·9
10 21	+11·2	−21·1	10 27	+11·0	−20·8
10 34	+11·3	−21·0	10 40	+11·1	−20·7
10 47	+11·4	−20·9	10 54	+11·2	−20·6
11 01	+11·5	−20·8	11 08	+11·3	−20·5
11 15	+11·6	−20·7	11 23	+11·4	−20·4
11 30	+11·7	−20·6	11 38	+11·5	−20·3
11 46	+11·8	−20·5	11 54	+11·6	−20·2
12 02	+11·9	−20·4	12 10	+11·7	−20·1
12 19	+12·0	−20·3	12 28	+11·8	−20·0
12 37	+12·1	−20·2	12 46	+11·9	−19·9
12 55	+12·2	−20·1	13 05	+12·0	−19·8
13 14	+12·3	−20·0	13 24	+12·1	−19·7
13 35	+12·4	−19·9	13 45	+12·2	−19·6
13 56	+12·5	−19·8	14 07	+12·3	−19·5
14 18	+12·6	−19·7	14 30	+12·4	−19·4
14 42	+12·7	−19·6	14 54	+12·5	−19·3
15 06	+12·8	−19·5	15 19	+12·6	−19·2
15 32	+12·9	−19·4	15 46	+12·7	−19·1
15 59	+13·0	−19·3	16 14	+12·8	−19·0
16 28	+13·1	−19·2	16 44	+12·9	−18·9
16 59	+13·2	−19·1	17 15	+13·0	−18·8
17 32	+13·3	−19·0	17 48	+13·1	−18·7
18 06	+13·4	−18·9	18 24	+13·2	−18·6
18 42	+13·5	−18·8	19 01	+13·3	−18·5
19 21	+13·6	−18·7	19 42	+13·4	−18·4
20 03	+13·7	−18·6	20 25	+13·5	−18·3
20 48	+13·8	−18·5	21 11	+13·6	−18·2
21 35	+13·9	−18·4	22 00	+13·7	−18·1
22 26	+14·0	−18·3	22 54	+13·8	−18·0
23 22	+14·1	−18·2	23 51	+13·9	−17·9
24 21	+14·2	−18·1	24 53	+14·0	−17·8
25 26	+14·3	−18·0	26 00	+14·1	−17·7
26 36	+14·4	−17·9	27 13	+14·2	−17·6
27 52	+14·5	−17·8	28 33	+14·3	−17·5
29 15	+14·6	−17·7	30 00	+14·4	−17·4
30 46	+14·7	−17·6	31 35	+14·5	−17·3
32 26	+14·8	−17·5	33 20	+14·6	−17·2
34 17	+14·9	−17·4	35 17	+14·7	−17·1
36 20	+15·0	−17·3	37 26	+14·8	−17·0
38 36	+15·1	−17·2	39 50	+14·9	−16·9
41 08	+15·2	−17·1	42 31	+15·0	−16·8
43 59	+15·3	−17·0	45 31	+15·1	−16·7
47 10	+15·4	−16·9	48 55	+15·2	−16·6
50 46	+15·5	−16·8	52 44	+15·3	−16·5
54 49	+15·6	−16·7	57 02	+15·4	−16·4
59 23	+15·7	−16·6	61 51	+15·5	−16·3
64 30	+15·8	−16·5	67 17	+15·6	−16·2
70 12	+15·9	−16·4	73 16	+15·7	−16·1
76 26	+16·0	−16·3	79 43	+15·8	−16·0
83 05	+16·1	−16·2	86 32	+15·9	−15·9
90 00			90 00		

STARS AND PLANETS

App. Alt.	Corrⁿ
9 56	−5·3
10 08	−5·2
10 20	−5·1
10 33	−5·0
10 46	−4·9
11 00	−4·8
11 14	−4·7
11 29	−4·6
11 45	−4·5
12 01	−4·4
12 18	−4·3
12 35	−4·2
12 54	−4·1
13 13	−4·0
13 33	−3·9
13 54	−3·8
14 16	−3·7
14 40	−3·6
15 04	−3·5
15 30	−3·4
15 57	−3·3
16 26	−3·2
16 56	−3·1
17 28	−3·0
18 02	−2·9
18 38	−2·8
19 17	−2·7
19 58	−2·6
20 42	−2·5
21 28	−2·4
22 19	−2·3
23 13	−2·2
24 11	−2·1
25 14	−2·0
26 22	−1·9
27 36	−1·8
28 56	−1·7
30 24	−1·6
32 00	−1·5
33 45	−1·4
35 40	−1·3
37 48	−1·2
40 08	−1·1
42 44	−1·0
45 36	−0·9
48 47	−0·8
52 18	−0·7
56 11	−0·6
60 28	−0·5
65 08	−0·4
70 11	−0·3
75 34	−0·2
81 13	−0·1
87 03	0·0
90 00	

Additional Corrⁿ — 1980

App. Alt.	Additional Corrⁿ
VENUS	
Jan. 1–Feb. 26	
42°	+0·1
Feb. 27–Apr. 13	
47°	+0·2
Apr. 14–May 9	
46°	+0·3
May 10–May 25	
11	+0·4
41	+0·5
May 26–June 3	
6	+0·5
20	+0·6
31	+0·7
June 4–June 26	
4	+0·6
12	+0·7
22	+0·8
June 27–July 6	
6	+0·5
20	+0·6
31	+0·7
July 7–July 21	
11	+0·4
41	+0·5
July 22–Aug. 17	
46	+0·3
Aug. 18–Oct. 2	
47	+0·2
Oct. 3–Dec. 31	
42	+0·1
MARS	
Jan. 1–Apr. 28	
41	+0·2
75	+0·1
Apr. 29–Dec. 31	
60	+0·1

DIP

Ht. of Eye (m)	Corrⁿ	Ht. of Eye (ft)
2·4	−2·8	8·0
2·6	−2·9	8·6
2·8		9·2
3·0	−3·0	9·8
3·2	−3·1	10·5
3·4		11·2
3·6	−3·3	11·9
3·8		12·6
4·0	−3·5	13·3
4·3	−3·6	14·1
4·5	−3·7	14·9
4·7		15·7
5·0	−3·9	16·5
5·2		17·4
5·5	−4·1	18·3
5·8		19·1
6·1	−4·3	20·1
6·3	−4·4	21·0
6·6	−4·5	22·0
6·9		22·9
7·2		23·9
7·5	−4·8	24·9
7·9		26·0
8·2	−5·0	27·1
8·5	−5·1	28·1
8·8	−5·2	29·2
9·2	−5·3	30·4
9·5	−5·4	31·5
9·9	−5·5	32·7
10·3	−5·6	33·9
10·6	−5·7	35·1
11·0	−5·8	36·3
11·4	−6·0	37·6
11·8	−6·1	38·9
12·2	−6·2	40·1
12·6	−6·3	41·5
13·0	−6·3	42·8
13·4	−6·4	44·2
13·8	−6·5	45·5
14·2	−6·6	46·9
14·7	−6·7	48·4
15·1	−6·9	49·8
15·5	−7·0	51·3
16·0	−7·0	52·8
16·5	−7·1	54·3
16·9	−7·2	55·8
17·4	−7·3	57·4
17·9	−7·4	58·9
18·4	−7·5	60·5
18·8	−7·6	62·1
19·3	−7·7	63·8
19·8	−7·8	65·4
20·4	−7·9	67·1
20·9	−8·0	68·8
21·4	−8·1	70·5

Ht. of Eye — Corrⁿ

Ht. of Eye (m)	Corrⁿ
1·0	− 1·8
1·5	− 2·2
2·0	− 2·5
2·5	− 2·8
3·0	− 3·0

See table →

Ht. of Eye (m)	Corrⁿ
20	− 7·9
22	− 8·3
24	− 8·6
26	− 9·0
28	− 9·3
30	− 9·6
32	−10·0
34	−10·3
36	−10·6
38	−10·8
40	−11·1
42	−11·4
44	−11·7
46	−11·9
48	−12·2

Ht. of Eye (ft)	Corrⁿ
2	− 1·4
4	− 1·9
6	− 2·4
8	− 2·7
10	− 3·1

See table →

Ht. of Eye (ft)	Corrⁿ
70	− 8·1
75	− 8·4
80	− 8·7
85	− 8·9
90	− 9·2
95	− 9·5
100	− 9·7
105	− 9·9
110	−10·2
115	−10·4
120	−10·6
125	−10·8
130	−11·1
135	−11·3
140	−11·5
145	−11·7
150	−11·9
155	−12·1

App. Alt. = Apparent altitude = Sextant altitude corrected for index error and dip.

Dec. 18h   N 12°52.8'
d corr.   −   0.4'
Dec.   N 12°52.4'

This gives you your tabulated declination of 12° and declination increment of 52.4' for *Pub 229*. Now convert GHA to LHA by establishing your AP at Lo 57°13.5' as this is the closest point to your DR longitude that will give you a whole degree of LHA:

Table 6-7. Daily Page Data for the Sun Sight. Note d Factor.

1980 AUGUST 16, 17, 18 (SAT., SUN., MON.)   163

G.M.T.	SUN G.H.A.	SUN Dec.	MOON G.H.A.	v	Dec.	d	H.P.
16 00	178 55.8	N13 45.7	123 24.7	15.5	S 3 54.5	9.8	54.1
01	193 55.9	44.9	137 59.2	15.5	4 04.3	9.7	54.1
02	208 56.1	44.1	152 33.7	15.5	4 14.0	9.7	54.1
03	223 56.2 ··	43.3	167 08.2	15.5	4 23.7	9.7	54.1
04	238 56.3	42.5	181 42.7	15.4	4 33.4	9.7	54.1
05	253 56.5	41.7	196 17.2	15.4	4 43.1	9.6	54.1
06	268 56.6	N13 40.9	210 51.6	15.5	S 4 52.7	9.7	54.1
07	283 56.7	40.1	225 26.1	15.4	5 02.4	9.6	54.1
S 08	298 56.8	39.3	240 00.5	15.4	5 12.0	9.6	54.1
A 09	313 57.0 ··	38.5	254 34.9	15.4	5 21.6	9.6	54.1
T 10	328 57.1	37.8	269 09.3	15.3	5 31.2	9.5	54.1
U 11	343 57.2	37.0	283 43.6	15.4	5 40.7	9.6	54.1
R 12	358 57.4	N13 36.2	298 18.0	15.3	S 5 50.3	9.5	54.2
D 13	13 57.5	35.4	312 52.3	15.4	5 59.8	9.5	54.2
A 14	28 57.6	34.6	327 26.7	15.3	6 09.3	9.4	54.2
Y 15	43 57.7 ··	33.8	342 01.0	15.2	6 18.7	9.5	54.2
16	58 57.9	33.0	356 35.2	15.3	6 28.2	9.4	54.2
17	73 58.0	32.2	11 09.5	15.2	6 37.6	9.4	54.2
18	88 58.1	N13 31.4	25 43.7	15.2	S 6 47.0	9.3	54.2
19	103 58.3	30.6	40 17.9	15.2	6 56.3	9.4	54.2
20	118 58.4	29.8	54 52.1	15.2	7 05.7	9.3	54.2
21	133 58.5 ··	29.0	69 26.3	15.1	7 15.0	9.3	54.2
22	148 58.7	28.2	84 00.4	15.1	7 24.3	9.2	54.2
23	163 58.8	27.4	98 34.5	15.1	7 33.5	9.3	54.2
17 00	178 58.9	N13 26.6	113 08.6	15.0	S 7 42.8	9.2	54.2
01	193 59.1	25.8	127 42.6	15.1	7 52.0	9.2	54.2
02	208 59.2	25.0	142 16.7	15.0	8 01.2	9.1	54.2
03	223 59.3 ··	24.2	156 50.7	14.9	8 10.3	9.1	54.3
04	238 59.5	23.4	171 24.6	15.0	8 19.4	9.1	54.3
05	253 59.6	22.6	185 58.6	14.9	8 28.5	9.0	54.3
06	268 59.7	N13 21.8	200 32.5	14.8	S 8 37.5	9.1	54.3
07	283 59.9	21.0	215 06.3	14.9	8 46.6	8.9	54.3
08	299 00.0	20.2	229 40.2	14.8	8 55.5	9.0	54.3
S 09	314 00.1 ··	19.4	244 14.0	14.8	9 04.5	8.9	54.3
U 10	329 00.3	18.6	258 47.8	14.7	9 13.4	8.9	54.3
N 11	344 00.4	17.8	273 21.5	14.7	9 22.3	8.8	54.3
D 12	359 00.5	N13 17.0	287 55.2	14.7	S 9 31.1	8.8	54.3
A 13	14 00.7	16.2	302 28.9	14.6	9 39.9	8.8	54.4
Y 14	29 00.8	15.4	317 02.5	14.6	9 48.7	8.7	54.4
15	44 00.9 ··	14.6	331 36.1	14.6	9 57.5	8.7	54.4
16	59 01.1	13.8	346 09.7	14.5	10 06.2	8.6	54.4
17	74 01.2	13.0	0 43.2	14.5	10 14.8	8.6	54.4
18	89 01.3	N13 12.2	15 16.7	14.4	S10 23.4	8.6	54.4
19	104 01.5	11.4	29 50.1	14.4	10 32.0	8.6	54.4
20	119 01.6	10.6	44 23.5	14.4	10 40.6	8.5	54.4
21	134 01.8 ··	09.8	58 56.9	14.3	10 49.1	8.4	54.5
22	149 01.9	09.0	73 30.2	14.3	10 57.5	8.4	54.5
23	164 02.0	08.2	88 03.5	14.2	11 05.9	8.4	54.5
18 00	179 02.2	N13 07.4	102 36.7	14.2	S11 14.3	8.3	54.5
01	194 02.3	06.6	117 09.9	14.1	11 22.6	8.3	54.5
02	209 02.4	05.8	131 43.0	14.1	11 30.9	8.3	54.5
03	224 02.6 ··	05.0	146 16.1	14.1	11 39.2	8.2	54.6
04	239 02.7	04.1	160 49.2	14.0	11 47.4	8.1	54.6
05	254 02.9	03.3	175 22.2	14.0	11 55.5	8.1	54.6
06	269 03.0	N13 02.5	189 55.2	13.9	S12 03.6	8.1	54.6
07	284 03.1	01.7	204 28.1	13.8	12 11.7	8.0	54.6
08	299 03.3	00.9	219 00.9	13.9	12 19.7	8.0	54.6
M 09	314 03.4	13 00.1	233 33.8	13.7	12 27.7	7.9	54.7
O 10	329 03.6	12 59.3	248 06.5	13.8	12 35.6	7.9	54.7
N 11	344 03.7	58.5	262 39.3	13.6	12 43.5	7.8	54.7
D 12	359 03.8	N12 57.7	277 11.9	13.6	S12 51.3	7.7	54.7
A 13	14 04.0	56.9	291 44.5	13.6	12 59.0	7.7	54.7
Y 14	29 04.1	56.0	306 17.1	13.5	13 06.7	7.7	54.7
15	44 04.3 ··	55.2	320 49.6	13.5	13 14.4	7.6	54.8
16	59 04.4	54.4	335 22.1	13.4	13 22.0	7.6	54.8
17	74 04.5	53.6	349 54.5	13.4	13 29.6	7.5	54.8
18	89 04.7	N12 52.8	4 26.9	13.3	S13 37.1	7.4	54.8
19	104 04.8	52.0	18 59.2	13.2	13 44.5	7.4	54.8
20	119 05.0	51.2	33 31.4	13.2	13 51.9	7.3	54.9
21	134 05.1 ··	50.4	48 03.6	13.1	13 59.2	7.3	54.9
22	149 05.3	49.5	62 35.7	13.1	14 06.5	7.2	54.9
23	164 05.4	48.7	77 07.8	13.1	14 13.7	7.2	54.9
	S.D. 15.8	d 0.8	S.D. 14.8		14.8		14.9

Lat.	Twilight Naut.	Twilight Civil	Sunrise	Moonrise 16	17	18	19
N 72	////	////	02 38	11 19	12 58	14 43	16 41
N 70	////	01 00	03 07	11 10	12 42	14 17	15 56
68	////	01 59	03 28	11 03	12 29	13 57	15 26
66	////	02 32	03 45	10 57	12 19	13 41	15 04
64	00 53	02 55	03 58	10 53	12 10	13 28	14 47
62	01 45	03 14	04 09	10 48	12 03	13 18	14 32
60	02 15	03 29	04 19	10 45	11 56	13 08	14 20
N 58	02 37	03 41	04 27	10 42	11 51	13 00	14 10
56	02 55	03 52	04 35	10 39	11 46	12 53	14 01
54	03 09	04 01	04 41	10 36	11 42	12 47	13 53
52	03 21	04 10	04 47	10 34	11 38	12 42	13 46
50	03 32	04 17	04 53	10 32	11 34	12 36	13 39
45	03 53	04 32	05 04	10 27	11 26	12 26	13 25
N 40	04 10	04 45	05 13	10 23	11 20	12 17	13 14
35	04 23	04 55	05 21	10 20	11 14	12 09	13 04
30	04 34	05 03	05 28	10 17	11 09	12 02	12 56
20	04 51	05 18	05 41	10 12	11 01	11 50	12 41
N 10	05 04	05 29	05 51	10 08	10 53	11 40	12 29
0	05 15	05 39	06 01	10 04	10 47	11 31	12 17
S 10	05 24	05 49	06 10	10 00	10 40	11 21	12 05
20	05 32	05 58	06 20	09 56	10 32	11 11	11 53
30	05 39	06 07	06 32	09 51	10 24	11 00	11 39
35	05 42	06 12	06 38	09 48	10 20	10 53	11 30
40	05 45	06 18	06 45	09 45	10 14	10 46	11 21
45	05 49	06 24	06 54	09 41	10 08	10 37	11 10
S 50	05 53	06 31	07 04	09 37	10 01	10 27	10 57
52	05 54	06 34	07 09	09 35	09 57	10 22	10 51
54	05 56	06 37	07 14	09 33	09 54	10 17	10 44
56	05 57	06 41	07 20	09 31	09 50	10 11	10 37
58	05 59	06 45	07 26	09 28	09 45	10 05	10 28
S 60	06 00	06 49	07 34	09 25	09 40	09 57	10 19

Lat.	Sunset	Twilight Civil	Twilight Naut.	Moonset 16	17	18	19
N 72	21 25	////	////	20 49	20 39	20 26	20 05
N 70	20 57	22 55	////	20 59	20 56	20 53	20 51
68	20 37	22 03	////	21 08	21 10	21 14	21 21
66	20 21	21 32	////	21 15	21 22	21 31	21 44
64	20 08	21 10	23 03	21 21	21 31	21 44	22 02
62	19 57	20 52	22 17	21 26	21 39	21 56	22 17
60	19 47	20 37	21 49	21 31	21 47	22 06	22 30
N 58	19 39	20 25	21 28	21 35	21 53	22 14	22 40
56	19 32	20 14	21 11	21 39	21 59	22 22	22 50
54	19 25	20 05	20 57	21 42	22 04	22 28	22 58
52	19 20	19 57	20 45	21 45	22 08	22 35	23 06
50	19 14	19 50	20 35	21 48	22 12	22 40	23 13
45	19 03	19 36	20 16	21 53	22 21	22 52	23 27
N 40	18 54	19 23	19 57	21 58	22 29	23 02	23 39
35	18 46	19 13	19 44	22 03	22 35	23 10	23 49
30	18 39	19 04	19 34	22 06	22 41	23 18	23 58
20	18 27	18 50	19 17	22 13	22 51	23 31	24 12
N 10	18 17	18 38	19 04	22 19	22 59	23 42	24 27
0	18 07	18 29	18 53	22 24	23 08	23 53	24 40
S 10	17 58	18 19	18 44	22 30	23 16	24 03	00 03
20	17 48	18 11	18 37	22 36	23 25	24 15	00 15
30	17 37	18 01	18 29	22 42	23 35	24 28	00 28
35	17 30	17 56	18 26	22 46	23 40	24 35	00 35
40	17 23	17 51	18 23	22 50	23 47	24 44	00 44
45	17 15	17 45	18 19	22 56	23 55	24 54	00 54
S 50	17 05	17 38	18 16	23 02	24 04	00 04	01 06
52	17 00	17 35	18 15	23 04	24 08	00 08	01 12
54	16 55	17 32	18 13	23 07	24 13	00 13	01 18
56	16 49	17 28	18 12	23 11	24 18	00 18	01 26
58	16 43	17 24	18 10	23 15	24 24	00 24	01 33
S 60	16 35	17 20	18 09	23 19	24 31	00 31	01 42

Day	SUN Eqn. of Time 00h	SUN Eqn. of Time 12h	SUN Mer. Pass.	MOON Mer. Pass. Upper	MOON Mer. Pass. Lower	Age	Phase
16	04 17	04 11	12 04	06 14	03 53	06	
17	04 05	03 58	12 04	16 57	04 35	07	
18	03 52	03 45	12 04	17 42	05 19	08	●

Table 6-8. GHA Increment for 28m 35s and d Correction for 0.8 Factor.

28	SUN PLANETS	ARIES	MOON	v or Corrn d	v or Corrn d	v or Corrn d
00	7 00·0	7 01·1	6 40·9	0·0 0·0	6·0 2·9	12·0 5·7
01	7 00·3	7 01·4	6 41·1	0·1 0·0	6·1 2·9	12·1 5·7
02	7 00·5	7 01·7	6 41·3	0·2 0·1	6·2 2·9	12·2 5·8
03	7 00·8	7 01·9	6 41·6	0·3 0·1	6·3 3·0	12·3 5·8
04	7 01·0	7 02·2	6 41·8	0·4 0·2	6·4 3·0	12·4 5·9
05	7 01·3	7 02·4	6 42·1	0·5 0·2	6·5 3·1	12·5 5·9
06	7 01·5	7 02·7	6 42·3	0·6 0·3	6·6 3·1	12·6 6·0
07	7 01·8	7 02·9	6 42·5	0·7 0·3	6·7 3·2	12·7 6·0
08	7 02·0	7 03·2	6 42·8	0·8 0·4	6·8 3·2	12·8 6·1
09	7 02·3	7 03·4	6 43·0	0·9 0·4	6·9 3·3	12·9 6·1
10	7 02·5	7 03·7	6 43·3	1·0 0·5	7·0 3·3	13·0 6·2
11	7 02·8	7 03·9	6 43·5	1·1 0·5	7·1 3·4	13·1 6·2
12	7 03·0	7 04·2	6 43·7	1·2 0·6	7·2 3·4	13·2 6·3
13	7 03·3	7 04·4	6 44·0	1·3 0·6	7·3 3·5	13·3 6·3
14	7 03·5	7 04·7	6 44·2	1·4 0·7	7·4 3·5	13·4 6·4
15	7 03·8	7 04·9	6 44·4	1·5 0·7	7·5 3·6	13·5 6·4
16	7 04·0	7 05·2	6 44·7	1·6 0·8	7·6 3·6	13·6 6·5
17	7 04·3	7 05·4	6 44·9	1·7 0·8	7·7 3·7	13·7 6·5
18	7 04·5	7 05·7	6 45·2	1·8 0·9	7·8 3·7	13·8 6·6
19	7 04·8	7 05·9	6 45·4	1·9 0·9	7·9 3·8	13·9 6·6
20	7 05·0	7 06·2	6 45·6	2·0 1·0	8·0 3·8	14·0 6·7
21	7 05·3	7 06·4	6 45·9	2·1 1·0	8·1 3·8	14·1 6·7
22	7 05·5	7 06·7	6 46·1	2·2 1·0	8·2 3·9	14·2 6·7
23	7 05·8	7 06·9	6 46·4	2·3 1·1	8·3 3·9	14·3 6·8
24	7 06·0	7 07·2	6 46·6	2·4 1·1	8·4 4·0	14·4 6·8
25	7 06·3	7 07·4	6 46·8	2·5 1·2	8·5 4·0	14·5 6·9
26	7 06·5	7 07·7	6 47·1	2·6 1·2	8·6 4·1	14·6 6·9
27	7 06·8	7 07·9	6 47·3	2·7 1·3	8·7 4·1	14·7 7·0
28	7 07·0	7 08·2	6 47·5	2·8 1·3	8·8 4·2	14·8 7·0
29	7 07·3	7 08·4	6 47·8	2·9 1·4	8·9 4·2	14·9 7·1
30	7 07·5	7 08·7	6 48·0	3·0 1·4	9·0 4·3	15·0 7·1
31	7 07·8	7 08·9	6 48·3	3·1 1·5	9·1 4·3	15·1 7·2
32	7 08·0	7 09·2	6 48·5	3·2 1·5	9·2 4·4	15·2 7·2
33	7 08·3	7 09·4	6 48·7	3·3 1·6	9·3 4·4	15·3 7·3
34	7 08·5	7 09·7	6 49·0	3·4 1·6	9·4 4·5	15·4 7·3
35	7 08·8	7 09·9	6 49·2	3·5 1·7	9·5 4·5	15·5 7·4
36	7 09·0	7 10·2	6 49·5	3·6 1·7	9·6 4·6	15·6 7·4
37	7 09·3	7 10·4	6 49·7	3·7 1·8	9·7 4·6	15·7 7·5
38	7 09·5	7 10·7	6 49·9	3·8 1·8	9·8 4·7	15·8 7·5
39	7 09·8	7 10·9	6 50·2	3·9 1·9	9·9 4·7	15·9 7·6
40	7 10·0	7 11·2	6 50·4	4·0 1·9	10·0 4·8	16·0 7·6
41	7 10·3	7 11·4	6 50·6	4·1 1·9	10·1 4·8	16·1 7·6
42	7 10·5	7 11·7	6 50·9	4·2 2·0	10·2 4·8	16·2 7·7
43	7 10·8	7 11·9	6 51·1	4·3 2·0	10·3 4·9	16·3 7·7
44	7 11·0	7 12·2	6 51·4	4·4 2·1	10·4 4·9	16·4 7·8
45	7 11·3	7 12·4	6 51·6	4·5 2·1	10·5 5·0	16·5 7·8
46	7 11·5	7 12·7	6 51·8	4·6 2·2	10·6 5·0	16·6 7·9
47	7 11·8	7 12·9	6 52·1	4·7 2·2	10·7 5·1	16·7 7·9
48	7 12·0	7 13·2	6 52·3	4·8 2·3	10·8 5·1	16·8 8·0
49	7 12·3	7 13·4	6 52·6	4·9 2·3	10·9 5·2	16·9 8·0
50	7 12·5	7 13·7	6 52·8	5·0 2·4	11·0 5·2	17·0 8·1
51	7 12·8	7 13·9	6 53·0	5·1 2·4	11·1 5·3	17·1 8·1
52	7 13·0	7 14·2	6 53·3	5·2 2·5	11·2 5·3	17·2 8·2
53	7 13·3	7 14·4	6 53·5	5·3 2·5	11·3 5·4	17·3 8·2
54	7 13·5	7 14·7	6 53·8	5·4 2·6	11·4 5·4	17·4 8·3
55	7 13·8	7 14·9	6 54·0	5·5 2·6	11·5 5·5	17·5 8·3
56	7 14·0	7 15·2	6 54·2	5·6 2·7	11·6 5·5	17·6 8·4
57	7 14·3	7 15·4	6 54·5	5·7 2·7	11·7 5·6	17·7 8·4
58	7 14·5	7 15·7	6 54·7	5·8 2·8	11·8 5·6	17·8 8·5
59	7 14·8	7 15·9	6 54·9	5·9 2·8	11·9 5·7	17·9 8·5
60	7 15·0	7 16·2	6 55·2	6·0 2·9	12·0 5·7	18·0 8·6

29	SUN PLANETS	ARIES	MOON	v or Corrn d	v or Corrn d	v or Corrn d
00	7 15·0	7 16·2	6 55·2	0·0 0·0	6·0 3·0	12·0 5·9
01	7 15·3	7 16·4	6 55·4	0·1 0·0	6·1 3·0	12·1 5·9
02	7 15·5	7 16·7	6 55·7	0·2 0·1	6·2 3·0	12·2 6·0
03	7 15·8	7 16·9	6 55·9	0·3 0·1	6·3 3·1	12·3 6·0
04	7 16·0	7 17·2	6 56·1	0·4 0·2	6·4 3·1	12·4 6·1
05	7 16·3	7 17·4	6 56·4	0·5 0·2	6·5 3·2	12·5 6·1
06	7 16·5	7 17·7	6 56·6	0·6 0·3	6·6 3·2	12·6 6·2
07	7 16·8	7 17·9	6 56·9	0·7 0·3	6·7 3·3	12·7 6·2
08	7 17·0	7 18·2	6 57·1	0·8 0·4	6·8 3·3	12·8 6·3
09	7 17·3	7 18·4	6 57·3	0·9 0·4	6·9 3·4	12·9 6·3
10	7 17·5	7 18·7	6 57·6	1·0 0·5	7·0 3·4	13·0 6·4
11	7 17·8	7 18·9	6 57·8	1·1 0·5	7·1 3·5	13·1 6·4
12	7 18·0	7 19·2	6 58·0	1·2 0·6	7·2 3·5	13·2 6·5
13	7 18·3	7 19·4	6 58·3	1·3 0·6	7·3 3·6	13·3 6·5
14	7 18·5	7 19·7	6 58·5	1·4 0·7	7·4 3·6	13·4 6·6
15	7 18·8	7 20·0	6 58·8	1·5 0·7	7·5 3·7	13·5 6·6
16	7 19·0	7 20·2	6 59·0	1·6 0·8	7·6 3·7	13·6 6·7
17	7 19·3	7 20·5	6 59·2	1·7 0·8	7·7 3·8	13·7 6·7
18	7 19·5	7 20·7	6 59·5	1·8 0·9	7·8 3·8	13·8 6·8
19	7 19·8	7 21·0	6 59·7	1·9 0·9	7·9 3·9	13·9 6·8
20	7 20·0	7 21·2	7 00·0	2·0 1·0	8·0 3·9	14·0 6·9
21	7 20·3	7 21·5	7 00·2	2·1 1·0	8·1 4·0	14·1 6·9
22	7 20·5	7 21·7	7 00·4	2·2 1·1	8·2 4·0	14·2 7·0
23	7 20·8	7 22·0	7 00·7	2·3 1·1	8·3 4·1	14·3 7·0
24	7 21·0	7 22·2	7 00·9	2·4 1·2	8·4 4·1	14·4 7·1
25	7 21·3	7 22·5	7 01·1	2·5 1·2	8·5 4·2	14·5 7·1
26	7 21·5	7 22·7	7 01·4	2·6 1·3	8·6 4·2	14·6 7·2
27	7 21·8	7 23·0	7 01·6	2·7 1·3	8·7 4·3	14·7 7·2
28	7 22·0	7 23·2	7 01·9	2·8 1·4	8·8 4·3	14·8 7·3
29	7 22·3	7 23·5	7 02·1	2·9 1·4	8·9 4·4	14·9 7·3
30	7 22·5	7 23·7	7 02·3	3·0 1·5	9·0 4·4	15·0 7·4
31	7 22·8	7 24·0	7 02·6	3·1 1·5	9·1 4·5	15·1 7·4
32	7 23·0	7 24·2	7 02·8	3·2 1·6	9·2 4·5	15·2 7·5
33	7 23·3	7 24·5	7 03·1	3·3 1·6	9·3 4·6	15·3 7·5
34	7 23·5	7 24·7	7 03·3	3·4 1·7	9·4 4·6	15·4 7·6
35	7 23·8	7 25·0	7 03·5	3·5 1·7	9·5 4·7	15·5 7·6
36	7 24·0	7 25·2	7 03·8	3·6 1·8	9·6 4·7	15·6 7·7
37	7 24·3	7 25·5	7 04·0	3·7 1·8	9·7 4·8	15·7 7·7
38	7 24·5	7 25·7	7 04·3	3·8 1·9	9·8 4·8	15·8 7·8
39	7 24·8	7 26·0	7 04·5	3·9 1·9	9·9 4·9	15·9 7·8
40	7 25·0	7 26·2	7 04·7	4·0 2·0	10·0 4·9	16·0 7·9
41	7 25·3	7 26·5	7 05·0	4·1 2·0	10·1 5·0	16·1 7·9
42	7 25·5	7 26·7	7 05·2	4·2 2·1	10·2 5·0	16·2 8·0
43	7 25·8	7 27·0	7 05·4	4·3 2·1	10·3 5·1	16·3 8·0
44	7 26·0	7 27·2	7 05·7	4·4 2·2	10·4 5·1	16·4 8·1
45	7 26·3	7 27·5	7 05·9	4·5 2·2	10·5 5·2	16·5 8·1
46	7 26·5	7 27·7	7 06·2	4·6 2·3	10·6 5·2	16·6 8·2
47	7 26·8	7 28·0	7 06·4	4·7 2·3	10·7 5·3	16·7 8·2
48	7 27·0	7 28·2	7 06·6	4·8 2·4	10·8 5·3	16·8 8·3
49	7 27·3	7 28·5	7 06·9	4·9 2·4	10·9 5·4	16·9 8·3
50	7 27·5	7 28·7	7 07·1	5·0 2·5	11·0 5·4	17·0 8·4
51	7 27·8	7 29·0	7 07·4	5·1 2·5	11·1 5·5	17·1 8·4
52	7 28·0	7 29·2	7 07·6	5·2 2·6	11·2 5·5	17·2 8·5
53	7 28·3	7 29·5	7 07·8	5·3 2·6	11·3 5·6	17·3 8·5
54	7 28·5	7 29·7	7 08·1	5·4 2·7	11·4 5·6	17·4 8·6
55	7 28·8	7 30·0	7 08·3	5·5 2·7	11·5 5·7	17·5 8·6
56	7 29·0	7 30·2	7 08·5	5·6 2·8	11·6 5·7	17·6 8·7
57	7 29·3	7 30·5	7 08·8	5·7 2·8	11·7 5·8	17·7 8·7
58	7 29·5	7 30·7	7 09·0	5·8 2·9	11·8 5·8	17·8 8·8
59	7 29·8	7 31·0	7 09·3	5·9 2·9	11·9 5·9	17·9 8·8
60	7 30·0	7 31·2	7 09·5	6·0 3·0	12·0 5·9	18·0 8·9

GHA   96°13.5′
AP Lo 57°13.5′
LHA   39°

With this information, you can draw your time diagram (Fig. 6-6). It can go right on the worksheet if there's no room for it. Note that the actual diagrams need not have the angles identified as "Lo," "GHA," "LHA," etc.,

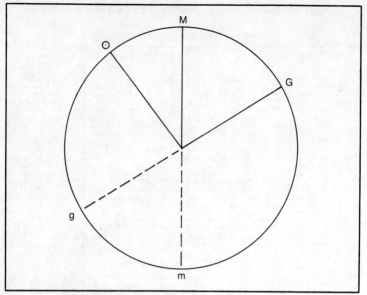

Fig. 6-6. Time diagram for the sun sight.

but you should be able to measure the angles—at least roughly—as a check on your calculations.

You have declination and LHA needed for the *Pub. 229* tables. The last item necessary is your AP latitude. You establish this as N 38° because this is the closest whole degree to your DR latitude.

In *Pub. 229* (Table 6-9), you'll use the left-hand page for 39° and 321° because latitude and declination are the same name—both north. For a declination of 12° at 38° latitude, you find:

Hc	d	Z
46°38.2′	+41.2	116.3

You also note that Z for 13° declination is 115.2.

At the back of *Pub. 229* (Table 6-10), the interpolation table for a declination increment of 52° shows 34.9′ under the "40" column, and 1.0′ for the 1.2′ balance of the d factor. Since d is positive, the sum of these figures will be added to the Tab Hc for 12°:

	34.9′
	+ 1.0′
d corr.	+35.9′

Tab Hc	46°38.2′
d corr.	+ 35.9′
Hc	47°14.1′

Table 6-9. *Pub. 229* Information for the Sight. Note That Z Figure Is Also Taken for the Next Higher Declination.

39°, 321° L.H.A. LATITUDE SAME NAME AS DECLINATION

N. Lat. { L.H.A. greater than 180°......Zn=Z / L.H.A. less than 180°........Zn=360°−Z }

Dec.	38° Hc	38° d	38° Z	39° Hc	39° d	39° Z	40° Hc	40° d	40° Z	41° Hc	41° d	41° Z	42° Hc	42° d	42° Z	43° Hc	43° d	43° Z	44° Hc	44° d	44° Z
0	37 45.8	+46.6	127.2	37 09.2	+47.2	127.7	36 32.2	+47.8	128.4	35 54.6	+48.5	129.0	35 16.6	+49.1	129.6	34 38.2	+49.6	130.1	33 59.3	+50.2	130.6
1	38 32.4	46.2	126.4	37 56.4	47.0	127.1	37 20.0	47.6	127.7	36 43.1	48.2	128.3	36 05.7	48.8	128.9	35 27.8	49.4	129.4	34 49.5	49.9	130.0
2	39 18.6	45.9	125.6	38 43.4	46.5	126.3	38 07.6	47.2	126.9	37 31.3	47.9	127.5	36 54.5	48.5	128.1	36 17.2	49.1	128.7	35 39.4	49.7	129.3
3	40 04.5	45.5	124.8	39 29.9	46.3	125.5	38 54.8	47.0	126.1	38 19.2	47.6	126.8	37 43.0	48.3	127.4	37 06.3	48.9	128.0	36 29.1	49.5	128.6
4	40 50.0	45.1	123.9	40 16.2	45.8	124.6	39 41.8	46.6	125.3	39 06.8	47.3	126.0	38 31.3	47.9	126.6	37 55.2	48.6	127.3	37 18.6	49.2	127.9
5	41 35.1	+44.7	123.1	41 02.0	+45.5	123.8	40 28.4	+46.2	124.5	39 54.1	+46.9	125.2	39 19.2	+47.6	125.9	38 43.8	+48.3	126.5	38 07.8	+48.9	127.2
6	42 19.8	44.3	122.2	41 47.5	45.1	122.9	41 14.6	45.8	123.7	40 41.0	46.6	124.4	40 06.8	47.3	125.1	39 32.1	48.0	125.8	38 56.7	48.7	126.4
7	43 04.1	43.8	121.2	42 32.6	44.7	122.0	42 00.4	45.5	122.8	41 27.6	46.2	123.5	40 54.1	47.0	124.3	40 20.1	47.6	125.0	39 45.4	48.3	125.7
8	43 47.9	43.4	120.3	43 17.3	44.2	121.1	42 45.9	45.0	121.9	42 13.8	45.9	122.7	41 41.1	46.6	123.4	41 07.7	47.3	124.2	40 33.7	48.0	124.9
9	44 31.3	42.8	119.3	44 01.5	43.7	120.2	43 30.9	44.6	121.0	42 59.7	45.4	121.8	42 27.7	46.2	122.6	41 55.0	47.0	123.4	41 21.7	47.7	124.1
10	45 14.1	+42.3	118.3	44 45.2	+43.3	119.2	44 15.5	+44.2	120.1	43 45.1	+45.0	120.9	43 13.9	+45.8	121.7	42 42.0	+46.6	122.5	42 09.4	+47.4	123.3
11	45 56.4	41.8	117.3	45 28.5	42.7	118.2	44 59.7	43.6	119.1	44 30.1	44.5	120.0	43 59.7	45.4	120.8	43 28.6	46.2	121.6	42 56.8	46.9	122.4
12	46 38.2	41.2	116.3	46 11.2	42.2	117.2	45 43.3	43.2	118.1	45 14.6	44.1	119.0	44 45.1	44.9	119.9	44 14.8	45.7	120.8	43 43.7	46.6	121.6
13	47 19.4	40.6	115.2	46 53.4	41.6	116.2	46 26.5	42.6	117.1	45 58.7	43.5	118.1	45 30.0	44.5	119.0	45 00.5	45.4	119.9	44 30.3	46.1	120.7
14	48 00.0	40.0	114.1	47 35.0	41.0	115.1	47 09.1	42.0	116.1	46 42.2	43.0	117.1	46 14.5	43.9	118.0	45 45.9	44.8	118.9	45 16.4	45.8	119.8
15	48 40.0	+39.3	113.0	48 16.0	+40.4	114.0	47 51.1	+41.4	115.1	47 25.2	+42.5	116.1	46 58.4	+43.4	117.0	46 30.7	+44.4	118.0	46 02.2	+45.2	118.9
16	49 19.3	38.6	111.9	48 56.4	39.7	112.9	48 32.5	40.8	113.9	48 07.7	41.8	115.0	47 41.8	42.9	116.0	47 15.1	43.8	117.0	46 47.4	44.8	117.9
17	49 57.9	37.8	110.7	49 36.1	39.0	111.8	49 13.3	40.2	112.9	48 49.5	41.2	113.9	48 24.7	42.3	115.0	47 58.9	43.3	116.0	47 32.2	44.2	116.9
18	50 35.7	37.1	109.5	50 15.1	38.3	110.6	49 53.5	39.4	111.7	49 30.7	40.6	112.8	49 07.0	41.6	113.9	48 42.2	42.7	114.9	48 16.4	43.7	115.9
19	51 12.8	36.2	108.2	50 53.4	37.5	109.4	50 32.9	38.7	110.5	50 11.3	39.9	111.7	49 48.6	41.0	112.8	49 24.9	42.1	113.8	49 00.1	43.2	114.9
20	51 49.0	+35.4	106.9	51 30.9	+36.7	108.1	51 11.6	+38.0	109.3	50 51.2	+39.2	110.5	50 29.6	+40.4	111.6	50 07.0	+41.4	112.7	49 43.3	+42.5	113.8
21	52 24.4	34.5	105.6	52 07.6	35.8	106.9	51 49.6	37.1	108.1	51 30.4	38.4	109.3	51 10.0	39.6	110.5	50 48.4	40.8	111.6	50 25.8	41.9	112.7
22	52 58.9	33.5	104.3	52 43.4	35.0	105.5	52 26.7	36.3	106.8	52 08.8	37.5	108.0	51 49.6	38.8	109.2	51 29.2	40.0	110.4	51 07.7	41.2	111.6
23	53 32.4	32.6	102.9	53 18.4	34.0	104.2	53 03.0	35.4	105.5	52 46.3	36.8	106.8	52 28.4	38.0	108.0	52 09.2	39.3	109.2	51 48.9	40.5	110.4
24	54 05.0	31.5	101.5	53 52.4	33.0	102.8	53 38.4	34.4	104.1	53 23.1	35.8	105.4	53 06.4	37.2	106.7	52 48.5	38.5	108.0	52 29.4	39.7	109.2
25	54 36.5	+30.4	100.0	54 25.4	+31.9	101.4	54 12.8	+33.5	102.7	53 58.9	+34.9	104.1	53 43.6	+36.3	105.4	53 27.0	+37.7	106.7	53 09.1	+39.0	108.0
26	55 06.9	29.4	98.5	54 57.3	30.9	99.9	54 46.3	32.4	101.3	54 33.8	33.9	102.7	54 19.9	35.4	104.1	54 04.7	36.7	105.4	53 48.1	38.1	106.7
27	55 36.3	28.1	97.0	55 28.2	29.8	98.4	55 18.7	31.3	99.8	55 07.7	32.9	101.3	54 55.3	34.3	102.7	54 41.4	35.8	104.0	54 26.2	37.2	105.4
28	56 04.4	26.9	95.4	55 58.0	28.6	96.9	55 50.0	30.2	98.3	55 40.6	31.8	99.8	55 29.6	33.4	101.2	55 17.2	34.9	102.6	55 03.4	36.3	104.0
29	56 31.3	25.6	93.8	56 26.6	27.3	95.3	56 20.2	29.0	96.8	56 12.4	30.6	98.3	56 03.0	32.2	99.7	55 52.1	33.8	101.2	55 39.7	35.3	102.6

Table 6-10. For Declination Increment of 52.4', Correction for a d Factor of 41.2' Is 34.9' Plus 1.0'.

INTERPOLATION TABLE

Left section — Altitude Difference (d)

Dec. Inc.	Tens 10'	20'	30'	40'	50'	Units 0	1	2	3	4	5	6	7	8	9
44.0	7.3	14.6	22.0	29.3	36.6	0.0	0.7	1.5	2.2	3.0	3.7	4.4	5.2	5.9	6.7
44.1	7.3	14.7	22.0	29.4	36.7	0.1	0.8	1.6	2.3	3.0	3.8	4.5	5.3	6.0	6.7
44.2	7.3	14.7	22.1	29.4	36.8	0.1	0.9	1.6	2.4	3.1	3.9	4.6	5.3	6.1	6.8
44.3	7.4	14.8	22.1	29.5	36.9	0.2	1.0	1.7	2.4	3.2	3.9	4.7	5.4	6.2	6.9
44.4	7.4	14.8	22.2	29.6	37.0	0.3	1.0	1.8	2.5	3.4	4.0	4.7	5.5	6.2	7.0
44.5	7.4	14.8	22.3	29.7	37.1	0.4	1.1	1.9	2.6	3.4	4.1	4.8	5.6	6.3	7.0
44.6	7.4	14.9	22.3	29.7	37.2	0.4	1.2	1.9	2.7	3.4	4.2	4.9	5.6	6.4	7.1
44.7	7.5	14.9	22.4	29.8	37.3	0.5	1.3	2.0	2.7	3.5	4.2	5.0	5.7	6.5	7.2
44.8	7.5	15.0	22.4	29.9	37.4	0.6	1.3	2.1	2.8	3.6	4.3	5.0	5.8	6.5	7.3
44.9	7.5	15.0	22.5	30.0	37.5	0.7	1.4	2.2	2.9	3.6	4.4	5.1	5.9	6.6	7.3
45.0	7.5	15.0	22.5	30.0	37.5	0.0	0.8	1.5	2.3	3.0	3.8	4.5	5.3	6.1	6.8
45.1	7.5	15.0	22.6	30.1	37.6	0.1	0.8	1.6	2.3	3.1	3.9	4.6	5.4	6.1	6.9
45.2	7.5	15.1	22.6	30.1	37.6	0.2	0.9	1.7	2.4	3.2	3.9	4.7	5.5	6.2	7.0
45.3	7.6	15.1	22.7	30.2	37.7	0.2	1.0	1.7	2.5	3.3	4.0	4.8	5.5	6.3	7.1
45.4	7.6	15.1	22.7	30.3	37.8	0.3	1.1	1.8	2.6	3.4	4.1	4.9	5.6	6.4	7.1
45.5	7.6	15.2	22.8	30.3	37.9	0.4	1.1	1.9	2.7	3.4	4.2	4.9	5.7	6.4	7.2
45.6	7.6	15.2	22.8	30.4	38.0	0.5	1.2	2.0	2.7	3.5	4.3	5.0	5.8	6.5	7.3
45.7	7.6	15.3	22.9	30.5	38.1	0.5	1.3	2.0	2.8	3.6	4.4	5.1	5.9	6.6	7.4
45.8	7.7	15.3	22.9	30.6	38.2	0.6	1.4	2.1	2.9	3.6	4.4	5.2	5.9	6.7	7.4
45.9	7.7	15.3	23.0	30.6	38.3	0.7	1.4	2.2	3.0	3.7	4.5	5.2	6.0	6.7	7.5
46.0	7.6	15.3	23.0	30.6	38.3	0.0	0.8	1.5	2.3	3.1	3.9	4.6	5.4	6.2	7.0
46.1	7.7	15.3	23.0	30.7	38.4	0.1	0.9	1.6	2.4	3.2	4.0	4.7	5.5	6.3	7.1
46.2	7.7	15.4	23.1	30.8	38.5	0.2	0.9	1.7	2.5	3.3	4.1	4.8	5.6	6.4	7.1
46.3	7.7	15.4	23.1	30.9	38.6	0.2	1.0	1.8	2.6	3.3	4.1	4.9	5.7	6.4	7.2
46.4	7.7	15.5	23.2	30.9	38.7	0.3	1.1	1.9	2.6	3.4	4.2	5.0	5.7	6.5	7.3

Left — Double Second Diff. and Corr.

Diff.	Corr.
1.1	0.1
3.2	0.2
5.3	0.3
7.5	0.4
9.6	0.5
11.7	0.6
13.9	0.7
16.0	0.8
18.1	0.9
20.3	1.0
22.4	1.1
24.5	1.2
26.7	1.3
28.8	1.4
30.9	1.5
33.1	1.6
35.2	
1.2	0.1
3.5	0.2
5.8	0.3
8.1	0.4
10.5	

Right section — Altitude Difference (d)

Dec. Inc.	Tens 10'	20'	30'	40'	50'	Units 0	1	2	3	4	5	6	7	8	9
52.0	8.6	17.3	26.0	34.6	43.3	0.0	0.9	1.7	2.6	3.5	4.4	5.2	6.1	7.0	7.9
52.1	8.7	17.3	26.0	34.7	43.4	0.1	1.0	1.8	2.7	3.6	4.5	5.3	6.2	7.1	8.0
52.2	8.7	17.4	26.1	34.8	43.5	0.1	1.0	1.9	2.8	3.7	4.5	5.4	6.3	7.2	8.0
52.3	8.7	17.4	26.1	34.8	43.6	0.3	1.1	2.0	2.9	3.8	4.6	5.5	6.4	7.3	8.1
52.4	8.7	17.5	26.2	34.9	43.7	0.3	1.2	2.1	3.0	3.8	4.7	5.6	6.5	7.3	8.2
52.5	8.8	17.5	26.3	35.0	43.8	0.5	1.4	2.3	3.1	3.9	4.8	5.7	6.6	7.4	8.3
52.6	8.8	17.5	26.3	35.1	43.8	0.5	1.4	2.3	3.1	4.0	4.9	5.8	6.6	7.5	8.4
52.7	8.8	17.6	26.4	35.2	43.9	0.6	1.5	2.4	3.2	4.1	5.0	5.9	6.7	7.6	8.5
52.8	8.8	17.6	26.4	35.2	44.0	0.7	1.6	2.4	3.3	4.2	5.1	5.9	6.8	7.7	8.6
52.9	8.9	17.7	26.5	35.3	44.1	0.8	1.7	2.5	3.4	4.3	5.2	6.0	6.9	7.8	8.7
53.0	8.8	17.6	26.5	35.3	44.1	0.0	0.9	1.8	2.7	3.6	4.5	5.3	6.2	7.1	8.0
53.1	8.8	17.7	26.5	35.4	44.2	0.1	1.0	1.9	2.8	3.7	4.6	5.4	6.3	7.2	8.1
53.2	8.8	17.7	26.6	35.4	44.3	0.2	1.1	2.0	2.9	3.7	4.6	5.5	6.4	7.3	8.2
53.3	8.9	17.8	26.6	35.5	44.4	0.2	1.2	2.1	2.9	3.8	4.7	5.6	6.5	7.4	8.3
53.4	8.9	17.8	26.7	35.6	44.5	0.4	1.2	2.1	3.0	3.9	4.8	5.7	6.6	7.5	8.4
53.5	8.9	17.8	26.8	35.7	44.6	0.4	1.3	2.2	3.1	4.0	4.9	5.8	6.7	7.6	8.5
53.6	9.0	17.9	26.8	35.7	44.7	0.5	1.4	2.3	3.2	4.1	5.0	5.9	6.8	7.7	8.6
53.7	9.0	17.9	26.9	35.8	44.8	0.6	1.5	2.4	3.3	4.2	5.1	6.0	6.8	7.8	8.6
53.8	9.0	18.0	26.9	35.9	44.9	0.7	1.6	2.5	3.4	4.3	5.2	6.1	6.9	7.8	8.7
53.9	9.0	18.0	27.0	36.0	45.0	0.8	1.7	2.6	3.5	4.3	5.3	6.2	7.0	7.9	8.8
54.0	9.0	18.0	27.0	36.0	45.0	0.0	0.9	1.8	2.7	3.6	4.5	5.4	6.3	7.3	8.2
54.1	9.0	18.0	27.1	36.1	45.1	0.1	1.0	1.9	2.8	3.7	4.6	5.5	6.4	7.4	8.3
54.2	9.0	18.1	27.1	36.1	45.1	0.2	1.1	2.0	2.9	3.8	4.7	5.6	6.5	7.4	8.4
54.3	9.1	18.1	27.1	36.2	45.2	0.3	1.2	2.1	3.0	3.9	4.8	5.7	6.6	7.5	8.4
54.4	9.1	18.1	27.2	36.3	45.3	0.4	1.3	2.2	3.1	4.0	4.9	5.8	6.7	7.6	8.5

Right — Double Second Diff. and Corr.

Diff.	Corr.
1.8	0.1
5.5	0.2
9.1	0.3
12.8	0.4
16.5	0.5
20.1	0.6
23.8	0.7
27.4	0.8
31.1	0.9
34.7	
2.1	0.1
6.2	0.2
10.4	0.3
14.5	0.4
18.6	0.5
22.8	0.6
26.9	0.7
31.1	0.8
35.2	
2.4	0.1
7.2	0.2
12.0	0.3
16.8	0.4

The intercept is the distance in nautical miles equal to the difference in minutes and tenths of minutes of arc between calculated altitude (Hc) and observed altitude (Ho):

$$\begin{array}{ll} \text{Ho} & 47°28.6' \\ \text{Hc} & \underline{47°14.1'} \\ \text{diff.} & 14.5' = 14.5 \text{ nautical miles} \end{array}$$

Since observed altitude is greater than Hc, the intercept is *toward* the body.

This leaves just azimuth north to be determined. Azimuth is decreasing from 116.3° at declination 12° to 115.2° at declination 13°. The difference factor is − 1.1°. Use this factor in the same interpolation table block as used for the declination increment (Table 6-10). The correction is − 1.0°, so:

$$\begin{array}{ll} \text{Tab Z} & 116.3° \\ \text{Z corr.} & \underline{- 1.0°} \\ \text{Z} & 115.3° \end{array}$$

We are in north latitudes and LHA is less than 180 degrees, so Z must be subtracted from 360 degrees to get Zn:

$$\begin{array}{lr} & 360.0° \\ \text{Z} & \underline{-115.3°} \\ \text{Zn} & 244.7° \end{array}$$

Table 6-11. Worksheet for the Sun Sight.

BODY-SUN LL	DR L 37°58.7'N, Lo 58°29.9'W	
		GHA
Hs 47° 19.7'	WT 15-29-52 18 AUG	18h 89°04.7'
IC − 2.8'	WE (F)-1- 17	28m 35s 7°08.8'
DIP − 3.4'		GHA 96°13.5'
Ha 47° 13.5'	ZT 15-28-35	AP Lo 57°13.5'
MAIN + 15.1'	ZD + 3	
		LHA 39°
Ho 47° 28.6'	GMT 18-28-35 18 AUG	
	DEC. 18h N 12°52.8'	TAB DEC N 12°
	d - 0.8'	DEC. INC. 52.4'
	d corr. − 0.4'	
		AP Lat N 38°
	DEC. N 12°52.4'	

PUB. 229:					
	TAB Hc 46°38.2'	d 41.2'	TAB Z 116.3°		(115.2° DEC. 13°)
d 34.9'	d corr. + 35.9'		Z corr. − 1.0°		Z diff − 1.1°
+ 1.0'	Hc 47°14.1'		Z 115.3°		Z corr. − 1.0°
d corr. 35.9'	Ho 47°28.6'		360		
	diff 14.5'		−115.3°		
	INT. 14.5 MILES TOWARD—Zm 244.7°				

95

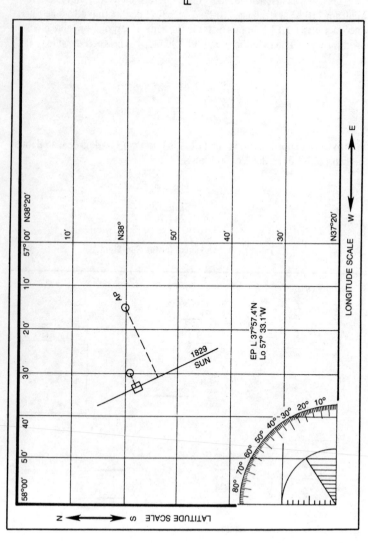

Fig. 6-7. Plot of the sun sight.

All of these computations are actually done on the worksheet shown complete in Table 6-11. Note the little box that encloses the information needed for *Pub. 229*. There are specially designed worksheets available to members of the U.S. Power Squadrons and there are commercially prepared sheets available through major navigation supply outlets. They're handy, but not necessary as long as you develop all the information you need, in any format you find convenient.

On your plotting sheet, plot both your DR position and the assumed position (AP). From the AP, draw the intercept as a dashed line in the direction of 244.7° for a distance of 14.5 nautical miles. Perpendicular to the end of the intercept draw the LOP so that it extends to both sides, and past the DR position.

From the DR position, drop a perpendicular to the LOP. At the intersection of this dashed line and the LOP, draw the little square to indicate that this is an EP. On the plotting sheet, you can determine the coordinates of the EP. Note these on the sheet itself and in the regular ship's log (not the sight log).

In this case, you're at EP L 37°57.4′N, Lo 57°33.1′W, which is less than four miles from your DR position (see Fig. 6-7). This position then can be transferred to the nautical chart on which you're keeping record of your track.

If you're in doubt or confused about any of the steps, review the material discussed in detail. Don't worry about the plotting sheets. These are covered fully in Chapter 7.

Chapter 7

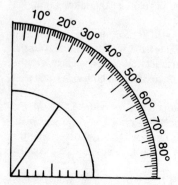

The Plotting Sheet

An ocean chart covers so much territory that you can't accurately draw the intercepts, lines of position, and other features needed in construction of a position plot. For plotting a sight you can use a small sheet of paper that is set up to cover just a small bit of sea territory: usually one degree of latitude and one degree of longitude. This is almost always enough area in which to develop the position plot.

The United States Power Squadrons have special plotting forms for use by their members. "Universal plotting sheets" printed by navigation equipment specialists such as Simex can be obtained at major chart supply firms. The government also sells universal plotting sheets in a 12″-×-12″ format, plus a series of larger forms that cover three-degree latitude ranges. These are discussed later in this chapter. In a pinch, you can construct your own plotting sheet.

MID-LATITUDE

Universal plotting sheets are printed with horizontal lines that represent latitude lines, usually at 10′ intervals. Since you want to develop your plot in the middle of the worksheet as much as possible, and you know the latitude of your DR position and assumed position for the time of a sight, it's best to select the 10′ interval of latitude that is closest to the midpoint between these two positions. Identify this line by its full degree plus 10′ interval and the other lines by the 10′ interval only, except for those that represent a full degree, plus the top and bottom lines (see Fig. 7-1). Be sure to designate north or south latitude wherever the full degree figures are used.

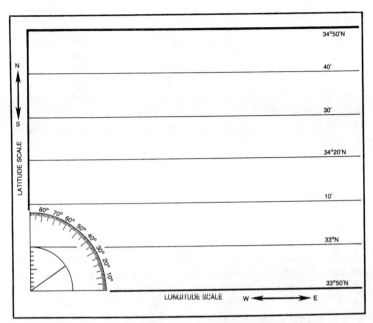

Fig. 7-1. Angle in bottom left corner is equal to the middle latitude.

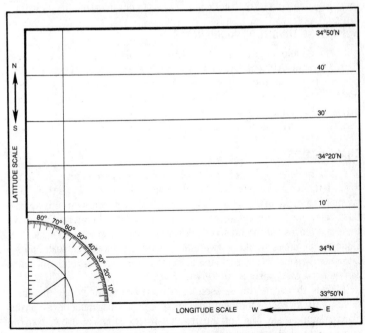

Fig. 7-2. First vertical establishes distance between lines of longitude.

LINES OF LONGITUDE

On a sheet such as the Simex-type illustrated, you next draw a line from the lower left corner out to the segment of a circle. This line must be at an angle equal to the mid-latitude you have designated. In this case "mid-lat" is 34°20'N. The line will be drawn at an angle slightly less than 34.5°, using the protractor arc on the sheet. It is difficult to precisely match the exact 10' interval of mid-lat, unless this is a whole degree of latitude, but try to work it as close as possible. It adds to the accuracy of your plot. This line is also shown in Fig. 7-1.

Now draw a line from top to bottom of the chart that intersects the point where your construction line touches the segment of the circle (Fig. 7-2). This is a line of longitude and it establishes the spacing for the balance of longitude lines on the sheet.

Place your ruler so that the edge intersects the bottom left corner of the chart and the point where the longitude line crosses the lowest latitude line above the bottom. Make a small tick mark across each of the remaining latitude lines, along the edge of the ruler (Fig. 7-3). By drawing a vertical line through each of these marks, you establish the balance of the longitude lines. You will wind up with seven longitude lines (including the preprinted heavy line at the left). Be sure the lines are exactly perpendicular, parallel, and evenly spaced. Note that the farther you are from the equator, the more vertical in shape will be the rectangles that are formed by the latitude and longitude lines. You would get squares only if your mid-lat is the equator.

NUMBERING LONGITUDE COORDINATES

The middle longitude can be the 10' interval that is closest to midway between your AP and DR position. This will help you keep the plot construction centered on the sheet. Unlike your mid-lat line, it can be numbered with just its 10' interval designation. The other longitude lines can also be numbered with 10' intervals, except those at the extreme left and right or any that represent a whole degree of longitude (Fig. 7-4).

CHART SCALES

A latitude scale in increments of one minute of arc is provided at the lower left, along the preprinted longitude line. Distances of less than 10' or arc or 10 nautical miles can be picked off with your dividers directly against this scale. For longer distances, measure from the 10' latitude line above the scale closest to total distance down into the the graduated scale. For example, 26 miles would be two full 10' intervals above the scale, plus 6 1-minute divisions of the scale itself. Tenths-of-minutes or tenths-of-miles can be estimated fairly accurately by eye.

While this scale can be used for latitude and nautical mile measurements, you can't use it for longitude measurements. You must develop your own longitude scale. Mark off the first construction line you drew, from the corner to the circle segment, at intervals equal to those of the latitude scale.

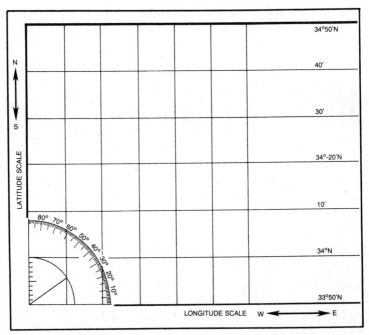

Fig. 7-3. Lines of longitude drawn through "tick" marks on latitude lines.

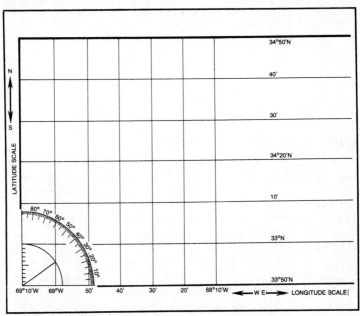

Fig. 7-4. Longitude designations are added at bottom of sheet.

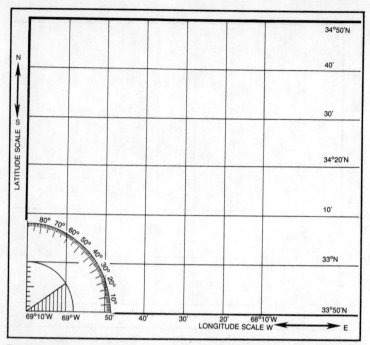

Fig. 7-5. Spacings for latitude minutes are transferred to the angle line at the lower left. Simply make tick marks along the edge of a sheet of paper held to the latitude scale, then hold this sheet along the angle line and transfer the space markings to it. Perpendiculars from the base line to these marks provide minutes of longitude.

From these marks drop perpendiculars to the bottom latitude line. This gives you your longitude scale (Fig. 7-5). This is the completed sheet ready for the sight plot.

DRAWING YOUR OWN PLOTTING SHEET

Start by drawing a line at the bottom of the page to represent latitude and a vertical line at the left to represent longitude. Be sure that they make an exact right angle (Fig. 7-6). Divide this line into six equal segments that will establish your 10′ intervals of latitude. Since you need to divide the lowest segment into tenths, to represent one minute intervals of arc, you might find it easiest to establish this distance first, mark off the ten units, and use the total distance to mark off the other 10′ intervals. With a drawing compass, draw the segment of circle at the lower left, with a radius of 10′ of latitude. With a ruler, draw in the remaining latitude lines (Fig. 7-7).

After you have decided on your mid-lat, place the center of a drawing protractor at the lower left corner of your chart and measure and mark the angle equal to the mid-lat. Be sure to measure up from the bottom latitude

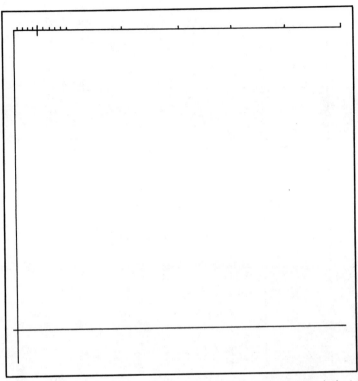

Fig. 7-6. Vertical line at left is marked for 10-minute latitude intervals, and minute intervals in lowest section.

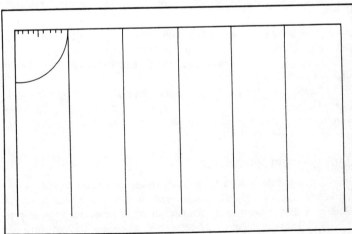

Fig. 7-7. Horizontal latitude lines are added, and arc with radius of 10 minutes of latitude.

103

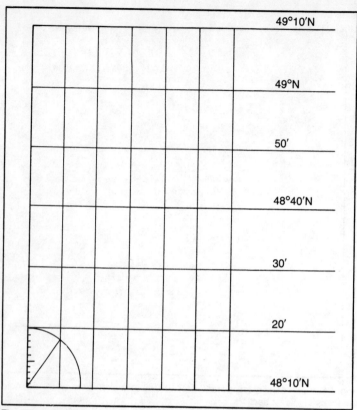

Fig. 7-8. Longitude lines are added in same manner as for pre-printed plotting sheets.

line and not down from the vertical longitude line at the left. Draw a line at this angle from the corner to the segment of a circle and proceed as with the Simex chart. Figure 7-8 shows a completed plotting sheet based on a mid-lat of 48°40′N.

A simple way to solve the problem of determining distance for single minutes of latitude, and the 10′ latitude intervals, is to use graph paper printed in light green or blue. Just be sure to ignore the printed vertical lines in any longitude determination.

LARGE AREA PLOTTING SHEETS

As noted earlier, there are plotting sheets available that cover three degrees of latitude. Printed by the government and available through major dealers for government charts and navigation publications, they measure 17″ × 19″. Those in the 0° to 12° range cover 4 degrees of longitude, those in the 12° to 33° range cover 3 degrees of longitude, and those that range from 33° to 49° range again cover 4 degrees of longitude.

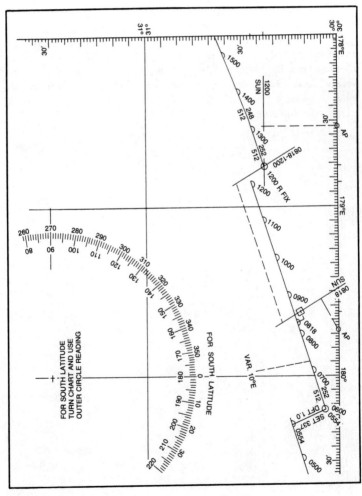

Fig. 7-9. Section of a government-printed plotting sheet covering three degrees of latitude and four degrees of longitude.

Latitude scales of one minute of arc are printed along the entire length of each side of the chart. Longitude scales to one minute of arc are printed along the top and bottom. The latitude lines are numbered. Since distance between latitude lines increases with distance from the equator, they are also numbered upside down so the chart can be reversed for use in southern latitudes. A large compass rose in the center of the sheet also has its numbers printed both right side up and upside down. The longitude lines are not numbered. You must identify these to suit your location.

An advantage of these sheets is that you not only plot individual sights on them, but you can plot your DR track for quite a distance. Figure 7-9 shows a section of the No. 970 chart for 30° to 33° latitudes, as used in the Southern Hemisphere. The solid course lines are DR tracks; each starts from a fix. The dashed line parallel to the course line was used in advancing the 0818 sun sight LOP. That procedure is explained later.

Note too that the square indicating estimated position (EP) at 0818 is offset from the DR position. This takes current set and drift into consideration as determined by the difference between the 0554 DR position and the fix at that time. Recording the DR position along the track at each hour is a worthwhile standard practice. Be sure to take zone boundaries into account when these are crossed. It's also good practice to indicate the magnetic variation for the area.

Chapter 8

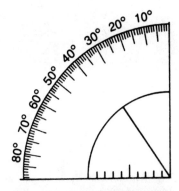

More Sun Sights

As useful as the noon sight is, you'll find there are often times when you need additional position information. Morning or afternoon sun sights can provide this. A morning sight, for example, can be combined with a dawn star sight in a running fix. The afternoon sun sight can be combined with the noon sight or morning and afternoon sights can provide a running fix when the sun was obscured at noon. How to establish a running fix is covered later in this chapter. First I'll show how to ensure the greatest degree of accuracy in the sights you use.

RUN OF SIGHTS AND SLOPE

If conditions permit, it is always advisable to take a run of at least three sights on a body. Then choose the one you feel is most accurate to reduce. It is even better to take a run of five or more sights to establish the *slope* for the body. From this, determine which sight is actually the most accurate.

Record the altitude and time for each sight. Select the time of a sight in the middle of the run, convert this to Greenwich mean time, and establish the GHA of the body for this time. Determine an assumed position in the way outlined in Chapter 6, and then determine the local hour angle. In the following example, watch error is (f) -0-06, index error is −1.5′, and eye height is 21 feet. Date is 14 September 1980, and the DR position is L 40°37.3′ N; Lo 73°16.7′ W.

Sight	Time	Sextant Altitude
1	12-37-45	51°04.2′
2	12-38-17	51°01.8′
3	12-38-52	50°59.9′

Sight	Time	Sextant Altitude
4	12-39-43	50°55.3′
5	12-40-30	50°53.0′

WT	12-38-52	(Middle sight of run)
WE (f)	− 0-06	
ZT	12-38-48	
ZD	+5	
GMT	17-38-48	

GHA Sun 17h	76°09.1′		Dec. N 3°10.0′
38m 48s	9°42.0′		d − 1.0′
GHA	85°51.1′		d corr. − 0.6′
AP Lo	72°51.1′		
LHA	13°		Dec. N 3°09.4′

With LHA and declination, you can determine the rate at which altitude is increasing or decreasing. To find this, use *Pub. 229* in the following manner. Set your AP latitude in the normal manner and as close as possible to your DR latitude; in this case it's N 41°. Take just the whole degree of declination for the sight and look up the calculated altitude, not for just the LHA established above, but for the LHAs that immediately precede and follow it. Here you have:

LHA 12°	Hc 50°29.5′
13°	Hc 50°14.2′
14°	Hc 49°57.7′

Determine the rate of change for each degree of LHA increase separately. Then average the two rates thus found:

LHA 12°	Hc 50°59.5′	15.3′
13°	Hc 50°14.2′	
14°	Hc 49°57.7′	16.4
		31.8′ 12 = 15.9′

This is the average rate of change for each degree of LHA. Since the GP of the sun moves west at exactly 1 degree each 4 minutes, you can plot this rate of change as a line (the slope) on a graph. Mark vertical lines on the graph as whole degrees of minutes, spanning the time of your run of sights. Mark horizontal lines as sextant altitude at intervals of 10′ of arc (see Fig. 8-1).

By plotting your run of sights on the graph, you can see the general direction that the slope will run. Near the top or bottom of the left hand time line, depending on slope direction, mark a point on a 10′ altitude line. Then mark a point on the time line four minutes to the right that corresponds to the rate of change found above. This mark will be above or below the one to the left, depending on slope direction. Draw the slope line between these two points.

In Fig. 8-1, you can see that the line of plotted sights is roughly parallel to the slope. Now determine the distance of each sight from the slope by drawing and measuring dashed vertical lines from sight to slope. Total this distance and take the average:

Sight	Distance	
1.	11.8′	
2.	13.0′	
3.	12.7′	$\frac{64.8}{5} = 12.5'$
4.	14.1′	
5.	13.2′	
	64.8′	

Now measure this distance down from each end of the sloped line (measure up if the slope is below the plotted sights) and draw a second line parallel to the slope. Now pick the sight that is closest to this line and reduce it in the normal manner. In this case, it's sight #3. Half your work is done already. You just need to determine the declination increment to the calculated altitude you found for the LHA of 13°, and determine Zn and intercept.

THE RUNNING FIX

A running fix based on celestial sights is exactly the same as one based on bearings of shoreside objects or aids to navigation. In each case, the LOP

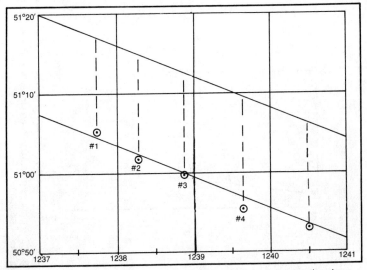

Fig. 8-1. Sights plotted on a graph to determine which is closest to the slope.

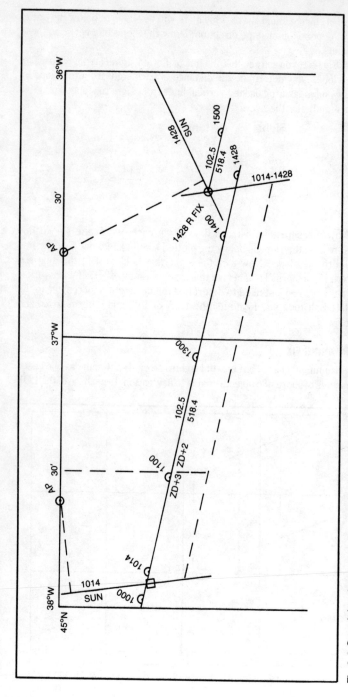

Fig. 8-2. Sun sight taken at 1014 is advanced in same direction and distance of DR course to provide a running fix when crossed with the 1428 sun sight line of position.

of the first sighted object is advanced in the distance and direction of the DR course for the period between the first and second sighting. The advanced LOP is identified by both times and the point at which it intersects the second LOP is the running fix—your position at the time of the second bearing or sight (see Fig. 8-2).

Although the EP is shown for the 1014 sight, the DR track is continued from the 1014 DR position. An EP is not considered to be definite enough to be a point from which a new DR track can be started. This is borne out in the plot of the running fix.

In this example, a time zone boundary is crossed shortly after 1100 and, since time moves ahead one hour, there is no 1200 position noted. A dashed line is drawn to indicate the boundary and the zones are indicated below the course line. Adding these details adds to accuracy. If you forgot you were in a new time zone, you would work the 1428 sight on the basis of 1328 as your time. The resulting LOP would be far from the mark.

At 1428, however, the sun is again shot and the sight is reduced and plotted. The LOP of the first sight is then advanced a distance equal to that between the 1014 and 1428 DR positions, in a direction parallel to the course. This is indicated by the dashed line drawn below the course line to keep it clear of the plot construction at 1428. Note that the advanced LOP is exactly parallel to the original. It is identified by the times of both sights. It is not necessary to again identify it as a sun sight.

The intersection of the advanced LOP with the 1428 sight LOP gives you your running fix, and it is identified as such on the plot. This is considered to be accurate enough to be the point where a new DR course line can be started.

Chapter 9

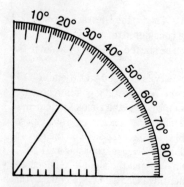

Sights on Other Bodies

While it is often easiest to take sun sights, you may want to take dawn or dusk sights of stars, planets, or the moon. A moon sight, taken when the body is visible in daylight, can be combined with a sun sight for a positive fix. The following steps are needed to reduce each type of sight to a line of position.

THE STAR SIGHT

For the star sight, we will use Sirius, the brightest star visible during the periods when the constellation Orion is in the morning or evening sky (star identification is covered in Chapter 12). Here we will take a morning sight on 28 September 1980. Watch time is 06-05-13 (DST) with an error of (S) +01-21, Hs is 31°46.7', index error is −1.1', and eye height is 14.2 feet above sea level. The DR position is L 38°34.4'N; Lo 63°27.8'W.

From the front of the *Almanac* (Table 9-1). Determine the observed altitude:

Hs	31°46.7'
IC	+ 1.1'
Dip	− 3.7'
Ha	31°44.1'
Main	− 1.5'
Ho	31°42.5'

Convert watch time to GMT. Note that we are using the ZD to the east of the actual zone to compensate for daylight saving time:

Table 9-1. Refraction Correction for the Sirius Sight.

A2 ALTITUDE CORRECTION TABLES 10°-90°—SUN, STARS, PLANETS

SUN

OCT.—MAR. App. Alt.	Lower Limb	Upper Limb	APR.—SEPT. App. Alt.	Lower Limb	Upper Limb
9 34	10·8	21 5	9 39	10·6	21 2
9 45	10·9	21 4	9 51	10·7	21 1
9 56	11·0	21 3	10 03	10·8	21 0
10 08	11·1	21 2	10 15	10·9	20 9
10 21	11·2	21 1	10 27	11·0	20 8
10 34	11·3	21 0	10 40	11·1	20 7
10 47	11·4	20 9	10 54	11·2	20 6
11 01	11·5	20 8	11 08	11·3	20 5
11 15	11·6	20 7	11 23	11·4	20 4
11 30	11·7	20 6	11 38	11·5	20 3
11 46	11·8	20 5	11 54	11·6	20 2
12 02	11·9	20 4	12 10	11·7	20 1
12 19	12·0	20 3	12 28	11·8	20 0
12 37	12·1	20 2	12 46	11·9	19 9
12 55	12·2	20 1	13 05	12·0	19 8
13 14	12·3	20 0	13 24	12·1	19 7
13 35	12·4	19 9	13 45	12·2	19 6
13 56	12·5	19 8	14 07	12·3	19 5
14 18	12·6	19 7	14 30	12·4	19 4
14 42	12·7	19 6	14 54	12·5	19 4
15 06	12·8	19 5	15 19	12·6	19 2
15 32	12·9	19 4	15 46	12·7	19 1
15 59	13·0	19 3	16 14	12·8	19 0
16 28	13·1	19 2	16 44	12·9	18 9
16 59	13·2	19 1	17 15	13·0	18 8
17 32	13·3	19 0	17 48	13·1	18 8
18 06	13·4	18 9	18 24	13·2	18 6
18 42	13·5	18 8	19 01	13·3	18 5
19 21	13·6	18 7	19 42	13·4	18 4
20 03	13·7	18 6	20 25	13·5	18 3
20 48	13·8	18 5	21 11	13·6	18 2
21 35	13·9	18 4	22 00	13·7	18 2
22 26	14·0	18 3	22 54	13·8	18 0
23 22	14·1	18 2	23 51	13·9	1 v
24 21	14·2	18 1	24 53	14·0	1 v
25 26	14·3	18 0	26 00	14·1	1 4
26 36	14·4	17 9	27 13	14·2	1 6
27 52	14·5	17 8	28 33	14·3	1 4
29 15	14·6	17 7	30 00	14·4	1 4
30 46	14·7	17 6	31 35	14·5	1 3
32 26	14·8	17 5	33 20	14·6	1 2
34 17	14·9	17 4	35 17	14·7	1 1
36 20	15·0	17 3	37 26	14·8	1 0
38 36	15·1	17 2	39 50	14·9	16 9
41 08	15·2	17 1	42 31	15·0	16 8
43 59	15·3	17 0	45 31	15·1	16 7
47 10	15·4	16 9	48 55	15·2	16 6
50 46	15·5	16 8	52 44	15·3	16 5
54 49	15·6	16 7	57 02	15·4	16 4
59 23	15·7	16 6	61 51	15·5	16 3
64 30	15·8	16 5	67 17	15·6	16 2
70 12	15·9	16 4	73 16	15·7	16 1
76 26	16·0	16 3	79 43	15·8	16 0
83 05	16·1	16 2	86 32	15·8	16 0
90 00			90 00	15·9	15 9

STARS AND PLANETS

App. Alt.	Corrn	App. Alt.	Additional Corrn
9 56	5·3		**1980**
10 08	5·2		**VENUS**
10 20	5·1		Jan. 1-Feb. 26
10 33	5·0		0 . 0·1
10 46	4·9		42
11 00	4·8		Feb. 27-Apr. 13
11 14	4·7		0 . 0·2
11 29	4·6		47
11 45	4·5		Apr. 14-May 9
12 01	4·4		0 . 0·3
12 18	4·3		46
12 35	4·2		May 10-May 25
12 54	4·1		0 . 0·4
13 13	4·0		11 . 0·5
13 33	3·9		41
13 54	3·8		May 26-June 3
14 16	3·7		0 . 0·5
14 40	3·6		6 . 0·6
15 04	3·5		20 . 0·7
15 30	3·4		31
15 57	3·3		June 4-June 26
16 26	3·2		0 . 0·6
16 56	3·1		4 . 0·7
17 28	3·0		12 . 0·7
18 02	2·9		22 . 0·8
18 38	2·8		
19 17	2·7		June 27-July 6
19 58	2·6		0 . 0·5
20 42	2·5		6 . 0·6
21 28	2·4		20 . 0·7
22 19	2·3		31
23 13	2·2		July 7-July 21
24 11	2·1		0 . 0·4
25 14	2·0		11 . 0·5
26 22	1·9		41
27 36	1·8		July 22-Aug. 17
28 56	1·7		0 . 0·3
30 24	1·6		46
32 00	1·5		Aug. 18-Oct. 2
33 45	1·4		0 . 0·2
35 40	1·3		47
37 48	1·2		Oct. 3-Dec. 31
40 08	1·1		0 . 0·1
42 44	1·0		42
45 36	0·9		
48 47	0·8		**MARS**
52 18	0·7		Jan. 1-Apr. 28
56 11	0·6		0 . 0·2
60 28	0·5		41 . 0·1
65 08	0·4		75
70 11	0·3		Apr. 29-Dec. 31
75 34	0·2		0 . 0·1
81 13	0·1		60
87 03	0·1		
90 00	0·0		

DIP

Ht. of Eye (m)	Corrn	Ht. of Eye (ft)	Ht. of Eye (m)	Corrn
2·4	2·8	8·0	1·0	1·8
2·6	2·9	8·6	1·5	2·2
2·8		9·2	2·0	2·5
3·0	3·0	9·8	2·5	2·8
3·2	3·1	10·5	3·0	3·0
3·4	3·2	11·2	See table	
3·6	3·3	11·9		
3·8	3·4	12·6	(m)	
4·0	3·5	13·3	20	7·9
4·3	3·6	14·1	22	8·3
4·5	3·7	14·9	24	8·6
4·7	3·8	15·7	26	9·0
5·0	3·9	16·5	28	9·3
5·2	4·0	17·4		
5·5	4·1	18·3	30	9·6
5·8	4·2	19·1	32	10·0
6·1	4·3	20·1	34	10·3
6·3	4·4	21·0	36	10·6
6·6	4·5	22·0	38	10·8
6·9	4·6	22·9		
7·2	4·7	23·9	40	11·1
7·5	4·8	24·9	42	11·4
7·9	4·9	26·0	44	11·7
8·2	5·0	27·1	46	11·9
8·5	5·1	28·1	48	12·2
8·8	5·2	29·2		
9·2	5·3	30·4	(ft.)	
9·5	5·4	31·5	2	1·4
9·9	5·5	32·7	4	1·9
10·3	5·6	33·9	6	2·4
10·6	5·7	35·1	8	2·7
11·0	5·8	36·3	10	3·1
11·4	5·9	37·6	See table	
11·8	6·0	38·9		
12·2	6·1	40·1	(ft.)	
12·6	6·2	41·5	70	8·1
13·0	6·3	42·8	75	8·4
13·4	6·4	44·2	80	8·7
13·8	6·5	45·5	85	8·9
14·2	6·6	46·9	90	9·2
14·7	6·7	48·4	95	9·5
15·1	6·8	49·8		
15·5	6·9	51·3	100	9·7
16·0	7·0	52·8	105	9·9
16·5	7·1	54·3	110	10·2
16·9	7·2	55·8	115	10·4
17·4	7·3	57·4	120	10·6
17·9	7·4	58·9	125	10·8
18·4	7·5	60·5		
18·8	7·6	62·1	130	11·1
19·3	7·7	63·8	135	11·3
19·8	7·8	65·4	140	11·5
20·4	7·9	67·1	145	11·7
20·9	8·0	68·8	150	11·9
21·4	8·1	70·5	155	12·1

App. Alt. Apparent altitude   Sextant altitude corrected for index error and dip.

Table 9-2. Daily Page and GHA Increment for the Sirius Sight.

190 — 1980 SEPTEMBER 27, 28, 29 (SAT., SUN., MON.)

The page shows a nautical almanac daily page (columns: ARIES, VENUS −3.7, MARS +1.5, JUPITER −1.2, SATURN +1.2, STARS) with a GHA Increments-and-Corrections strip (the "6ᵐ" page) overlaid across the MARS/JUPITER/SATURN area.

Daily page — ARIES and VENUS

G.M.T.	ARIES G.H.A.	VENUS G.H.A.	VENUS Dec.
27 00	5 56.1	222 42.2	N14 1.
01	20 58.6	237 41.9	1.
02	36 01.0	252 41.5	1.
03	51 03.5	267 41.2 ··	1.
04	66 06.0	282 40.8	0
05	81 08.4	297 40.5	0
06	96 10.9	312 40.1 N14	0
07	111 13.4	327 39.8	0
S 08	126 15.8	342 39.4	0
A 09	141 18.3	357 39.1 ··	0
T 10	156 20.8	12 38.8	0
U 11	171 23.2	27 38.4	0
R 12	186 25.7	42 38.1 N14	0
D 13	201 28.1	57 37.7	0
A 14	216 30.6	72 37.4	0
Y 15	231 33.1	87 37.0 ··	0
16	246 35.5	102 36.7	0
17	261 38.0	117 36.3 14	0
18	276 40.5	132 36.0 N13	5
19	291 42.9	147 35.7	5
20	306 45.4	162 35.3	5
21	321 47.9	177 35.0 ··	5
22	336 50.3	192 34.6	5
23	351 52.8	207 34.3	5
28 00	6 55.3	222 33.9 N13	54
01	21 57.7	237 33.6	54
02	37 00.2	252 33.2	53
03	52 02.6	267 32.9 ··	53
04	67 05.1	282 32.6	52
05	82 07.6	297 32.2	51
06	97 10.0	312 31.9 N13	50
07	112 12.5	327 31.5	49
08	127 15.0	342 31.2	49
S 09	142 17.4	357 30.8 ··	48
U 10	157 19.9	12 30.5	47
N 11	172 22.4	27 30.1	46
D 12	187 24.8	42 29.8 N13	46
A 13	202 27.3	57 29.5	45
Y 14	217 29.7	72 29.1	44
15	232 32.2	87 28.8 ··	43
16	247 34.7	102 28.4	43
17	262 37.1	117 28.1	42
18	277 39.6	132 27.7 N13	41
19	292 42.1	147 27.4	40
20	307 44.5	162 27.0	40
21	322 47.0	177 26.7 ··	39
22	337 49.5	192 26.4	38
23	352 51.9	207 26.0	37
29 00	7 54.4	222 25.7 N13	37
01	22 56.9	237 25.3	36
02	37 59.3	252 25.0	36
03	53 01.8	267 24.6 ··	34
04	68 04.2	282 24.3	33
05	83 06.7	297 23.9	33
06	98 09.2	312 23.6 N13	32
07	113 11.6	327 23.2	31
08	128 14.1	342 22.9	30.9
M 09	143 16.6	357 22.5	30.6
O 10	158 19.0	12 22.2	29.4
N 11	173 21.5	27 21.9	28.6
D 12	188 24.0	42 21.5 N13	27.8
A 13	203 26.4	57 21.2	27.1
Y 14	218 28.9	72 20.8	26.3
15	233 31.4	87 20.5 ··	25.5
16	248 33.8	102 20.1	24.8
17	263 36.3	117 19.8	24.0
18	278 38.7	132 19.5 N13	23.2
19	293 41.2	147 19.1	22.5
20	308 43.7	162 18.8	21.7
21	323 46.1	177 18.4	20.9
22	338 48.6	192 18.1	20.2
23	353 51.1	207 17.7	19.4
Mer. Pass.	23 28.5	v −0.3	d 0.7

Daily page — MARS, JUPITER, SATURN (lower rows, SEPT 29)

G.M.T.	MARS G.H.A. Dec.	JUPITER G.H.A. Dec.	SATURN G.H.A. Dec.
06	244 56.4 37.7	298 07.0 16.5	291 28.7 30.8
07	259 57.1 38.2	313 09.0 16.3	306 30.9 30.6
08	274 57.8 ·· 38.6	328 11.0 ·· 16.1	321 33.1 ·· 30.5
09	289 58.5 39.1	343 13.0 15.9	336 35.2 30.4
10	304 59.2 39.6	358 15.0 15.7	351 37.4 30.3
11	319 59.9 S18 40.1	13 16.9 N 3 15.5	6 39.6 N 1 30.1
12	335 00.6 40.6	28 18.9 15.3	21 41.8 30.0
13	350 01.3 41.0	43 20.8 15.1	36 44.0 29.9
14	5 02.0 ·· 41.5	58 22.8 ·· 14.9	51 46.1 ·· 29.8
15	20 02.7 42.0	73 24.8 14.7	66 48.3 29.7
16	35 03.4 42.5	88 26.7 14.4	81 50.5 29.5
17	50 04.1 S18 43.0	103 28.7 N 3 14.2	96 52.7 N 1 29.4
18	65 04.8 43.4	118 30.7 14.0	111 54.9 29.3
19	80 05.5 43.9	133 32.7 13.8	126 57.0 29.2
20	95 06.2 ·· 44.4	148 34.6 ·· 13.6	141 59.2 ·· 29.1
21	110 06.9 44.9	163 36.6 13.4	157 01.4 28.9
22	125 07.6 45.4	178 38.6 13.2	172 03.6 28.8
Mer. Pass.	v 0.7 d 0.5	v 2.0 d 0.2	v 2.2 d 0.1

STARS

Name	S.H.A.	Dec.
Acamar	315 36.9	S40 22.8
Achernar	335 44.7	S57 20.0
Acrux	173 37.9	S62 59.4
Adhara	255 32.2	S28 56.5
Aldebaran	291 17.9	N16 28.2
Alioth	166 43.1	N56 04.1
Alkaid	153 19.0	N49 24.8
Al Na'ir	28 14.6	S47 03.4
Alnilam	276 11.6	S 1 12.8
Alphard	218 20.8	S 8 34.3
Alphecca	126 32.4	N26 47.1
Alpheratz	358 08.9	N28 59.1
Altair	62 32.4	N 8 49.2
Ankaa	353 39.8	S42 24.6
Antares	112 57.0	S26 23.3
Arcturus	146 18.8	N19 17.2
Atria	108 21.5	S68 59.7
Avior	234 28.5	S59 26.6
Bellatrix	278 58.7	N 6 20.0
Betelgeuse	271 28.2	N 7 24.3
Canopus	264 07.2	S52 40.8
Capella	281 11.2	N45 58.6
Deneb	49 48.2	N45 13.0
Denebola	182 59.4	N14 41.0
Diphda	349 20.6	S18 05.5
Dubhe	194 22.8	N61 51.3
Elnath	278 44.0	N28 35.4
Eltanin	90 57.8	N51 29.9
Enif	34 11.3	N 9 47.3
Fomalhaut	15 51.1	S29 43.5
Gacrux	172 29.3	S57 00.2
Gienah	176 18.3	S17 25.9
Hadar	149 23.9	S60 16.8
Hamal	328 28.6	N23 22.3
Kaus Aust.	84 16.9	S34 23.7
Kochab	137 20.0	N74 14.4
Markab	14 02.9	N15 06.2
Menkar	314 40.9	N 4 00.9
Menkent	148 37.4	S36 14.4
Miaplacidus	221 45.5	S69 38.0
Mirfak	309 15.7	N49 47.4
Nunki	76 29.1	S26 19.2
Peacock	53 58.1	S56 48.0
Pollux	243 58.3	N28 04.3
Procyon	245 25.9	N 5 16.6
Rasalhague	96 29.7	N12 34.7
Regulus	208 10.3	N12 03.8
Rigel	281 35.9	S 8 13.3
Rigil Kent.	140 26.3	S60 45.3
Sabik	102 41.2	S15 42.0
Schedar	350 08.4	N56 25.9
Shaula	96 55.8	S37 05.4
Sirius	258 55.7	S16 41.2
Spica	158 57.9	S11 03.4
Suhail	223 11.0	S43 21.0
Vega	80 55.8	N38 46.3
Zuben'ubi	137 33.3	S15 57.5

	S.H.A.	Mer. Pass.
Venus	215 38.7	9 10
Mars	132 39.2	14 41
Jupiter	185 10.7	11 10
Saturn	178 25.9	11 37

INCREMENTS AND CORRECTIONS — 6ᵐ

s	SUN PLANETS	ARIES	MOON	v or d Corrⁿ	v or d Corrⁿ	v or d Corrⁿ
00	1 30.0	1 30.2	1 25.9	0.0 0.0	6.0 0.7	12.0 1.3
01	1 30.3	1 30.5	1 26.1	0.1 0.0	6.1 0.7	12.1 1.3
02	1 30.5	1 30.7	1 26.4	0.2 0.0	6.2 0.7	12.2 1.3
03	1 30.8	1 31.0	1 26.6	0.3 0.0	6.3 0.7	12.3 1.3
04	1 31.0	1 31.2	1 26.9	0.4 0.0	6.4 0.7	12.4 1.3
05	1 31.3	1 31.5	1 27.1	0.5 0.1	6.5 0.7	12.5 1.4
06	1 31.5	1 31.8	1 27.3	0.6 0.1	6.6 0.7	12.6 1.4
07	1 31.8	1 32.0	1 27.6	0.7 0.1	6.7 0.7	12.7 1.4
08	1 32.0	1 32.3	1 27.8	0.8 0.1	6.8 0.7	12.8 1.4
09	1 32.3	1 32.5	1 28.0	0.9 0.1	6.9 0.7	12.9 1.4
10	1 32.5	1 32.8	1 28.3	1.0 0.1	7.0 0.8	13.0 1.4
11	1 32.8	1 33.0	1 28.5	1.1 0.1	7.1 0.8	13.1 1.4
12	1 33.0	1 33.3	1 28.8	1.2 0.1	7.2 0.8	13.2 1.4
13	1 33.3	1 33.5	1 29.0	1.3 0.1	7.3 0.8	13.3 1.4
14	1 33.5	1 33.8	1 29.2	1.4 0.2	7.4 0.8	13.4 1.5
15	1 33.8	1 34.0	1 29.5	1.5 0.2	7.5 0.8	13.5 1.5
16	1 34.0	1 34.3	1 29.7	1.6 0.2	7.6 0.8	13.6 1.5
17	1 34.3	1 34.5	1 30.0	1.7 0.2	7.7 0.8	13.7 1.5
18	1 34.5	1 34.8	1 30.2	1.8 0.2	7.8 0.8	13.8 1.5
19	1 34.8	1 35.0	1 30.4	1.9 0.2	7.9 0.9	13.9 1.5
20	1 35.0	1 35.3	1 30.7	2.0 0.2	8.0 0.9	14.0 1.5
21	1 35.3	1 35.5	1 30.9	2.1 0.2	8.1 0.9	14.1 1.5
22	1 35.5	1 35.8	1 31.1	2.2 0.2	8.2 0.9	14.2 1.5
23	1 35.8	1 36.0	1 31.4	2.3 0.2	8.3 0.9	14.3 1.5
24	1 36.0	1 36.3	1 31.6	2.4 0.3	8.4 0.9	14.4 1.6
25	1 36.3	1 36.5	1 31.9	2.5 0.3	8.5 0.9	14.5 1.6
26	1 36.5	1 36.8	1 32.1	2.6 0.3	8.6 0.9	14.6 1.6
27	1 36.8	1 37.0	1 32.3	2.7 0.3	8.7 0.9	14.7 1.6
28	1 37.0	1 37.3	1 32.6	2.8 0.3	8.8 1.0	14.8 1.6
29	1 37.3	1 37.5	1 32.8	2.9 0.3	8.9 1.0	14.9 1.6
30	1 37.5	1 37.8	1 33.1	3.0 0.3	9.0 1.0	15.0 1.6
31	1 37.8	1 38.0	1 33.3	3.1 0.3	9.1 1.0	15.1 1.6
32	1 38.0	1 38.3	1 33.5	3.2 0.3	9.2 1.0	15.2 1.6
33	1 38.3	1 38.5	1 33.8	3.3 0.4	9.3 1.0	15.3 1.7
34	1 38.5	1 38.8	1 34.0	3.4 0.4	9.4 1.0	15.4 1.7
35	1 38.8	1 39.0	1 34.3	3.5 0.4	9.5 1.0	15.5 1.7
36	1 39.0	1 39.3	1 34.5	3.6 0.4	9.6 1.0	15.6 1.7
37	1 39.3	1 39.5	1 34.7	3.7 0.4	9.7 1.1	15.7 1.7
38	1 39.5	1 39.8	1 35.0	3.8 0.4	9.8 1.1	15.8 1.7
39	1 39.8	1 40.0	1 35.2	3.9 0.4	9.9 1.1	15.9 1.7

Time reduction

```
WT        06-05-13   28 Sept
WE   (s)  +01-21
ZT        06-06-34
ZD    +       3
GMT       09-06-34   28 Sept
```

Now determine the GP of Sirius. For GHA, you'll need its sidereal hour angle plus GHA of Aries. Table 9-2 shows elements of the daily and correction pages used for the calculations:

```
SHA *           258°55.7'  Dec. S 16°41.2'
GHA Aries
09h             142°17.4'
06m 34s           1°38.8'
                402°51.9'
            −   360°
GHA *            42°51.9'
```

Declination of a star is considered constant so no correction to the given figure is needed. For *Pub. 229*, tabulated declination is S 16° and the declination increment is 41.2°.

Establish your AP latitude and longitude, and then the LHA of Sirius. N 38° is the closest whole degree of DR latitude. For LHA you have:

```
GHA*            42°51.9'
               +360°
               402°51.9'
AP Lo           63°51.9'
LHA            339°
```

Latitude is contrary in name to declination so in *Pub. 229* use the right hand page for LHA 21°, 339°. For 16° declination at Lat. 38° you have (Table 9-3):

Tab Hc	d	Z	Z -- Dec. 17°
32°30.7'	−56.6'	155.9°	156.3°

For the d correction, use the interpolation table inside the back cover (Table 9-4). For a declination increment of 41.2' and a d factor of −56.6' you have:

```
        D       34.3'
      +          4.6'
d corr. +       38.9'
```

As this is positive, it is added to Tab Hc:

```
Tab Hc      32°30.7'
d corr.       +38.9'
Hc          33°09.6'
```

Compare this with Ho to get intercept distance:

```
Hc          33°09.6'
Ho          31°42.5'
Intercept       9.3   miles, away
```

Table 9-3. Tabulated Altitude, d Factor and Azimuth Information for the Sirius Sight, from Pub. 229.

LATITUDE CONTRARY NAME TO DECLINATION   L.H.A.

Dec.	38° Hc	d	Z	39° Hc	d	Z	40° Hc	d	Z	41° Hc	d	Z	42° Hc	d	Z	43° Hc	d	Z	44° Hc	d	Z	Hc
0	47 21.8	−54.6	148.1	46 30.8	−55.0	148.6	45 39.4	55.3	149.2	44 47.7	55.5	149.7	43 55.8	55.8	150.2	43 03.7	56.1	150.6	42 11.3	56.3	151.1	41 18.6
1	46 27.2	54.8	148.7	45 35.8	55.1	149.2	44 44.1	55.4	149.7	43 52.2	55.7	150.2	43 00.0	55.9	150.7	42 07.6	56.2	151.1	41 15.0	56.5	151.5	40 22.1
2	45 32.4	55.0	149.2	44 40.7	55.3	149.8	43 48.7	55.5	150.2	42 56.5	55.8	150.7	42 04.1	56.1	151.2	41 11.4	56.3	151.6	40 18.5	56.5	152.0	39 25.5
3	44 37.4	55.2	149.8	43 45.4	55.4	150.3	42 53.2	55.7	150.7	42 00.7	55.9	151.2	41 08.0	56.2	151.6	40 15.1	56.4	152.0	39 22.0	56.6	152.4	38 28.7
4	43 42.2	55.3	150.4	42 50.0	55.6	150.8	41 57.5	55.9	151.3	41 04.8	56.1	151.7	40 11.8	56.3	152.1	39 18.7	56.5	152.5	38 25.4	56.7	152.9	37 31.9
5	42 46.9	−55.4	150.9	41 54.4	−55.7	151.3	41 01.6	55.9	151.8	40 08.7	56.2	152.2	39 15.5	56.3	152.5	38 22.2	56.6	152.9	37 28.7	56.8	153.3	36 35.0
6	41 51.5	55.6	151.4	40 58.7	55.8	151.8	40 05.7	56.1	152.2	39 12.5	56.3	152.6	38 19.2	56.5	153.0	37 25.6	56.7	153.3	36 31.9	56.8	153.7	35 38.1
7	40 55.9	55.7	151.9	40 02.9	56.0	152.3	39 09.6	56.1	152.7	38 16.2	56.3	153.1	37 22.7	56.6	153.4	36 28.9	56.7	153.7	35 35.1	57.0	154.1	34 41.0
8	40 00.2	55.8	152.4	39 06.9	56.0	152.8	38 13.5	56.3	153.1	37 19.9	56.5	153.5	36 26.1	56.7	153.8	35 32.2	56.9	154.1	34 38.1	57.0	154.4	33 43.9
9	39 04.4	56.0	152.9	38 10.9	56.2	153.2	37 17.2	56.4	153.6	36 23.4	56.6	153.9	35 29.4	56.7	154.2	34 35.3	56.9	154.5	33 41.1	57.1	154.8	32 46.7
10	38 08.4	−56.0	153.3	37 14.7	−56.3	153.7	36 20.8	56.4	154.0	35 26.8	56.6	154.3	34 32.7	56.8	154.6	33 38.4	57.0	154.9	32 44.0	57.1	155.2	31 49.5
11	37 12.4	56.2	153.8	36 18.4	56.3	154.1	35 24.4	56.5	154.4	34 30.2	56.7	154.7	33 35.9	56.9	155.0	32 41.4	57.0	155.3	31 46.9	57.2	155.6	30 52.2
12	36 16.2	56.2	154.2	35 22.1	56.4	154.5	34 27.9	56.5	154.8	33 33.5	56.8	155.1	32 39.0	57.0	155.4	31 44.4	57.1	155.7	30 49.7	57.3	155.9	29 54.8
13	35 20.0	56.4	154.7	34 25.7	56.6	155.0	33 31.2	56.7	155.2	32 36.7	56.9	155.5	31 42.0	57.0	155.8	30 47.3	57.2	156.0	29 52.4	57.3	156.3	28 57.4
14	34 23.6	56.4	155.1	33 29.1	56.6	155.4	32 34.6	56.8	155.6	31 39.8	56.9	155.9	30 45.0	57.0	156.1	29 50.1	57.2	156.4	28 55.1	57.3	156.6	28 00.0
15	33 27.2	−56.5	155.5	32 32.6	−56.7	155.8	31 37.8	56.8	156.0	30 42.9	57.0	156.3	29 48.0	57.2	156.5	28 52.9	57.3	156.7	27 57.7	57.4	156.9	27 02.5
16	32 30.7	56.6	155.9	31 35.9	56.7	156.1	30 41.0	56.9	156.4	29 45.9	57.0	156.6	28 50.8	57.2	156.8	27 55.6	57.3	157.1	27 00.3	57.4	157.3	26 04.9
17	31 34.1	56.6	156.3	30 39.2	56.8	156.5	29 44.1	57.0	156.8	28 48.9	57.1	157.0	27 53.6	57.2	157.2	26 58.3	57.4	157.4	26 02.9	57.5	157.6	25 07.4
18	30 37.5	56.7	156.7	29 42.4	56.9	156.9	28 47.1	57.0	157.1	27 51.8	57.1	157.3	26 56.4	57.3	157.5	26 00.9	57.4	157.7	25 05.4	57.6	157.9	24 09.7
19	29 40.8	56.8	157.0	28 45.5	56.9	157.3	27 50.1	57.1	157.5	26 54.7	57.2	157.7	25 59.1	57.3	157.9	25 03.5	57.4	158.0	24 07.8	57.5	158.2	23 12.1
20	28 44.0	−56.8	157.4	27 48.6	−57.0	157.6	26 53.0	57.1	157.8	25 57.5	57.3	158.0	25 01.8	57.4	158.2	24 06.1	57.5	158.4	23 10.3	57.7	158.5	22 14.4
21	27 47.2	56.9	157.8	26 51.6	57.0	158.0	25 55.9	57.2	158.2	25 00.2	57.3	158.3	24 04.4	57.4	158.5	23 08.6	57.6	158.7	22 12.6	57.6	158.8	21 16.7
22	26 50.3	56.9	158.1	25 54.6	57.1	158.3	24 58.7	57.2	158.5	24 02.9	57.3	158.7	23 07.0	57.5	158.8	22 11.0	57.6	159.0	21 15.0	57.7	159.1	20 18.9
23	25 53.3	56.9	158.5	24 57.5	57.1	158.7	24 01.6	57.3	158.8	23 05.6	57.4	159.0	22 09.6	57.5	159.1	21 13.5	57.6	159.3	20 17.3	57.7	159.4	19 21.1
24	24 56.4	57.1	158.8	24 00.4	57.1	159.0	23 04.3	57.3	159.2	22 08.2	57.4	159.3	21 12.1	57.6	159.4	20 15.9	57.6	159.6	19 19.6	57.7	159.7	18 23.3
25	23 59.3	−57.1	159.2	23 03.2	−57.2	159.3	22 07.0	57.3	159.5	21 10.8	57.4	159.6	20 14.6	57.5	159.7	19 18.2	57.6	159.9	18 21.9	57.8	159.9	17 25.5
26	23 02.2	57.1	159.5	22 06.0	57.2	159.7	21 09.7	57.3	159.8	20 13.4	57.5	159.9	19 17.0	57.6	160.0	18 20.6	57.7	160.2	17 24.1	57.8	160.3	16 27.6
27	22 05.1	57.2	159.8	21 08.8	57.3	160.0	20 12.4	57.4	160.1	19 15.9	57.5	160.2	18 19.4	57.6	160.3	17 22.9	57.7	160.5	16 26.4	57.9	160.6	15 29.8
28	21 07.9	57.2	160.2	20 11.5	57.3	160.3	19 15.0	57.5	160.4	18 18.4	57.5	160.4	17 21.8	57.6	160.6	16 25.2	57.7	160.7	15 28.5	57.8	160.8	14 31.9
29	20 10.7	57.2	160.5	19 14.2	57.4	160.6	18 17.5	57.4	160.7	17 20.9	57.6	160.8	16 24.2	57.7	160.9	15 27.5	57.8	161.0	14 30.7	57.8	161.1	13 33.9

Table 9-4. The d Factor and Z Corrections for the Sirius Sight.

Left table

40'	50'		.		0'	1'	2'	3'	4'	5'	6'	7'	8'	9'		and Corr.
18.6	23.3															
18.7	23.4															0.8 0.1
18.8	23.5															2.4 0.2
18.9	23.6															5.6 0.4
	23.7															7.2

40'	50'		.		0'	1'	2'	3'	4'	5'	6'	7'	8'	9'		and Corr.
22.0	27.5		.0		0.0	0.6	1.1	1.7	2.2	2.8	3.3	3.9	4.5	5.0		18.6 1.2
22.1	27.6		.1		0.0	0.5	1.1	1.6	2.2	2.7	3.3	3.8	4.4	5.0		20.2 1.3
22.2	27.7		.2		0.1	0.7	1.2	1.8	2.3	2.9	3.4	4.0	4.6	5.1		21.8 1.4
22.3	27.7		.3		0.1	0.7	1.3	1.8	2.4	3.0	3.5	4.1	4.6	5.2		23.4 1.5
22.3	27.8		.4		0.2	0.8	1.3	1.9	2.5	3.0	3.6	4.1	4.7	5.2		25.1 1.6
22.3	27.9		.5		0.3	0.8	1.4	2.0	2.5	3.1	3.6	4.2	4.7	5.3		26.7 1.7
22.4	28.0		.6		0.3	0.9	1.5	2.0	2.6	3.1	3.7	4.3	4.8	5.4		28.3 1.8
22.5	28.1		.7		0.4	1.0	1.5	2.1	2.7	3.2	3.8	4.3	4.9	5.4		29.9 1.9
22.6	28.2		.8		0.4	1.0	1.6	2.1	2.7	3.3	3.8	4.4	4.9	5.5		31.5 2.0
22.6	28.3		.9		0.5	1.1	1.6	2.2	2.8	3.3	3.9	4.4	5.0	5.5		33.1
																34.7

40'	50'		.		0'	1'	2'	3'	4'	5'	6'	7'	8'	9'		and Corr.
22.6	28.3		.0		0.0	0.6	1.1	1.7	2.3	2.9	3.4	4.0	4.6	5.2		0.8 0.1
22.7	28.4		.1		0.1	0.7	1.2	1.8	2.4	2.9	3.5	4.1	4.7	5.2		2.4 0.2
22.8	28.5		.2		0.1	0.7	1.3	1.8	2.4	3.0	3.6	4.2	4.8	5.3		4.1 0.3
22.9	28.6		.3		0.2	0.8	1.3	1.9	2.5	3.0	3.6	4.2	4.8	5.4		5.8 0.4
23.0	28.7		.4		0.2	0.8	1.4	2.0	2.5	3.1	3.7	4.3	4.8	5.4		7.4 0.5
23.0	28.8		.5		0.3	0.9	1.4	2.0	2.6	3.2	3.7	4.3	4.9	5.5		9.1 0.6
23.1	28.8		.6		0.3	0.9	1.5	2.1	2.7	3.2	3.8	4.4	4.9	5.5		10.7 0.7
23.2	29.0		.7		0.4	1.0	1.6	2.2	2.7	3.3	3.9	4.5	5.0	5.6		12.3 0.8
23.2	29.0		.8		0.5	1.0	1.6	2.2	2.8	3.3	3.9	4.5	5.1	5.6		14.0 0.9
23.3	29.1		.9		0.5	1.1	1.7	2.3	2.8	3.4	4.0	4.5	5.1	5.7		15.6 1.0
																17.3 1.1
																18.9

40'	50'		.		0'	1'	2'	3'	4'	5'	6'	7'	8'	9'		and Corr.
23.3	29.1		.0		0.0	0.6	1.2	1.8	2.4	3.0	3.5	4.1	4.7	5.3		20.6 1.2
23.4	29.2		.1		0.1	0.7	1.2	1.8	2.4	3.0	3.6	4.2	4.8	5.4		22.2 1.3
23.4	29.3		.2		0.1	0.7	1.3	1.9	2.5	3.1	3.7	4.3	4.9	5.5		23.9 1.4
23.5	29.4		.3		0.2	0.8	1.4	2.0	2.6	3.1	3.7	4.3	4.9	5.5		25.5 1.5
23.6	29.5		.4		0.2	0.8	1.4	2.0	2.6	3.2	3.8	4.4	5.0	5.6		27.2 1.6
23.7	29.6		.5		0.3	0.9	1.5	2.1	2.7	3.3	3.8	4.4	5.0	5.6		28.8 1.7
23.7	29.7		.6		0.4	0.9	1.5	2.1	2.7	3.3	3.9	4.5	5.1	5.7		30.4 1.8
23.8	29.8		.7		0.4	1.0	1.6	2.2	2.8	3.4	4.0	4.6	5.1	5.7		32.1 2.0
23.9	29.9		.8		0.5	1.1	1.7	2.2	2.8	3.4	4.0	4.6	5.2	5.8		33.7
24.0	30.0		.9		0.5	1.1	1.7	2.3	2.9	3.5	4.1	4.7	5.3	5.9		35.4 2.1

Right table

Inc.	10'	20'	30'	40'	50'		.		0'	1'	2'	3'	4'	5'	6'	7'	8'	9'		and Corr.
36.0	6.0	12.0	18.0	24.0	30.0															
36.1	6.0	12.0	18.0	24.0	30.1															0.8 0.1
36.2	6.0	12.1	18.1	24.1	30.1															2.5 0.2
36.3	6.0	12.1	18.1	24.2	30.2															4.2 0.2
36.3	6.1	12.1	18.2	24.3	30.3															5.9 0.3

Inc.	10'	20'	30'	40'	50'		.		0'	1'	2'	3'	4'	5'	6'	7'	8'	9'		and Corr.
41.0	6.8	13.6	20.5	27.3	34.1		.0		0.0	0.7	1.4	2.1	2.8	3.5	4.2	5.0	5.7	6.4		17.6 1.0
41.1	6.8	13.7	20.5	27.4	34.2		.1		0.1	0.7	1.5	2.1	2.8	3.5	4.2	4.9	5.6	6.3		19.4 1.1
41.2	6.8	13.7	20.6	27.4	34.3		.2		0.1	0.8	1.5	2.2	2.9	3.6	4.3	5.0	5.7	6.4		21.3 1.2
41.3	6.9	13.8	20.6	27.5	34.4		.3		0.2	0.9	1.6	2.3	3.0	3.7	4.4	5.1	5.8	6.5		23.1
41.4	6.9	13.8	20.7	27.6	34.5		.4		0.3	1.0	1.7	2.4	3.0	3.7	4.4	5.1	5.8	6.5		25.0 1.3
41.5	6.9	13.8	20.8	27.7	34.6		.5		0.3	1.0	1.7	2.4	3.1	3.8	4.5	5.2	5.9	6.6		26.8 1.4
41.6	6.9	13.9	20.8	27.7	34.7		.6		0.4	1.1	1.8	2.5	3.2	3.9	4.6	5.3	5.9	6.6		28.7 1.5
41.7	7.0	13.9	20.9	27.8	34.8		.7		0.5	1.2	1.9	2.6	3.3	4.0	4.7	5.4	6.0	6.7		30.5 1.6
41.8	7.0	13.9	20.9	27.9	34.8		.8		0.6	1.3	1.9	2.6	3.3	4.0	4.7	5.4	6.1	6.8		32.3 1.7
41.9	7.0	14.0	21.0	28.0	35.0		.9		0.6	1.3	2.0	2.7	3.4	4.1	4.8	5.5	6.2	6.8		34.2 1.8

Inc.	10'	20'	30'	40'	50'		.		0'	1'	2'	3'	4'	5'	6'	7'	8'	9'		and Corr.
42.0	7.0	14.0	21.0	28.0	35.0		.0		0.0	0.7	1.4	2.1	2.8	3.5	4.2	5.0	5.7	6.4		1.0 0.1
42.1	7.0	14.0	21.0	28.0	35.1		.1		0.1	0.8	1.5	2.2	2.9	3.6	4.3	5.0	5.7	6.5		3.0 0.2
42.2	7.0	14.1	21.1	28.1	35.1		.2		0.1	0.8	1.5	2.2	3.0	3.6	4.3	5.1	5.8	6.5		4.9 0.3
42.3	7.0	14.1	21.2	28.2	35.2		.3		0.2	0.9	1.6	2.3	3.0	3.8	4.5	5.2	5.9	6.6		6.9 0.4
42.4	7.1	14.1	21.2	28.3	35.3		.4		0.3	1.0	1.7	2.4	3.1	3.8	4.5	5.2	5.9	6.7		8.9 0.5
42.5	7.1	14.2	21.3	28.3	35.4		.5		0.4	1.1	1.8	2.5	3.2	3.9	4.6	5.3	6.0	6.7		9.1 0.6
42.6	7.1	14.2	21.3	28.4	35.5		.6		0.4	1.1	1.8	2.6	3.3	4.0	4.7	5.4	6.1	6.8		10.8 0.7
42.7	7.1	14.3	21.4	28.5	35.6		.7		0.5	1.2	1.9	2.6	3.3	4.0	4.7	5.5	6.2	6.9		12.8 0.7
42.8	7.2	14.3	21.4	28.6	35.7		.8		0.6	1.3	2.0	2.7	3.4	4.1	4.8	5.5	6.2	6.9		14.8 0.8
42.9	7.2	14.3	21.5	28.6	35.8		.9		0.6	1.3	2.1	2.8	3.5	4.2	4.9	5.6	6.3	7.0		16.7 0.9
																				18.7 1.0

Inc.	10'	20'	30'	40'	50'		.		0'	1'	2'	3'	4'	5'	6'	7'	8'	9'		and Corr.
43.0	7.1	14.3	21.5	28.6	35.8		.0		0.0	0.7	1.4	2.2	2.9	3.6	4.3	5.1	5.8	6.5		17.0 1.0
43.1	7.2	14.3	21.5	28.7	35.9		.1		0.1	0.8	1.5	2.2	3.0	3.7	4.4	5.1	5.9	6.6		22.7 1.1
43.2	7.2	14.4	21.6	28.8	36.0		.2		0.1	0.9	1.6	2.3	3.0	3.8	4.5	5.2	5.9	6.7		24.6 1.2
43.3	7.2	14.4	21.6	28.8	36.1		.3		0.2	1.0	1.7	2.4	3.1	3.8	4.5	5.3	6.0	6.7		26.6 1.3
43.4	7.2	14.5	21.7	28.9	36.2		.4		0.3	1.0	1.7	2.5	3.2	3.9	4.6	5.4	6.1	6.8		28.6 1.4
43.5	7.3	14.5	21.8	29.0	36.3		.5		0.4	1.1	1.8	2.5	3.3	4.0	4.7	5.4	6.2	6.9		30.5 1.5
43.6	7.3	14.5	21.8	29.1	36.3		.6		0.4	1.2	1.9	2.6	3.4	4.1	4.8	5.5	6.2	7.0		32.5 1.6
43.7	7.3	14.6	21.9	29.1	36.4		.7		0.5	1.3	2.0	2.7	3.4	4.1	4.9	5.6	6.3	7.0		34.5 1.7
43.8	7.3	14.6	21.9	29.2	36.5		.8		0.6	1.3	2.0	2.8	3.5	4.2	4.9	5.7	6.4	7.1		
43.9	7.4	14.7	22.0	29.3	36.6		.9		0.7	1.4	2.1	2.8	3.6	4.3	5.0	5.7	6.5	7.2		

Computed altitude is greater than observed altitude so the intercept will be plotted away from the body. Now determine azimuth north (Zn). The Z of 155.9° at declination 16° increases by 0.4° to 156.3° at declination 17° as noted above. Use the same interpolation table as for the declination increment. For +0.4° the correction is +0.3°:

$$\begin{array}{ll} \text{Tab Z} & 155.9° \\ \text{Z corr.} & +\ 0.3° \\ \hline \text{Z} & 156.2° \end{array}$$

In this example, LHA is greater than 180 degrees, so Z = Zn. With this information you can do your plot. Figure 9-1 shows the complete worksheet (which includes the time diagram). Note that the position of the sun is shown about 15 degrees below your horizon. It's at this point, in clear weather, that you first get a distinct horizon. Figure 9-2 shows the plot for this sight.

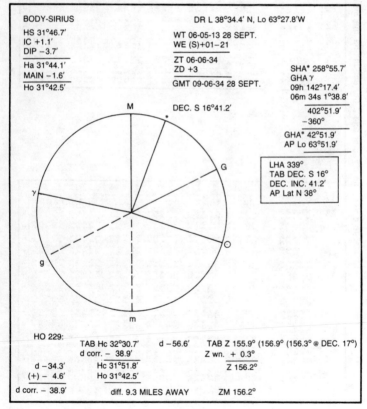

Fig. 9-1. Worksheet for the Sirius sight. Note that the time diagram shows the GHA of the sun at the time of the sight, as well as Aries and the star.

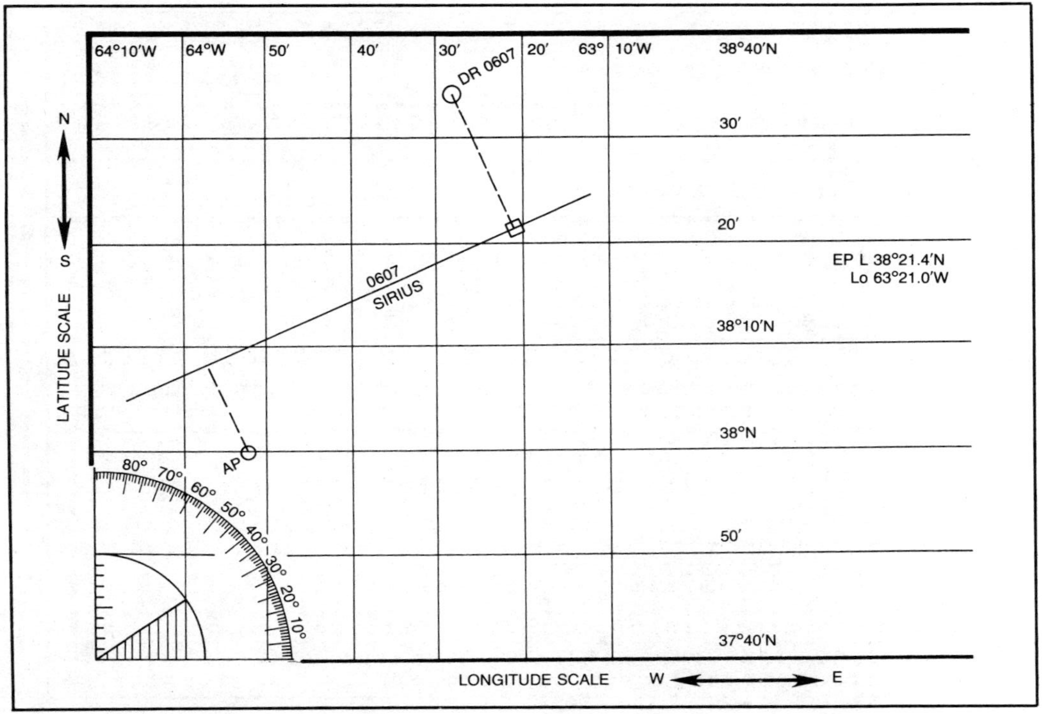

Fig. 9-2. The Sirius sight plot.

A SHOT OF JUPITER

Here's a plan sight: It's 16 June 1980 and you're at DR Lat. 30°41.6'N; Lo 164°37.1'W, with your watch set to standard time. Your watch error is (s) +01-04; index error is – 2.4'; your eye height is 8.7 feet above sea level.

Table 9-5. Dip and Refraction Corrections for the Jupiter Sight.

A2 ALTITUDE CORRECTION TABLES 10°-90°—SUN, STARS, PLANETS

SUN OCT.—MAR. App. Alt.	Lower Limb	Upper Limb	SUN APR.—SEPT. App. Alt.	Lower Limb	Upper Limb	STARS AND PLANETS App. Alt.	Corr^n	DIP Ht. of Eye	Corr^n	Ht. of Eye
° ′	′	′	° ′	′	′	° ′	′	m	′	ft.
9 34			9 39			9 56		2.4		8.0
	+10.8	-21.5		+10.6	-21.2		-5.3		-2.8	
9 45			9 51			10 08		2.6		8.6
	+10.9	-21.4		+10.7	-21.1		-5.2		-2.9	
9 56			10 03			10 20		2.8		9.2
	+11.0	-21.3		+10.8	-21.0		-5.1		-3.0	
10 08			10 15			10 33		3.0		9.8
	+11.1	-21.2		+10.9	-20.9		-5.0		-3.1	
10 21			10 27			10 46		3.2		10.5
	+11.2	-21.1		+11.0	-20.8		-4.9		-3.2	
10 34			10 40			11 00		3.4		11.2
	+11.3	-21.0		+11.1	-20.7		-4.8		-3.3	
10 47			10 54			11 14		3.6		11.9
	+11.4	-20.9		+11.2	-20.6		-4.7		-3.4	
11 01			11 08			11 29		3.8		12.6
	+11.5	-20.8		+11.3	-20.5		-4.6		-3.5	
11 15			11 23			11 45		4.0		13.3
	+11.6	-20.7		+11.4	-20.4		-4.5		-3.6	
11 30			11 38			12 01		4.3		14.1
	+11.7	-20.6		+11.5	-20.3		-4.4		-3.7	
11 46			11 54			12 18		4.5		14.9
	+11.8	-20.5		+11.6	-20.2		-4.3		-3.8	
12 02			12 10			12 35		4.7		15.7
	+11.9	-20.4		+11.7	-20.1		-4.2		-3.9	
12 19			12 28			12 54		5.0		16.5
	+12.0	-20.3		+11.8	-20.0		-4.1		-4.0	
12 37			12 46			13 13		5.2		17.4
	+12.1	-20.2		+11.9	-19.9		-4.0		-4.1	
12 55			13 05			13 33		5.5		18.3
	+12.2	-20.1		+12.0	-19.8		-3.9		-4.2	
13 14			13 24			13 54		5.8		19.1
	+12.3	-20.0		+12.1	-19.7		-3.8		-4.3	
13 35			13 45			14 16		6.1		20.1
	+12.4	-19.9		+12.2	-19.6		-3.7		-4.4	
13 56			14 07			14 40		6.3		21.0
	+12.5	-19.8		+12.3	-19.5		-3.6		-4.5	
14 18			14 30			15 04		6.6		22.0
	+12.6	-19.7		+12.4	-19.4		-3.5		-4.6	
14 42			14 54			15 30		6.9		22.9
	+12.7	-19.6		+12.5	-19.3		-3.4		-4.7	
15 06			15 19			15 57		7.2		23.9
	+12.8	-19.5		+12.6	-19.2		-3.3		-4.8	
15 32			15 46			16 26		7.5		24.9
	+12.9	-19.4		+12.7	-19.1		-3.2		-4.9	
15 59			16 14			16 56		7.9		26.0
	+13.0	-19.3		+12.8	-19.0		-3.1		-5.0	
16 28			16 44			17 28		8.2		27.1
	+13.1	-19.2		+12.9	-18.9		-3.0		-5.1	
16 59			17 15			18 02		8.5		28.1
	+13.2	-19.1		+13.0	-18.8		-2.9		-5.2	
17 32			17 48			18 38		8.8		29.2
	+13.3	-19.0		+13.1	-18.7		-2.8		-5.3	
18 06			18 24			19 17		9.2		30.4
	+13.4	-18.9		+13.2	-18.6		-2.7		-5.4	
18 42			19 01			19 58		9.5		31.5
	+13.5	-18.8		+13.3	-18.5		-2.6		-5.5	
19 21			19 42			20 42		9.9		32.7
	+13.6	-18.7		+13.4	-18.4		-2.5		-5.6	
20 03			20 25			21 28		10.3		33.9
	+13.7	-18.6		+13.5	-18.3		-2.4		-5.7	
20 48			21 11			22 19		10.6		35.1
	+13.8	-18.5		+13.6	-18.2		-2.3		-5.8	
21 35			22 00			23 13		11.0		36.3
	+13.9	-18.4		+13.7	-18.1		-2.2		-5.9	
22 26			22 54			24 11		11.4		37.6
	+14.0	-18.3		+13.8	-18.0		-2.1		-6.0	
23 22			23 51			25 14		11.8		38.9
	+14.1	-18.2		+13.9	-17.9		-2.0		-6.1	
24 21			24 53			26 22		12.2		40.1
	+14.2	-18.1		+14.0	-17.8		-1.9		-6.2	
25 26			26 00			27 36		12.6		41.5
	+14.3	-18.0		+14.1	-17.7		-1.8		-6.3	
26 36			27 13			28 56		13.0		42.8
	+14.4	-17.9		+14.2	-17.6		-1.7		-6.4	
27 52			28 33			30 24		13.4		44.2
	+14.5	-17.8		+14.3	-17.5		-1.6		-6.5	
29 15			30 00			32 00		13.8		45.5
	+14.6	-17.7		+14.4	-17.4		-1.5		-6.6	
30 46			31 35			33 45		14.2		46.9
	+14.7	-17.6		+14.5	-17.3		-1.4		-6.7	
32 26			33 20			35 40		14.7		48.4
	+14.8	-17.5		+14.6	-17.2		-1.3		-6.8	
34 17			35 17			37 48		15.1		49.8
	+14.9	-17.4		+14.7	-17.1		-1.2		-6.9	
36 20			37 26			40 08		15.5		51.3
	+15.0	-17.3		+14.8	-17.0		-1.1		-7.0	
38 36			39 50			42 44		16.0		52.8
	+15.1	-17.2		+14.9	-16.9		-1.0		-7.1	
41 08			42 31			45 36		16.5		54.3
	+15.2	-17.1		+15.0	-16.8		-0.9		-7.2	
43 59			45 31			48 47		16.9		55.8
	+15.3	-17.0		+15.1	-16.7		-0.8		-7.3	
47 10			48 55			52 18		17.4		57.4
	+15.4	-16.9		+15.2	-16.6		-0.7		-7.4	
50 46			52 44			56 11		17.9		58.9
	+15.5	-16.8		+15.3	-16.5		-0.6		-7.5	
54 49			57 02			60 28		18.4		60.5
	+15.6	-16.7		+15.4	-16.4		-0.5		-7.6	
59 23			61 51			65 08		18.8		62.1
	+15.7	-16.6		+15.5	-16.3		-0.4		-7.7	
64 30			67 17			70 11		19.3		63.8
	+15.8	-16.5		+15.6	-16.2		-0.3		-7.8	
70 12			73 16			75 34		19.8		65.4
	+15.9	-16.4		+15.7	-16.1		-0.2		-7.9	
76 26			79 43			81 13		20.4		67.1
	+16.0	-16.3		+15.8	-16.0		-0.1		-8.0	
83 05			86 32			87 03		20.9		68.8
	+16.1	-16.2		+15.9	-15.9		0.0		-8.1	
90 00			90 00			90 00		21.4		70.5

STARS AND PLANETS — App. Alt. / Additional Corr^n

1980

VENUS

Jan. 1-Feb. 26
0° +0.1'
42

Feb. 27-Apr. 13
0° +0.2'
47

Apr. 14-May 9
0° +0.3'
46

May 10-May 25
0° +0.4'
11 +0.5
41

May 26-June 3
0° +0.5'
6 +0.6
20 +0.7
31

June 4-June 26
0° +0.6'
4 +0.7
12 +0.8
22

June 27-July 6
0° +0.5'
6 +0.6
20 +0.7
31

July 7-July 21
0° +0.4'
11 +0.5
41

July 22-Aug. 17
0° +0.3'
46

Aug. 18-Oct. 2
0° +0.2'
47

Oct. 3-Dec. 31
0° +0.1'
42

MARS

Jan. 1-Apr. 28
0° +0.2'
41 +0.1
75

Apr. 29-Dec. 31
0° +0.1'
60

DIP

Ht. of Eye	Corr^n
m	′
1.0	1.8
1.5	2.2
2.0	2.5
2.5	2.8
3.0	3.0
See table ←	
m	′
20	7.9
22	8.3
24	8.6
26	9.0
28	9.3
30	9.6
32	10.0
34	10.3
36	10.6
38	10.8
40	11.1
42	11.4
44	11.7
46	11.9
48	12.2
ft.	′
2	1.4
4	1.9
6	2.4
8	2.7
10	3.1
See table ←	
ft.	′
70	8.1
75	8.4
80	8.7
85	8.9
90	9.2
95	9.5
100	9.7
105	9.9
110	10.2
115	10.4
120	10.6
125	10.8
130	11.1
135	11.3
140	11.5
145	11.7
150	11.9
155	12.1

App. Alt. = Apparent altitude = Sextant altitude corrected for index error and dip.

At 18-47-18 you record a sextant altitude of 54°35.1' for Jupiter. Use the front of the *Almanac* to establish Ho (see Table 9-5.)

Hs	54°35.1'
IC	+ 2.4'
Dip	− 2.9'
Ha	54°34.6'
Main	− 0.7'
Ho	54°33.9'

Determine GMT for the sight:

WT	18-47-18	16 Jun
WE (s)	+ 1-04	
ZT	18-48-22	
ZD	+11	
GMT	29-48-22	16 Jun
GMT	05-48-22	17 Jun

Note that the date in your area is 16 June, but you will use the daily page for 17 June, at 0500, for the Jupiter data. Find the GHA of the body and its declination and determine the LHA. Table 9-6 combines information from the daily page and the increments and corrections page for this sight.

GHA		Dec.	
05h	184°17.7'	05h	N 11°04.4'
48m 22s	12°05.5'	d +0.1'	
v = 2.1'		D corr.	+ 0.1'
v corr.	+ 1.7'	Dec.	N 11°04.5'
GHA	196°24.9'		
AP Lo	164°24.9' W		
LHA	32°		

Tabulated declination for *Pub. 229* is N 11°; declination increment is 04.5'. Use the left-hand page for LHA 32°, 328° as AP latitude and declination are both north. Your AP latitude is 31° closest to actual DR latitude. For declination 11° you find:

Tab Hc	d	Tab Z
54°16.6'	+37.2'	117.0°

You note that Z is 115.7° at declination 12°. Use the front interpolation table to get the d and Z correction figures:

	d	2.3'	Z diff.	−1.3°
		+ 0.5'	Z corr.	−0.1°
d corr.	+	2.8'		

121

Table 9-6. Daily Page Data and the Increment and Corrections for the Jupiter Sight.

122   1980 JUNE 17, 18, 19 (TUES., WED., THURS.)

G.M.T.	ARIES G.H.A.	VENUS −3.1 G.H.A.	Dec.	MARS +1.0 G.H.A.	Dec.	JUPITER −1.5 G.H.A.	Dec.	SATURN +1.3 G.H.A.	Dec.	STARS Name	S.H.A.	Dec.
17 00	265 24.0	182 37.7 N21 42.9		96 34.2 N 5 42.4		109 07.0 N11 05.1		93 01.7 N 5 42.2		Acamar	315 37.7 S40 22.9
01	280 26.4	197 41.8	42.2	111 35.6	41.9	124 09.2	04.9	108 04.1	42.1	Achernar	335 45.8 S57 20.0
02	295 28.9	212 45.9	41.6	126 36.9	41.3	139 11.3	04.8	123 06.4	42.1	Acrux	173 37.2 S62 59.7
03	310 31.4	227 50.0	40.9	141 38.2 ··	40.8	154 13.5 ··	04.7	138 08.8 ··	42.1	Adhara	255 32.6 S28 56.8
04	325 33.8	242 54.1	40.3	156 39.6	40.3	169 15.6	04.6	153 11.2	42.0	Aldebaran	291 18.5 N16 28.1
05	340 36.3	257 58.2	39.6	171 40.9	39.7	184 17.7	04.4	168 13.5	42.0	
06	355 38.8	273 02.3 N21 39.0		186 42.3 N 5 39.2		199 19.9 N11 04.3		183 15.9 N 5 41.9		Alioth	166 42.6 N56 04.3
T 07	10 41.2	288 06.4	38.3	201 43.6	38.7	214 22.0	04.2	198 18.3	41.9	Alkaid	153 18.4 N49 25.0
U 08	25 43.7	303 10.5	37.7	216 45.0	38.2	229 24.2	04.0	213 20.6	41.8	Al Na'ir	28 15.0 S47 03.2
E 09	40 46.2	318 14.6 ··	37.1	231 46.3 ··	37.6	244 26.3 ··	03.9	228 23.0 ··	41.8	Alnilam	276 12.2 S 1 13.0
S 10	55 48.6	333 18.7	36.4	246 47.7	37.1	259 28.5	03.8	243 25.4	41.7	Alphard	218 20.9 S 8 34.5
D 11	70 51.1	348 22.8	35.8	261 49.0	36.6	274 30.6	03.7	258 27.7	41.7	
12											N26 47.0

INCREMENTS AND CORRECTIONS

48	SUN PLANETS	ARIES	MOON	v or Corr d	v or Corr d	v or Corr d
00	12 00·0	12 02·0	11 27·2	0·0 0·0	6·0 4·9	12·0 9·7
01	12 00·3	12 02·2	11 27·4	0·1 0·1	6·1 4·9	12·1 9·8
02	12 00·5	12 02·5	11 27·7	0·2 0·2	6·2 5·0	12·2 9·9
03	12 00·8	12 02·7	11 27·9	0·3 0·2	6·3 5·1	12·3 9·9
04	12 01·0	12 03·0	11 28·2	0·4 0·3	6·4 5·2	12·4 10·0
05	12 01·3	12 03·2	11 28·4	0·5 0·4	6·5 5·3	12·5 10·1
06	12 01·5	12 03·5	11 28·6	0·6 0·5	6·6 5·4	12·6 10·2
07	12 01·8	12 03·7	11 28·9	0·7 0·6	6·7 5·4	12·7 10·3
08	12 02·0	12 04·0	11 29·1	0·8 0·6	6·8 5·5	12·8 10·3
09	12 02·3	12 04·2	11 29·3	0·9 0·7	6·9 5·6	12·9 10·4
10	12 02·5	12 04·5	11 29·6	1·0 0·8	7·0 5·7	13·0 10·5
11	12 02·8	12 04·7	11 29·8	1·1 0·9	7·1 5·7	13·1 10·6
12	12 03·0	12 05·0	11 30·1	1·2 1·0	7·2 5·8	13·2 10·7
13	12 03·3	12 05·2	11 30·3	1·3 1·1	7·3 5·9	13·3 10·8
14	12 03·5	12 05·5	11 30·5	1·4 1·1	7·4 6·0	13·4 10·8
15	12 03·8	12 05·7	11 30·8	1·5 1·2	7·5 6·1	13·5 10·9
16	12 04·0	12 06·0	11 31·0	1·6 1·3	7·6 6·1	13·6 11·0
17	12 04·3	12 06·2	11 31·3	1·7 1·4	7·7 6·2	13·7 11·1
18	12 04·5	12 06·5	11 31·5	1·8 1·5	7·8 6·3	13·8 11·2
19	12 04·8	12 06·7	11 31·7	1·9 1·5	7·9 6·4	13·9 11·2
20	12 05·0	12 07·0	11 32·0	2·0 1·6	8·0 6·5	14·0 11·3
21	12 05·3	12 07·2	11 32·2	2·1 1·7	8·1 6·5	14·1 11·4
22	12 05·5	12 07·5	11 32·4	2·2 1·8	8·2 6·6	14·2 11·5
23	12 05·8	12 07·7	11 32·7	2·3 1·9	8·3 6·7	14·3 11·5
24	12 06·0	12 08·0	11 32·9	2·4 1·9	8·4 6·8	14·4 11·6
25	12 06·3	12 08·2	11 33·2	2·5 2·0	8·5 6·9	14·5 11·7
26	12 06·5	12 08·5	11 33·4	2·6 2·1	8·6 7·0	14·6 11·8
27	12 06·8	12 08·7	11 33·6	2·7 2·2	8·7 7·0	14·7 11·9
28	12 07·0	12 09·0	11 33·9	2·8 2·3	8·8 7·1	14·8 12·0
29	12 07·3	12 09·2	11 34·1	2·9 2·3	8·9 7·2	14·9 12·0

49	SUN PLANETS	ARIES	MOON	v or Corr d	v or Corr d	v or Corr d
00	12 15·0	12 17·0	11 41·5	0·0 0·0	6·0 5·0	12·0 9·9
01	12 15·3	12 17·3	11 41·8	0·1 0·1	6·1 5·0	12·1 10·0
02	12 15·5	12 17·5	11 42·0	0·2 0·2	6·2 5·1	12·2 10·1
03	12 15·8	12 17·8	11 42·2	0·3 0·2	6·3 5·2	12·3 10·1
04	12 16·0	12 18·0	11 42·5	0·4 0·3	6·4 5·3	12·4 10·2
05	12 16·3	12 18·3	11 42·7	0·5 0·4	6·5 5·4	12·5 10·3
06	12 16·5	12 18·5	11 42·9	0·6 0·5	6·6 5·4	12·6 10·4
07	12 16·8	12 18·8	11 43·2	0·7 0·6	6·7 5·5	12·7 10·5
08	12 17·0	12 19·0	11 43·4	0·8 0·7	6·8 5·6	12·8 10·6
09	12 17·3	12 19·3	11 43·7	0·9 0·7	6·9 5·7	12·9 10·6
10	12 17·5	12 19·5	11 43·9	1·0 0·8	7·0 5·8	13·0 10·7
11	12 17·8	12 19·8	11 44·1	1·1 0·9	7·1 5·9	13·1 10·8
12	12 18·0	12 20·0	11 44·4	1·2 1·0	7·2 5·9	13·2 10·9
13	12 18·3	12 20·3	11 44·6	1·3 1·1	7·3 6·0	13·3 11·0
14	12 18·5	12 20·5	11 44·9	1·4 1·2	7·4 6·1	13·4 11·1
15	12 18·8	12 20·8	11 45·1	1·5 1·2	7·5 6·2	13·5 11·1
16	12 19·0	12 21·0	11 45·3	1·6 1·3	7·6 6·3	13·6 11·2
17	12 19·3	12 21·3	11 45·6	1·7 1·4	7·7 6·4	13·7 11·3
18	12 19·5	12 21·5	11 45·8	1·8 1·5	7·8 6·4	13·8 11·4
19	12 19·8	12 21·8	11 46·1	1·9 1·6	7·9 6·5	13·9 11·5
20	12 20·0	12 22·0	11 46·3	2·0 1·7	8·0 6·6	14·0 11·6
21	12 20·3	12 22·3	11 46·5	2·1 1·7	8·1 6·7	14·1 11·6
22	12 20·5	12 22·5	11 46·8	2·2 1·8	8·2 6·8	14·2 11·7
23	12 20·8	12 22·8	11 47·0	2·3 1·9	8·3 6·8	14·3 11·8
24	12 21·0	12 23·0	11 47·2	2·4 2·0	8·4 6·9	14·4 11·9
25	12 21·3	12 23·3	11 47·5	2·5 2·1	8·5 7·0	14·5 12·0
26	12 21·5	12 23·5	11 47·7	2·6 2·1	8·6 7·1	14·6 12·0
27	12 21·8	12 23·8	11 48·0	2·7 2·2	8·7 7·2	14·7 12·1
28	12 22·0	12 24·0	11 48·2	2·8 2·3	8·8 7·3	14·8 12·2
29	12 22·3	12 24·3	11 48·4	2·9 2·4	8·9 7·3	14·9 12·3

				STARS	S.H.A.	Dec.
07	12 39·5	291 21·3	07·5	202 47.8	13.1	216 04.8 57.9
T 08	27 42.0	306 25.3	06.9	217 49.1	12.5	231 06.9 57.8	Regulus	208 10.4 N12 03.8
H 09	42 44.4	321 29.3 ··	06.3	232 50.5 ··	12.0	246 09.1 ·· 57.7	Rigel	281 36.5 S 8 13.5
U 10	57 46.9	336 33.4	05.6	247 51.8	11.5	261 11.2 57.5	Rigil Kent.	140 25.5 S60 45.3
R 11	72 49.4	351 37.4	05.0	262 53.1	10.9	276 13.3 57.4	Sabik	102 41.0 S15 42.0
S 12	87 51.8	6 41.4 N21 04.4		277 54.5 N 5 10.4		291 15.5 N10 57.3	Schedar	350 09.3 N56 25.5
D 13	102 54.3	21 45.4	03.7	292 55.8	09.9	306 17.6 57.1	Shaula	96 55.6 S37 05.3
A 14	117 56.8	36 49.4	03.1	307 57.1	09.3	321 19.7 57.0	Sirius	258 56.2 S16 41.5
Y 15	132 59.2	51 53.4 ··	02.5	322 58.4 ··	08.8	336 21.9 ·· 56.9	Spica	158 57.6 S11 03.5
16	148 01.7	66 57.4	01.9	337 59.8	08.2	351 24.0 56.8	Suhail	223 11.2 S43 21.4
17	163 04.1	82 01.4	01.2	353 01.1	07.7	6 26.1 56.6	
18	178 06.6	97 05.4 N21 00.6		8 02.4 N 5 07.2		21 28.3 N10 56.5	Vega	80 55.5 N38 46.0
19	193 09.1	112 09.4 21 00.0		23 03.7	06.6	36 30.4 56.4	Zuben'ubi	137 33.0 S15 57.6
20	208 11.5	127 13.4 20 59.3		38 05.1	06.1	51 32.6 56.2	
21	223 14.0	142 17.4 ·· 58.7		53 06.4 ·· 05.6		66 34.7 ·· 56.1			Mer. Pass.
22	238 16.5	157 21.4	58.1	68 07.7	05.0	81 36.8 56.0	Venus	277 52.7 11 40
23	253 18.9	172 25.4	57.5	83 09.1	04.5	96 39.0 55.8	Mars	190 43.3 17 30
Mer. Pass.	6 13.4	v 4.1 d 0.6		v 1.3 d 0.5		v 2.1 d 0.1	v 2.4 d 0.1	Jupiter	203 35.4 16 38
								Saturn	187 35.4 17 41

Table 9-7 combines the necessary information from the LHA page and the interpolation table. Correct Tab Hc and Tab Z:

Tab Hc	54°16.6′	Tab Z	117.0°
d corr.	+ 2.8′	Z corr.	− 0.1′
Hc	54°19.4′	Z	116.9°

Since LHA is less than 180 degrees, subtract Z from 360 degrees to get Zn:

$$
\begin{array}{rr}
 & 360° \\
Z & -116.9° \\
\hline
Zn & 243.1° \\
\end{array}
$$

The difference between Ho and Hc gives intercept distance:

$$
\begin{array}{rl}
Hc & 54°19.4' \\
Ho & 54°33.9' \\
\hline
\text{Intercept} & 14.5 \text{ miles toward} \\
\end{array}
$$

With this information, you can plot the sight. Figure 9-3 shows the work-sheet. Figure 9-4 shows the plot.

A RUN OF MOON SIGHTS

Other than the sun, the moon is the easiest body to spot for a sight. It is, however the most complicated sight to reduce because of the HP factor that often requires interpolation. Here we will go through the entire procedure, including selecting the sight from a run of five. To add to the fun, we are south of the equator in east longitude.

The date is 17 November 1980; DR position L 39°35.6'S; Lo 137°42.7' E, which puts us in ZD −9 (see Fig. 4-1). Watch error is (s) +03-17; index error is +0.8'; and eye height is 18.6 feet above sea level. With good visibility, the following sights are recorded:

	WT	Hs
1.	18-30-16	55°18.3'
2.	18-31-10	55°21.5'
3.	18-32-09	55°23.5'
4.	18-32-55	55°29.7'
5.	18-33-46	55°31.3'

Determine LHA for the middle sight of the run, draw the slope, and plot the sights on a graph (Fig. 9-5). Choose the sight closest to the line representing average distance from sights to slope:

WT	18-32-09	17 Nov	GHA		
WE (s)	+3-17		9h	202°15.6'	
ZT	18-35-26		35m 26s	7°44.3'	
ZD	−9		v + 9.5'		
GMT	09-35-26	17 Nov	v corr.	+ 5.1'	
			GHA	210°05.0'	
Dec. GMT	9h S 7°21.8'		AP Lo	137°55.0'	
			LHA	348°	

123

Table 9-7. Pub. 229 Tables Used for the Jupiter Sight Reduction.

32°, 328° L.H.A. LATITUDE SAME NAME AS DECLINATION

N. Lat { L.H.A. greater than 180° ..... Zn=Z ; L.H.A. less than 180° ..... Zn= }

Dec.	30° Hc	d	Z	31° Hc	d	Z	32° Hc	d	Z	33° Hc	d	Z	34° Hc	d	Z	35° Hc	d	Z	36° Hc	d	Z	37° Hc	d	Z
0	47 15.6	43.9	128.7	46 37.7	44.8	129.5	45 59.2	45.6	130.3	45 20.1	46.3	131.1	44 40.4	47.0	131.8	44 00.1	47.7	132.5	43 19.3	48.3	133.2	42 37.9	48.9	133.9
1	47 59.5	43.4	127.7	47 22.5	44.2	128.5	46 44.8	45.0	129.4	46 06.4	45.8	130.2	45 27.4	46.5	130.9	44 47.8	47.2	131.7	44 07.6	47.9	132.4	43 26.8	48.6	133.1
2	48 42.9	42.8	126.6	48 06.7	43.7	127.5	47 29.8	44.6	128.4	46 52.2	45.4	129.2	46 13.9	46.2	130.0	45 35.0	46.9	130.8	44 55.5	47.6	131.6	44 15.4	48.2	132.3
3	49 25.7	42.1	125.4	48 50.4	43.1	126.4	48 14.4	44.0	127.4	47 37.6	44.8	128.3	47 00.1	45.6	129.1	46 21.9	46.3	129.9	45 43.1	47.1	130.7	45 03.6	47.9	131.5
4	50 07.9	41.2	124.4	49 33.5	42.6	125.4	48 58.4	43.4	126.4	48 22.4	44.3	127.3	47 45.7	45.2	128.1	47 08.3	46.0	129.0	46 30.2	46.7	129.8	45 51.5	47.4	130.6
5	50 49.5	40.9	123.3	50 16.1	41.9	124.3	49 41.8	42.9	125.3	49 06.7	43.8	126.2	48 30.9	44.6	127.2	47 54.3	45.4	128.0	47 16.9	46.3	128.9	46 38.9	47.0	129.7
6	51 30.4	40.0	122.1	50 58.0	41.2	123.1	50 24.7	42.2	124.1	49 50.5	43.2	125.2	49 15.5	44.1	126.1	48 39.7	45.0	127.1	48 03.2	45.8	128.0	47 25.9	46.6	128.8
7	52 10.5	39.4	120.9	51 39.2	40.5	122.0	51 06.9	41.5	123.0	50 33.7	42.5	124.1	49 59.6	43.5	125.1	49 24.7	44.4	126.1	48 49.0	45.3	127.0	48 12.5	46.1	127.9
8	52 49.9	38.7	119.7	52 19.7	39.8	120.8	51 48.4	40.9	121.9	51 16.2	41.9	123.0	50 43.1	42.9	124.0	50 09.1	43.9	125.0	49 34.3	44.7	126.0	48 58.6	45.6	126.9
9	53 28.6	37.7	118.4	52 59.5	38.9	119.6	52 29.3	40.1	120.7	51 58.1	41.2	121.8	51 26.0	42.2	122.9	50 53.0	43.2	123.9	50 19.0	44.2	124.9	49 44.2	45.1	125.9
10	54 06.3	36.9	117.1	53 38.4	38.2	118.3	53 09.4	39.3	119.5	52 39.3	40.5	120.6	52 08.2	41.6	121.8	51 36.2	42.6	122.8	51 03.2	43.6	123.9	50 29.3	44.5	124.9
11	54 43.2	36.0	115.8	54 16.6	37.2	117.0	53 48.7	38.5	118.2	53 19.8	39.7	119.4	52 49.8	40.8	120.6	52 18.8	41.9	121.7	51 46.8	42.9	122.8	51 13.8	43.9	123.8
12	55 19.2	34.9	114.4	54 53.8	36.3	115.7	54 27.2	37.6	116.9	53 59.5	38.8	118.2	53 30.6	40.0	119.4	53 00.7	41.1	120.5	52 29.7	42.2	121.6	51 57.7	43.3	122.7
13	55 54.1	34.0	112.9	55 30.1	35.4	114.3	55 04.8	36.7	115.6	54 38.3	38.0	116.8	54 10.7	39.2	118.1	53 41.8	40.3	119.3	53 12.0	41.5	120.5	52 41.0	42.6	121.6
14	56 05.5	34.3	112.8	56 05.5	34.3	112.8	55 41.5	35.7	114.2	55 16.3	37.1	115.5	54 49.9	38.3	116.8	54 22.3	39.6	118.0	53 53.5	40.8	119.2	53 23.6	41.9	120.4
15	57 00.9	31.7	109.9	56 39.8	33.2	111.3	56 17.3	34.6	112.7	55 53.4	36.1	114.1	55 28.2	37.5	115.4	55 01.9	38.7	116.7	54 34.3	40.0	118.0	54 05.5	41.2	119.2
16	57 32.6	30.5	108.3	57 13.0	32.1	109.8	56 51.9	33.6	111.3	56 29.5	35.0	112.7	56 05.7	36.5	114.1	55 40.6	37.8	115.4	55 14.3	39.1	116.7	54 46.7	40.4	118.0
17	58 03.1	29.2	106.7	57 45.1	30.8	108.2	57 25.5	32.5	109.7	57 04.6	34.0	111.2	56 42.2	35.4	112.6	56 18.4	34.9	114.0	55 53.4	38.2	115.4	55 27.1	39.5	116.7

Interpolation tables (bottom)

Dec.	0	1	2	3	4
4.0	0.6	1.3	2.0	3.3	3.8
4.1	0.7	1.3	2.0	3.4	3.8
4.2	0.7	1.4	2.1	3.5	3.9
4.3	0.7	1.4	2.1	3.6	4.0
4.4	0.7	1.5	2.2	3.7	4.1
4.5	0.8	1.5	2.3	3.8	4.2
4.6	0.8	1.6	2.3	3.8	4.3
4.7	0.8	1.6	2.4	3.9	4.3
4.8	0.8	1.6	2.4	4.0	4.3
4.9	0.9	1.7	2.5	4.1	4.3

	0	1	2	3	4
.0	0.0	0.0	0.1	0.1	0.2
.1	0.0	0.0	0.1	0.2	0.2
.2	0.0	0.1	0.2	0.2	0.3
.3	0.1	0.1	0.2	0.3	0.3
.4	0.1	0.1	0.2	0.3	0.4
.5	0.1	0.2	0.3	0.4	0.5
.6	0.1	0.2	0.3	0.4	0.5
.7	0.1	0.2	0.3	0.5	0.6
.8	0.1	0.2	0.4	0.5	0.6
.9	0.1	0.2	0.4	0.6	0.7

2.9	0.1	
8.6	0.2	
14.4	0.3	
20.2	0.4	
25.9	0.5	
31.7	0.6	
37.5	0.7	

			6	8	10
12.0	2.0	4.0	6.0	8.0	10.0
12.1	2.0	4.0	6.0	8.0	10.1
12.2	2.0	4.1	6.1	8.1	10.1
12.3	2.1	4.1	6.1	8.2	10.2
12.4	2.1	4.1	6.2	8.3	10.3
12.5	2.1	4.2	6.3	8.3	10.4
12.6	2.1	4.2	6.3	8.4	10.5
12.7	2.2	4.3	6.4	8.5	10.6
12.8	2.2	4.3	6.4	8.6	10.7
12.9	2.2	4.3	6.5	8.6	10.8

	5	6	7	8	9
.0	0.8	1.0	1.2	1.5	1.7
.1	0.8	1.0	1.3	1.5	1.7
.2	0.9	1.1	1.3	1.5	1.7
.3	0.9	1.1	1.3	1.5	1.7
.4	0.9	1.1	1.3	1.5	1.7
.5	0.9	1.1	1.4	1.6	1.8
.6	1.0	1.2	1.4	1.6	1.8
.7	1.0	1.2	1.4	1.6	1.8
.8	1.0	1.2	1.4	1.6	1.8
.9	1.0	1.2	1.4	1.6	1.9

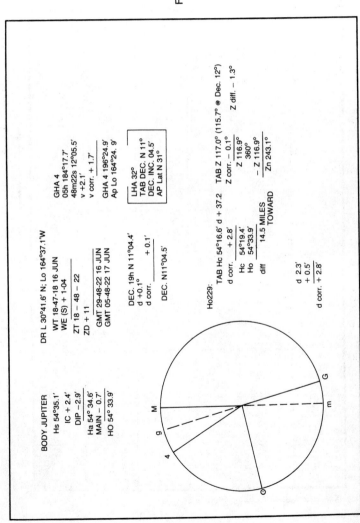

BODY JUPITER
Hs 54°35.1'
IC + 2.4'
DIP −2.9'
Ha 54° 34.6'
MAIN − 0.7'
HO 54° 33.9'

DR L 30°41.6' N; Lo 164°37.1'W
WT 18-47-18 16 JUN
WE (S) + 1-04
ZT 18 − 48 − 22
ZD + 11
GMT 29-48-22 16 JUN
GMT 05-48-22 17 JUN

DEC. 19h N 11°04.4'
d +0.1°   + 0.1'
d corr.   _____
DEC. N11°04.5'

GHA 4
05h 184°17.7'
48m22s 12°05.5'
v +2.1'
v corr. + 1.7'
_____
GHA 4 196°24.9'
Ap Lo 164°24. 9'

LHA 32°
TAB DEC. N 11°
DEC. INC. 04.5'
AP Lat N 31°

Ho229:   TAB Hc 54°16.6' d + 37.2   TAB Z 117.0° (115.7° @ Dec. 12°)
                               d corr. + 2.8'   Z corr. − 0.1°   Z diff. − 1.3°
                               Hc   54°19.4'   Z 116.9°
                               Ho   54°33.9'   360°
                                    _____   − Z 116.9°
                               diff   14.5 MILES   _____
                                       TOWARD   Zn 243.1°

                                       d 2.3'
                                      + 0.5'
                                      _____
                                      d corr. + 2.8'

Fig. 9-3. Jupiter sight worksheet.

125

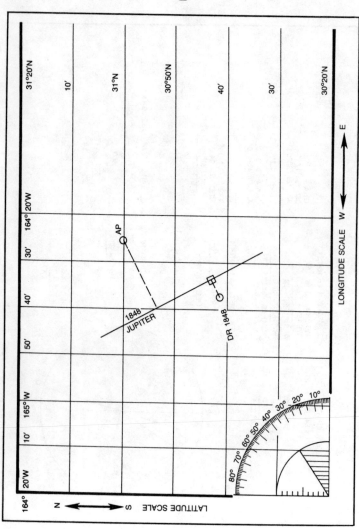

Fig. 9-4. The Jupiter sight plot.

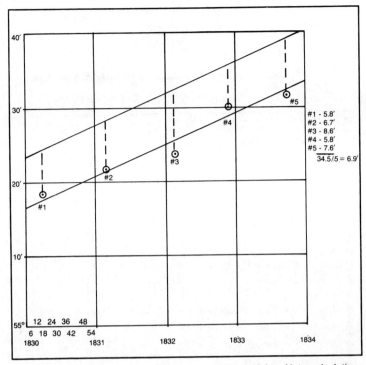

Fig. 9-5. Slope plotted on a graph for a run of moon sights. Note calculations above used to determine average distance from the slope line.

In *Pub. 229* at tabulated declination 7°:

$$
\begin{array}{ll}
\text{LHA } 347° & \text{Hc } 55°00.2' \\
& \qquad\qquad 17.3' \\
\text{LHA } 348° & \text{Hc } 55°17.5' \\
& \qquad\qquad 16.0' \\
\text{LHA } 349° & \text{Hc } 55°33.5' \quad \overline{} \\
& \qquad\qquad \dfrac{33.3°}{2} = 16.6°
\end{array}
$$

Hc changes an average of 16.6' for the four minutes covering the time of the run of sights. This line is drawn on the graph and the sight closest to it is chosen; in this case it's sight #2. Convert sextant altitude to observed altitude using the information on the inside back cover of the *Almanac* (Table 9-8):

$$
\begin{array}{ll}
\text{Hs} & 55°21.5' \\
\text{IC} & -0.8' \\
\text{Dip} & -4.2' \\
\text{Ha} & \overline{55°16.5'}
\end{array}
$$

127

Main + 42.8'
HP + 6.4'
Ho 56°05.7'

Table 9-8. Refraction and H.P. Corrections for the Selected Moon Sight.

ALTITUDE CORRECTION TABLES 35°–90°—MOON

App. Alt.	35°–39° Corrⁿ	40°–44° Corrⁿ	45°–49° Corrⁿ	50°–54° Corrⁿ	55°–59° Corrⁿ	60°–64° Corrⁿ	65°–69° Corrⁿ	70°–74° Corrⁿ	75°–79° Corrⁿ	80°–84° Corrⁿ	85°–89° Corrⁿ	App. Alt.
00	35 56.5	40 53.7	45 50.5	50 46.9	55 43.1	60 38.9	65 34.6	70 30.1	75 25.3	80 20.5	85 15.6	00
10	56.4	53.6	50.4	46.8	42.9	38.8	34.4	29.9	25.2	20.4	15.5	10
20	56.3	53.5	50.2	46.7	42.8	38.7	34.3	29.7	25.0	20.2	15.3	20
30	56.2	53.4	50.1	46.5	42.7	38.5	34.1	29.6	24.9	20.0	15.1	30
40	56.2	53.3	50.0	46.4	42.5	38.4	34.0	29.4	24.7	19.9	15.0	40
50	56.1	53.2	49.9	46.3	42.4	38.2	33.8	29.3	24.5	19.7	14.8	50
00	36 56.0	41 53.1	46 49.8	51 46.2	56 42.3	61 38.1	66 33.7	71 29.1	76 24.4	81 19.6	86 14.6	00
10	55.9	53.0	49.7	46.0	42.1	37.9	33.5	29.0	24.2	19.4	14.5	10
20	55.8	52.8	49.5	45.9	42.0	37.8	33.4	28.8	24.1	19.2	14.3	20
30	55.7	52.7	49.4	45.8	41.8	37.7	33.2	28.7	23.9	19.1	14.1	30
40	55.6	52.6	49.3	45.7	41.7	37.5	33.1	28.5	23.8	18.9	14.0	40
50	55.5	52.5	49.2	45.5	41.6	37.4	32.9	28.3	23.6	18.7	13.8	50
00	37 55.4	42 52.4	47 49.1	52 45.4	57 41.4	62 37.2	67 32.8	72 28.2	77 23.4	82 18.6	87 13.7	00
10	55.3	52.3	49.0	45.3	41.3	37.1	32.6	28.0	23.3	18.4	13.5	10
20	55.2	52.2	48.8	45.2	41.2	36.9	32.5	27.9	23.1	18.2	13.3	20
30	55.1	52.1	48.7	45.0	41.0	36.8	32.3	27.7	22.9	18.1	13.2	30
40	55.0	52.0	48.6	44.9	40.9	36.6	32.2	27.6	22.8	17.9	13.0	40
50	55.0	51.9	48.5	44.8	40.8	36.5	32.0	27.4	22.6	17.8	12.8	50
00	38 54.9	43 51.8	48 48.4	53 44.6	58 40.6	63 36.4	68 31.9	73 27.2	78 22.5	83 17.6	88 12.7	00
10	54.8	51.7	48.2	44.5	40.5	36.2	31.7	27.1	22.3	17.4	12.5	10
20	54.7	51.6	48.1	44.4	40.3	36.1	31.6	26.9	22.1	17.3	12.3	20
30	54.6	51.5	48.0	44.2	40.2	35.9	31.4	26.8	22.0	17.1	12.2	30
40	54.5	51.4	47.9	44.1	40.1	35.8	31.3	26.6	21.8	16.9	12.0	40
50	54.4	51.2	47.8	44.0	39.9	35.6	31.1	26.5	21.7	16.8	11.8	50
00	39 54.3	44 51.1	49 47.6	54 43.9	59 39.8	64 35.5	69 31.0	74 26.3	79 21.5	84 16.6	89 11.7	00
10	54.2	51.0	47.5	43.7	39.6	35.3	30.8	26.1	21.3	16.5	11.5	10
20	54.1	50.9	47.4	43.6	39.5	35.2	30.7	26.0	21.2	16.3	11.4	20
30	54.0	50.8	47.3	43.5	39.4	35.0	30.5	25.8	21.0	16.1	11.2	30
40	53.9	50.7	47.2	43.3	39.2	34.9	30.4	25.7	20.9	16.0	11.0	40
50	53.8	50.6	47.0	43.2	39.1	34.7	30.2	25.5	20.7	15.8	10.9	50

H.P.	L U	L U	L U	L U	L U	L U	L U	L U	L U	L U	L U	H.P.
54.0	1.1 1.7	1.3 1.9	1.5 2.1	1.7 2.4	2.0 2.6	2.3 2.9	2.6 3.2	2.9 3.5	3.2 3.8	3.5 4.1	3.8 4.5	54.0
54.3	1.4 1.8	1.6 2.0	1.8 2.2	2.0 2.5	2.3 2.7	2.5 3.0	2.8 3.2	3.0 3.5	3.3 3.8	3.6 4.1	3.9 4.4	54.3
54.6	1.7 2.0	1.9 2.2	2.1 2.4	2.3 2.6	2.5 2.8	2.7 3.0	3.0 3.3	3.2 3.5	3.5 3.8	3.7 4.1	4.0 4.3	54.6
54.9	2.0 2.2	2.2 2.3	2.3 2.5	2.5 2.7	2.7 2.9	2.9 3.1	3.2 3.3	3.4 3.5	3.6 3.8	3.9 4.0	4.1 4.3	54.9
55.2	2.3 2.3	2.5 2.4	2.6 2.6	2.8 2.8	3.0 2.9	3.2 3.1	3.4 3.3	3.6 3.5	3.8 3.7	4.0 4.0	4.2 4.2	55.2
55.5	2.7 2.5	2.8 2.6	2.9 2.7	3.1 2.9	3.2 3.0	3.4 3.2	3.6 3.4	3.7 3.5	3.9 3.7	4.1 3.9	4.3 4.1	55.5
55.8	3.0 2.6	3.1 2.7	3.2 2.8	3.3 3.0	3.5 3.1	3.6 3.3	3.8 3.4	3.9 3.6	4.1 3.7	4.3 3.9	4.4 4.0	55.8
56.1	3.3 2.8	3.4 2.9	3.5 3.0	3.6 3.1	3.7 3.2	3.8 3.3	4.0 3.4	4.1 3.6	4.2 3.7	4.4 3.8	4.5 4.0	56.1
56.4	3.6 2.9	3.7 3.0	3.8 3.1	3.9 3.2	3.9 3.3	4.0 3.4	4.1 3.5	4.3 3.6	4.4 3.7	4.5 3.8	4.6 3.9	56.4
56.7	3.9 3.1	4.0 3.1	4.1 3.2	4.1 3.3	4.2 3.3	4.3 3.4	4.3 3.5	4.4 3.6	4.5 3.7	4.6 3.8	4.7 3.8	56.7
57.0	4.3 3.2	4.3 3.3	4.3 3.3	4.4 3.4	4.4 3.4	4.5 3.5	4.5 3.5	4.6 3.6	4.7 3.6	4.7 3.7	4.8 3.8	57.0
57.3	4.6 3.4	4.6 3.4	4.6 3.4	4.7 3.5	4.7 3.5	4.7 3.6	4.8 3.6	4.8 3.6	4.8 3.7	4.9 3.7	4.9 3.7	57.3
57.6	4.9 3.6	4.9 3.6	4.9 3.6	4.9 3.6	4.9 3.6	4.9 3.6	4.9 3.6	5.0 3.6	5.0 3.6	5.0 3.6	5.0 3.6	57.6
57.9	5.2 3.7	5.2 3.7	5.2 3.7	5.2 3.7	5.1 3.6	5.1 3.6	5.1 3.6	5.1 3.6	5.1 3.6	5.1 3.6	5.1 3.6	57.9
58.2	5.5 3.9	5.5 3.8	5.5 3.8	5.4 3.8	5.4 3.7	5.3 3.7	5.3 3.7	5.3 3.6	5.2 3.6	5.2 3.5	5.2 3.5	58.2
58.5	5.9 4.0	5.8 4.0	5.8 3.9	5.7 3.9	5.6 3.8	5.6 3.8	5.5 3.7	5.5 3.6	5.4 3.6	5.3 3.5	5.3 3.4	58.5
58.8	6.2 4.2	6.1 4.1	6.0 4.1	6.0 4.0	5.9 3.9	5.8 3.8	5.7 3.7	5.6 3.6	5.5 3.5	5.4 3.5	5.3 3.4	58.8
59.1	6.5 4.3	6.4 4.3	6.3 4.2	6.2 4.1	6.1 4.0	6.0 3.9	5.9 3.8	5.8 3.6	5.6 3.5	5.5 3.4	5.3 3.3	59.1
59.4	6.8 4.5	6.7 4.4	6.6 4.3	6.5 4.2	6.4 4.1	6.2 3.9	6.1 3.8	6.0 3.7	5.8 3.5	5.7 3.4	5.5 3.2	59.4
59.7	7.1 4.6	7.0 4.5	6.9 4.4	6.8 4.3	6.6 4.1	6.5 4.0	6.3 3.8	6.2 3.7	6.0 3.5	5.8 3.3	5.6 3.2	59.7
60.0	7.5 4.8	7.3 4.7	7.2 4.5	7.0 4.4	6.9 4.2	6.7 4.0	6.5 3.9	6.3 3.7	6.1 3.5	5.9 3.3	5.7 3.1	60.0
60.3	7.8 5.0	7.6 4.8	7.5 4.7	7.3 4.5	7.1 4.3	6.9 4.1	6.7 3.9	6.5 3.7	6.3 3.5	6.0 3.2	5.8 3.0	60.3
60.6	8.1 5.1	7.9 5.0	7.7 4.8	7.5 4.6	7.3 4.4	7.1 4.2	6.9 3.9	6.7 3.7	6.4 3.4	6.2 3.2	5.9 2.9	60.6
60.9	8.4 5.3	8.2 5.1	8.0 4.9	7.8 4.7	7.6 4.5	7.3 4.2	7.1 4.0	6.8 3.7	6.6 3.4	6.3 3.2	6.0 2.9	60.9
61.2	8.7 5.4	8.5 5.2	8.3 5.0	8.1 4.8	7.8 4.5	7.6 4.3	7.3 4.0	7.0 3.7	6.7 3.4	6.4 3.1	6.1 2.8	61.2
61.5	9.1 5.6	8.8 5.4	8.6 5.1	8.3 4.9	8.1 4.6	7.8 4.3	7.5 4.0	7.2 3.7	6.9 3.4	6.5 3.1	6.2 2.7	61.5

Obviously this run of sights was based on the lower limb of the moon. If an upper limb sight had been reduced, we would have subtracted 30.0′ after adding the Main and HP corrections. Note that we were lucky with the HP factor; the exact figure was shown in the table and there was no need for interpolation. Work out GMT for the sight, and determine the moon's GP for this time (see Table 9-9).

WT	18-31-10 17 Nov
WE (s)	+ 3-17
ZT	18-34-27
ZD	−9
GMT	09-34-27 17 Nov

GHA		**Dec**	
9h	202°15.6′	9h	S 7°33.2′
34m 27s	8°13.2′	d −11.4′	
v +9.6°		d corr.	−6.6′
v corr.	+5.5′	Dec.	S 7°26.6′
GHA	210°34.3′		

This gives you your tabulated declination of S 7°, and declination increment of 26.6′. Establish LHA of the moon:

GHA	210°34.3′
AP Lo	137°25.7′ E
LHA	348°

Your AP latitude will be S 40°, as this is the closest whole degree to the DR latitude. In *Pub. 229* for LHA 12°, 348° and declination 7°, with latitude and declination the same name, you find:

Tab Hc	**d**	**Tab Z**
55°17.5′	+57.5′	158.8°

You also note that Z for declination 8° is 158.2°, so the Z correction factor is −0.6′. Use the front of the book to correct Tab Hc and Tab Z (Table 9-10):

d	22.2′	Z corr.	− 0.3°
	+ 3.3′		
d corr.	+25.5′		
Tab Hc	55°17.5′	Tab Z	158.8°
d corr.	+ 25.5′	Z corr.	− 0.3°
Hc	55°43.0′	Z	158.5°

Since we are in southern latitudes and LHA is greater than 180 degrees, Zn = 180° −Z:

Table 9-9. Daily Page Data and the Increment and Correction Figures for the Moon Sight.

G.M.T.	SUN G.H.A.	SUN Dec.	MOON G.H.A.	v	MOON Dec.	d	H.P.
17 00	183 45.6	S18 58.1	71 58.5	9.6	S 9 13.8	10.9	59.1
01	198 45.4	58.7	86 27.1	9.6	9 02.9	11.0	59.1
02	213 45.3	59.3	100 55.7	9.5	8 51.9	11.1	59.2
03	228 45.2	18 59.9	115 24.2	9.6	8 40.8	11.1	59.2
04	243 45.1	19 00.5	129 52.8	9.6	8 29.7	11.2	59.2
05	258 44.9	01.1	144 21.4	9.6	8 18.5	11.2	59.3
06	273 44.8	S19 01.7	158 50.0	9.5	S 8 07.3	11.3	59.3
07	288 44.7	02.3	173 18.5	9.6	7 56.0	11.4	59.3
08	303 44.6	03.0	187 47.1	9.5	7 44.6	11.4	59.3
M 09	318 44.4 ··	03.6	202 15.6	9.6	7 33.2	11.4	59.4
O 10	333 44.3	04.2	216 44.2	9.5	7 21.8	11.6	59.4
N 11	348 44.2	04.8	231 12.7	9.5	7 10.2	11.5	59.4
D 12	3 44.1	S19 05.4	245 41.2	9.5	S 6 58.7	11.7	59.5
A 13	18 43.9	06.0	260 09.7	9.6	6 47.0	11.7	59.5
Y 14	33 43.8	06.6	274 38.3	9.4	6 35.3	11.7	59.5
15	48 43.7 ··	07.2	289 06.7	9.5	6 23.6	11.8	59.6
16	63 43.6	07.8	303 35.2	9.5	6 11.8	11.8	59.6
17	78 43.4	08.4	318 03.7	9.5	6 00.0	11.9	59.6
18	93 43.3	S19 09.0	332 32.2	9.4	S 5 48.1	12.0	59.7
19	108 43.2	09.6	347 00.6	9.5	5 36.1	11.9	59.7
20	123 43.0	10.2	1 29.1	9.4	5 24.2	12.1	59.7
21	138 42.9 ··	10.8	15 57.5	9.4	5 12.1	12.0	59.7
22	153 42.8	11.4	30 25.9	9.4	5 00.1	12.1	59.8
23	168 42.7	12.0	44 54.3	9.4	4 48.0	12.2	59.8
18 00	183 42.5	S19 12.6	59 22.7	9.3	S 4 35.8	12.2	59.8
01	198 42.4	13.2	73 51.0	9.4	4 23.6	12.2	59.9
02	213 42.3	13.8	88 19.4	9.3	4 11.4	12.3	59.9
03	228 42.1 ··	14.4	102 47.7	9.3	3 59.1	12.3	59.9
04	243 42.0	15.0	117 16.0	9.3	3 46.8	12.3	59.9
05	258 41.9	15.6	131 44.3	9.3	3 34.5	12.4	60.0
06	273 41.7	S19 16.2	146 12.6	9.3	S 3 22.1	12.4	60.0
07	288 41.6	16.8	160 40.9	9.2	3 09.7	12.4	60.0
T 08	303 41.5	17.3	175 09.1	9.2	2 57.3	12.5	60.1
U 09	318 41.3 ··	17.9	189 37.3	9.2	2 44.8	12.5	60.1
E 10	333 41.2	18.5	204 05.5	9.2	2 32.3	12.5	60.1
S 11	348 41.1	19.1	218 33.7	9.2	2 19.8	12.5	60.1
D 12	3 40.9	S19 19.7	233 01.9	9.1	S 2 07.3	12.6	60.2
A 13	18 40.8	20.3	247 30.0	9.1	1 54.7	12.6	60.2
Y 14	33 40.7	20.9	261 58.1	9.1	1 42.1	12.6	60.2
15	48 40.5 ··	21.5	276 26.2	9.0	1 29.5	12.6	60.2
16	63 40.4	22.1	290 54.2	9.1	1 16.9	12.7	60.3
17	78 40.3	22.6	305 22.3	9.0	1 04.2	12.7	60.3
18	93 40.1	S19 23.2	319 50.3	9.0	S 0 51.5	12.7	60.3
19	108 40.0	23.8	334 18.3	8.9	0 38.8	12.7	60.3
20	123 39.9	24.4	348 46.2	8.9	0 26.1	12.7	60.4
21	138 39.7 ··	25.0	3 14.1	8.9	0 13.4	12.7	60.4
22	153 39.6	25.6	17 42.0	8.9	S 0 00.7	12.8	60.4
23	168 39.4	26.2	32 09.9	8.9	N 0 12.1	12.7	60.4
19 00	183 39.3	S19 26.7	46 37.8	8.8	N 0 24.8	12.8	60.5
01	198 39.2	27.3	61 05.6	8.8	0 37.6	12.7	60.5
02	213 39.0	27.9	75 33.4	8.7	0 50.3	12.8	60.5
03	228 38.9 ··	28.5	90 01.1	8.7	1 03.1	12.8	60.5
04	243 38.7	29.1	104 28.8	8.7	1 15.9	12.8	60.5
05	258 38.6	29.6	118 56.5	8.7	1 28.7	12.7	60.6
06	273 38.5	S19 30.2	133 24.2	8.6	N 1 41.4	12.8	60.6
W 07	288 38.3	30.8	147 51.8	8.6	1 54.2	12.8	60.6
E 08	303 38.2	31.4	162 19.4	8.5	2 07.0	12.8	60.6
D 09	318 38.0 ··	32.0	176 46.9	8.5	2 19.8	12.7	60.6
N 10	333 37.9	32.5	191 14.4	8.5	2 32.5	12.8	60.7
E 11	348 37.8	33.1	205 41.9	8.5	2 45.3	12.7	60.7
S 12	3 37.6	S19 33.7	220 09.4	8.4	N 2 58.0	12.8	60.7
D 13	18 37.5	34.3	234 36.8	8.4	3 10.8	12.7	60.7
A 14	33 37.3	34.8	249 04.2	8.3	3 23.5	12.7	60.7
Y 15	48 37.2 ··	35.4	263 31.5	8.3	3 36.2	12.7	60.7
16	63 37.0	36.0	277 58.8	8.3	3 48.9	12.7	60.8
17	78 36.9	36.6	292 26.1	8.2	4 01.6	12.7	60.8
18	93 36.7	S19 37.1	306 53.3	8.2	N 4 14.3	12.6	60.8
19	108 36.6	37.7	321 20.5	8.1	4 26.9	12.7	60.8
20	123 36.5	38.3	335 47.6	8.1	4 39.6	12.6	60.8
21	138 36.3 ··	38.8	350 14.7	8.1	4 52.2	12.6	60.8
22	153 36.2	39.4	4 41.8	8.0	5 04.8	12.5	60.8
23	168 36.0	40.0	19 08.8	8.0	5 17.3	12.6	60.9
	S.D. 16.2	d 0.6	S.D. 16.2		16.4		16.5

34	SUN PLANETS	ARIES	MOON	v or Corrⁿ d		v or Corrⁿ d		v or Corrⁿ d	
00	8 30.0	8 31.4	8 06.8	0.0	0.0	6.0	3.5	12.0	6.9
01	8 30.3	8 31.6	8 07.0	0.1	0.1	6.1	3.5	12.1	7.0
02	8 30.5	8 31.9	8 07.2	0.2	0.1	6.2	3.6	12.2	7.0
03	8 30.8	8 32.1	8 07.5	0.3	0.2	6.3	3.6	12.3	7.1
04	8 31.0	8 32.4	8 07.7	0.4	0.2	6.4	3.7	12.4	7.1
05	8 31.3	8 32.6	8 08.0	0.5	0.3	6.5	3.7	12.5	7.2
06	8 31.5	8 32.9	8 08.2	0.6	0.3	6.6	3.8	12.6	7.2
07	8 31.8	8 33.2	8 08.4	0.7	0.4	6.7	3.9	12.7	7.3
08	8 32.0	8 33.4	8 08.7	0.8	0.5	6.8	3.9	12.8	7.4
09	8 32.3	8 33.7	8 08.9	0.9	0.5	6.9	4.0	12.9	7.4
10	8 32.5	8 33.9	8 09.2	1.0	0.6	7.0	4.0	13.0	7.5
11	8 32.8	8 34.2	8 09.4	1.1	0.6	7.1	4.1	13.1	7.5
12	8 33.0	8 34.4	8 09.6	1.2	0.7	7.2	4.1	13.2	7.6
13	8 33.3	8 34.7	8 09.9	1.3	0.7	7.3	4.2	13.3	7.6
14	8 33.5	8 34.9	8 10.1	1.4	0.8	7.4	4.3	13.4	7.7
15	8 33.8	8 35.2	8 10.3	1.5	0.9	7.5	4.3	13.5	7.8
16	8 34.0	8 35.4	8 10.6	1.6	0.9	7.6	4.4	13.6	7.8
17	8 34.3	8 35.7	8 10.8	1.7	1.0	7.7	4.4	13.7	7.9
18	8 34.5	8 35.9	8 11.1	1.8	1.0	7.8	4.5	13.8	7.9
19	8 34.8	8 36.2	8 11.3	1.9	1.1	7.9	4.5	13.9	8.0
20	8 35.0	8 36.4	8 11.5	2.0	1.2	8.0	4.6	14.0	8.1
21	8 35.3	8 36.7	8 11.8	2.1	1.2	8.1	4.7	14.1	8.1
22	8 35.5	8 36.9	8 12.0	2.2	1.3	8.2	4.7	14.2	8.2
23	8 35.8	8 37.2	8 12.3	2.3	1.3	8.3	4.8	14.3	8.2
24	8 36.0	8 37.4	8 12.5	2.4	1.4	8.4	4.8	14.4	8.3
25	8 36.3	8 37.7	8 12.7	2.5	1.4	8.5	4.9	14.5	8.3
26	8 36.5	8 37.9	8 13.0	2.6	1.5	8.6	4.9	14.6	8.4
27	8 36.8	8 38.2	8 13.2	2.7	1.6	8.7	5.0	14.7	8.5
28	8 37.0	8 38.4	8 13.4	2.8	1.6	8.8	5.1	14.8	8.5
29	8 37.3	8 38.7	8 13.7	2.9	1.7	8.9	5.1	14.9	8.6
30	8 37.5	8 38.9	8 13.9	3.0	1.7	9.0	5.2	15.0	8.6
31	8 37.8	8 39.2	8 14.2	3.1	1.8	9.1	5.2	15.1	8.7
32	8 38.0	8 39.4	8 14.4	3.2	1.8	9.2	5.3	15.2	8.7
33	8 38.3	8 39.7	8 14.6	3.3	1.9	9.3	5.3	15.3	8.8
34	8 38.5	8 39.9	8 14.9	3.4	2.0	9.4	5.4	15.4	8.9
35	8 38.8	8 40.2	8 15.1	3.5	2.0	9.5	5.5	15.5	8.9
36	8 39.0	8 40.4	8 15.4	3.6	2.1	9.6	5.5	15.6	9.0
37	8 39.3	8 40.7	8 15.6	3.7	2.1	9.7	5.6	15.7	9.0
38	8 39.5	8 40.9	8 15.8	3.8	2.2	9.8	5.6	15.8	9.1
39	8 39.8	8 41.2	8 16.1	3.9	2.2	9.9	5.7	15.9	9.1
40	8 40.0	8 41.4	8 16.3	4.0	2.3	10.0	5.8	16.0	9.2
41	8 40.3	8 41.7	8 16.5	4.1	2.4	10.1	5.8	16.1	9.3
42	8 40.5	8 41.9	8 16.8	4.2	2.4	10.2	5.9	16.2	9.3
43	8 40.8	8 42.2	8 17.0	4.3	2.5	10.3	5.9	16.3	9.4
44	8 41.0	8 42.4	8 17.3	4.4	2.5	10.4	6.0	16.4	9.4
45	8 41.3	8 42.7	8 17.5	4.5	2.6	10.5	6.0	16.5	9.5
46	8 41.5	8 42.9	8 17.7	4.6	2.6	10.6	6.1	16.6	9.5
47	8 41.8	8 43.2	8 18.0	4.7	2.7	10.7	6.2	16.7	9.6
48	8 42.0	8 43.4	8 18.2	4.8	2.8	10.8	6.2	16.8	9.7
49	8 42.3	8 43.7	8 18.5	4.9	2.8	10.9	6.3	16.9	9.7
50	8 42.5	8 43.9	8 18.7	5.0	2.9	11.0	6.3	17.0	9.8
51	8 42.8	8 44.2	8 18.9	5.1	2.9	11.1	6.4	17.1	9.8
52	8 43.0	8 44.4	8 19.2	5.2	3.0	11.2	6.4	17.2	9.9
53	8 43.3	8 44.7	8 19.4	5.3	3.0	11.3	6.5	17.3	9.9
54	8 43.5	8 44.9	8 19.7	5.4	3.1	11.4	6.6	17.4	10.0
55	8 43.8	8 45.2	8 19.9	5.5	3.2	11.5	6.6	17.5	10.1
56	8 44.0	8 45.4	8 20.1	5.6	3.2	11.6	6.7	17.6	10.1
57	8 44.3	8 45.7	8 20.4	5.7	3.3	11.7	6.7	17.7	10.2
58	8 44.5	8 45.9	8 20.6	5.8	3.3	11.8	6.8	17.8	10.2
59	8 44.8	8 46.2	8 20.8	5.9	3.4	11.9	6.8	17.9	10.3
60	8 45.0	8 46.4	8 21.1	6.0	3.5	12.0	6.9	18.0	10.4

$$\begin{array}{rr} & 180° \\ Z & -158.5° \\ \hline Zn & 021.5° \end{array}$$

Compare Hc and Ho to get the intercept distance and direction in relation to the moon:

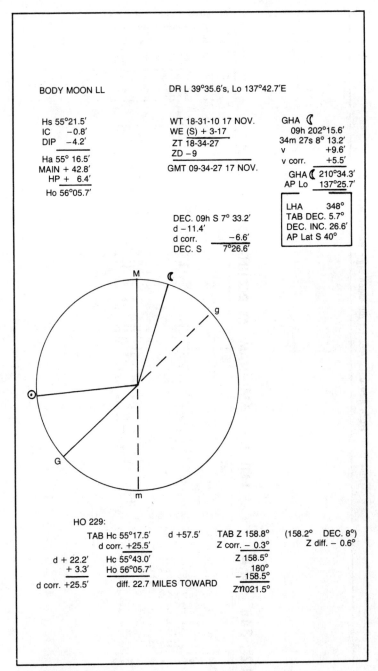

BODY MOON LL   DR L 39°35.6's, Lo 137°42.7'E

Hs 55°21.5'   WT 18-31-10 17 NOV.   GHA ☽
IC   -0.8'   WE (S) + 3-17   09h 202°15.6'
DIP   -4.2'   ZT 18-34-27   34m 27s 8° 13.2'
                     ZD -9   v   +9.6'
Ha 55° 16.5'   ─────────────   v corr.   +5.5'
MAIN + 42.8'   GMT 09-34-27 17 NOV.
HP + 6.4'   GHA ☽ 210°34.3'
                                                 AP Lo 137°25.7'
Ho 56°05.7'
                                              ┌──────────────────────┐
                                              │ LHA   348°   │
                     DEC. 09h S 7° 33.2'   │ TAB DEC. 5.7°   │
                     d -11.4'   │ DEC. INC. 26.6'   │
                     d corr.   -6.6'   │ AP Lat S 40°   │
                     DEC. S   7°26.6'   └──────────────────────┘

HO 229:
                    TAB Hc 55°17.5'   d +57.5'   TAB Z 158.8°   (158.2° DEC. 8°)
                    d corr. +25.5'   Z corr. - 0.3°   Z diff. - 0.6°
        d + 22.2'   Hc 55°43.0'   Z 158.5°
        + 3.3'   Ho 56°05.7'   180°
        ────────   - 158.5°
d corr. +25.5'   diff. 22.7 MILES TOWARD   Zn021.5°

Table 9-10. *Pub.* 229 Data Used for the Moon Sight Reduction.

12°, 348° L.H.A.   LATITUDE SAME NAME AS DECLINATION

N. Lat. { L.H.A. greater than 180°........Zn=Z
N. Lat. { L.H.A. less than 180°........Zn=360°−Z

Dec.	38° Hc	d	Z	39° Hc	d	Z	40° Hc	d	Z	41° Hc	d	Z
0	50 25.5	+57.9	161.0	49 28.7	+58.1	161.3	48 31.8	+58.2	161.7	47 34.8	+58.3	162.0
1	51 23.4	57.9	160.5	50 26.8	58.0	160.9	49 30.0	58.1	161.3	48 33.1	58.1	161.7
2	52 21.3	57.7	160.1	51 24.8	57.9	160.5	50 28.1	58.1	160.9	49 31.4	58.1	161.3
3	53 19.0	57.7	159.7	52 22.7	57.8	160.1	51 26.2	58.0	160.5	50 29.5	58.2	161.0
4	54 16.7	57.5	159.2	53 20.5	57.7	159.7	52 24.2	57.8	160.1	51 27.7	58.0	160.6
5	55 14.2	+57.4	158.7	54 18.2	+57.6	159.2	53 22.0	+57.8	159.7	52 25.7	+57.9	160.1
6	56 11.6	57.3	158.2	55 15.8	57.5	158.7	54 19.8	57.7	159.2	53 23.6	57.8	159.7
7	57 08.9	57.1	157.6	56 13.3	57.3	158.2	55 17.5	57.5	158.8	54 21.4	57.8	159.3
8	58 06.0	57.0	157.1	57 10.6	57.2	157.7	56 15.0	57.4	158.2	55 19.2	57.6	158.8
9	59 03.0	56.8	156.5	58 07.8	57.1	157.1	57 12.4	57.3	157.7	56 16.8	57.5	158.3

42° Hc	d	Z	43° Hc	d	Z	44° Hc	d	Z	45° Hc	d	Z	Dec.
46 37.7	+58.4	162.4	45 40.4	+58.4	162.7	44 43.1	+58.6	163.0	43 45.7	+58.7	163.3	0
47 36.1	58.4	162.0	46 39.0	58.4	162.4	45 41.7	58.6	162.7	44 44.4	58.7	163.0	1
48 34.5	58.3	161.7	47 37.4	58.5	162.0	46 40.3	58.5	162.4	45 43.1	58.6	162.7	2
49 32.8	58.2	161.3	48 35.9	58.3	161.7	47 38.8	58.5	162.0	46 41.7	58.6	162.4	3
50 31.0	58.2	161.0	49 34.2	58.3	161.3	48 37.3	58.4	161.7	47 40.3	58.5	162.1	4
51 29.2	+58.1	160.6	50 32.5	+58.2	161.0	49 35.7	+58.4	161.4	48 38.8	+58.5	161.7	5
52 27.3	57.9	160.2	51 30.7	58.2	160.6	50 34.1	58.2	161.0	49 37.3	58.4	161.4	6
53 25.2	57.9	159.7	52 28.9	58.0	160.2	51 32.3	58.2	160.6	50 35.7	58.3	161.0	7
54 23.1	57.8	159.3	53 26.9	58.0	159.8	52 30.5	58.2	160.2	51 34.0	58.3	160.7	8
55 20.9	57.7	158.8	54 24.9	57.9	159.3	53 28.7	58.0	159.8	52 32.3	58.2	160.3	9

INTERPOLATION TABLE

Altitude Difference (d) — Dec. Inc. 16.0–19.5

Dec. Inc.	10'	20'	Tens 30'	40'	50'	Decimals	0'	1'	2'	3'	Units 4'	5'	6'	7'	8'	9'	Double Second Diff. and Corr.
16.0	2.6	5.3	8.0	10.6	13.3	.0	0.0	0.3	0.5	0.8	1.1	1.4	1.6	1.9	2.2	2.5	
16.1	2.7	5.3	8.0	10.7	13.4	.1	0.0	0.3	0.6	0.9	1.1	1.4	1.7	2.0	2.2	2.5	1.0 0.1
16.2	2.7	5.4	8.1	10.8	13.5	.2	0.1	0.3	0.6	0.9	1.2	1.4	1.7	2.0	2.3	2.6	3.0 0.2
16.3	2.7	5.4	8.1	10.9	13.6	.3	0.1	0.4	0.6	0.9	1.2	1.5	1.7	2.0	2.3	2.6	4.9 0.2
16.4	2.7	5.5	8.2	10.9	13.7	.4	0.1	0.4	0.7	0.9	1.2	1.5	1.8	2.0	2.3	2.6	
18.5	3.1	6.2	9.3	12.3	15.4	.5	0.2	0.5	0.8	1.1	1.4	1.7	2.0	2.3	2.6	2.9	8.3 0.4
18.6	3.1	6.2	9.3	12.4	15.5	.6	0.2	0.5	0.8	1.1	1.4	1.7	2.0	2.3	2.7	3.0	10.2 0.5
18.7	3.1	6.3	9.4	12.5	15.6	.7	0.2	0.5	0.8	1.1	1.5	1.8	2.1	2.4	2.7	3.0	12.0 0.6
18.8	3.2	6.3	9.4	12.6	15.7	.8	0.2	0.6	0.9	1.2	1.5	1.8	2.1	2.4	2.7	3.0	13.9 0.7
18.9	3.2	6.3	9.5	12.6	15.8	.9	0.3	0.6	0.9	1.2	1.5	1.8	2.1	2.4	2.7	3.1	15.7 0.8
19.0	3.1	6.3	9.5	12.6	15.8	.0	0.0	0.3	0.6	1.0	1.3	1.6	1.9	2.3	2.6	2.9	17.6 0.9
19.1	3.2	6.3	9.5	12.7	15.9	.1	0.0	0.4	0.7	1.0	1.3	1.7	2.0	2.3	2.6	3.0	19.4 1.0
19.2	3.2	6.4	9.6	12.8	16.0	.2	0.1	0.4	0.7	1.0	1.4	1.7	2.0	2.3	2.7	3.0	21.3 1.1
19.3	3.2	6.4	9.6	12.9	16.1	.3	0.1	0.4	0.7	1.1	1.4	1.7	2.0	2.4	2.7	3.0	23.1 1.3
19.4	3.2	6.5	9.7	12.9	16.2	.4	0.1	0.5	0.8	1.1	1.4	1.8	2.1	2.4	2.7	3.1	26.8 1.4
19.5	3.3	6.5	9.8	13.0	16.3	.5	0.2	0.5	0.8	1.1	1.5	1.8	2.1	2.4	2.8	3.1	28.7 1.5
																	30.5 1.6

Altitude Difference (d) — Dec. Inc. 24.0–27.5

Dec. Inc.	10'	20'	Tens 30'	40'	50'	Decimals	0'	1'	2'	3'	Units 4'	5'	6'	7'	8'	9'	Double Second Diff. and Corr.
24.0	4.0	8.0	12.0	16.0	20.0	.0	0.0	0.4	0.8	1.2	1.6	2.0	2.4	2.9	3.3	3.7	0.8 0.1
24.1	4.0	8.0	12.0	16.0	20.1	.1	0.0	0.4	0.9	1.3	1.7	2.1	2.5	2.9	3.3	3.7	2.5 0.2
24.2	4.0	8.0	12.1	16.1	20.1	.2	0.1	0.5	0.9	1.3	1.7	2.1	2.5	2.9	3.3	3.8	4.1 0.3
24.3	4.0	8.1	12.1	16.2	20.2	.3	0.1	0.5	0.9	1.3	1.8	2.2	2.6	3.0	3.4	3.8	5.8 0.4
24.4	4.1	8.1	12.2	16.3	20.3	.4	0.2	0.6	1.0	1.4	1.8	2.2	2.6	3.0	3.4	3.8	7.4 0.5
26.5	4.4	8.8	13.3	17.7	22.1	.5	0.2	0.7	1.1	1.5	2.0	2.4	2.9	3.3	3.8	4.2	8.9 0.6
26.6	4.4	8.9	13.3	17.7	22.2	.6	0.3	0.7	1.1	1.6	2.0	2.5	2.9	3.4	3.8	4.2	10.5 0.7
26.7	4.5	8.9	13.4	17.8	22.3	.7	0.3	0.8	1.2	1.6	2.1	2.5	3.0	3.4	3.9	4.3	12.1 0.8
26.8	4.5	9.0	13.4	17.9	22.4	.8	0.4	0.8	1.2	1.7	2.1	2.6	3.0	3.5	3.9	4.3	13.7 0.9
26.9	4.5	9.0	13.5	18.0	22.5	.9	0.4	0.8	1.3	1.7	2.2	2.6	3.0	3.5	3.9		15.4 1.0
27.0	4.5	9.0	13.5	18.0	22.5	.0	0.0	0.5	0.9	1.4	1.8	2.3	2.7	3.2	3.7	4.1	17.0 1.1
27.1	4.5	9.0	13.5	18.0	22.6	.1	0.0	0.5	1.0	1.4	1.9	2.3	2.8	3.3	3.7	4.2	18.6 1.2
27.2	4.5	9.0	13.6	18.1	22.6	.2	0.1	0.6	1.0	1.5	1.9	2.4	2.8	3.3	3.8	4.2	20.2 1.3
27.3	4.5	9.1	13.6	18.2	22.7	.3	0.1	0.6	1.1	1.5	2.0	2.4	2.9	3.3	3.8	4.3	21.8 1.4
27.4	4.6	9.1	13.7	18.3	22.8	.4	0.2	0.6	1.1	1.6	2.0	2.5	2.9	3.4	3.8	4.3	23.4 1.5
27.5	4.6	9.2	13.8	18.3	22.9	.5	0.2	0.7	1.1	1.6	2.1	2.5	3.0	3.4	3.9	4.4	25.1 1.6
																	26.7 1.6
																	28.3 1.7

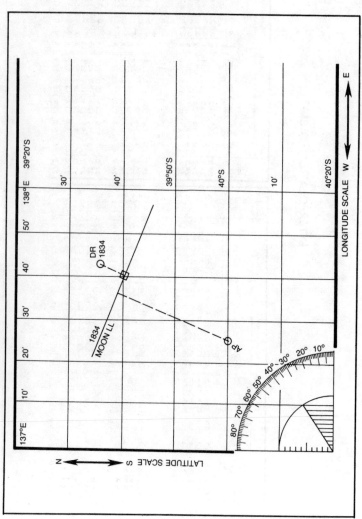

Fig. 9-7. Plot of the moon sight.

Hc 55°43.0′
Ho 56°05.7′
Intercept   22.7 miles toward

Figure 9-6 shows the complete worksheet. Figure 9-7 shows the plot of the sight. Note that latitude numbers increase from top to bottom of the sheet and that longitude numbers increase from left to right.

Chapter 10

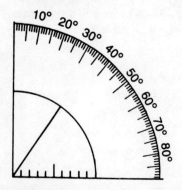

Multiple Body Fixes

A single sight results in a line of position and shows where your boat is on this line. Presumably you are close to your DR position at the time of the sight. Advancing an LOP and combining it with a new sight provides a running fix, and while this is usually reasonably accurate, it is not as positive as a fix based on two or more bodies, with sights taken at the same time.

For practical purposes, sights taken within five minutes of each other can be considered "simultaneous," although each sight would be reduced to an LOP based on actual time of the sight. In five minutes the average cruising yacht is not going to travel very far, and you can use the DR position at the time of the second sight for plotting purposes.

However, if time between sights is more than five minutes, you can ensure a higher degree of accuracy by plotting the AP for each sight at the time of the sight. Then advance each AP in the direction and distance of the DR course until the time of the last sight. Also, the sights should be within a span of about 15 minutes or so; if more time elapses, you're really dealing with a running fix.

As an example of advancing an AP for a two-body fix, you are working with a 1754 sight of Venus and an 1800 sight of Canopus. Your course is 248° True, at a speed of 12 knots, and your DR position at 1800 is L 30°49.2′S, Lo 178°31.6′W. Your AP at 1754 is L 31°S, Lo 178°24.9′W; and at 1800 your AP is L 31° S, Lo 178°55.8′W. Remember, the assumed positions are selected to provide whole degrees of LHA and they have no relation to boat speed and course. Both APs are drawn on your plotting sheet in the correct positions and labeled AP_1 and AP_2 respectively, in time of sight sequence.

Before plotting the intercept for the Venus sight, determine the DR distance and direction your boat has traveled in the six minutes between sights. At a speed through the water of 12 knots, distance is: speed × time, so:

$$\frac{\text{speed} \times \text{time}}{60}$$

$$12 \times 6 = \frac{72}{60} = 1.2 \text{ mile}$$

From AP_1 mark this distance off in the same direction (248°) as your DR course. Mark this point, draw a small circle around it to identify it as the advanced AP, and draw a line to it from AP_1. Now draw the intercept from the advanced AP, and proceed as with a standard sight. Plot the Canopus intercept directly from AP_2. Your fix is the point where the two LOPs intersect (see Fig. 10-1).

Note that the DR line is shown here. On a plotting sheet of this type it is not really necessary, but it is included here so you can see the short line connecting AP_1 to its advanced position is parallel to the DR course.

THREE-BODY FIXES

When sights are taken on three bodies, the first and second APs are advanced distances corresponding to distanced traveled along the DR track in the interval before the final sight. Here's an example, illustrated in Fig. 10-2:

Your DR course is 326° True, at a speed of 18 knots. You get a sight of Venus at 0533, a sight of Sirius at 0539, and a sight of Mars at 0546. For Venus, AP_1 Lo at 0533 is 63°01.3′W; Sirius AP_2 Lo is 63°06.8′W at 0539; and at 0546, AP_3 Lo is 63°08.9′ for Mars. Latitude for all three N 41°. Advance AP_1 3.9 miles (13 minutes × 18 knots + 60) and AP_2 2.1 miles (7 minutes × 18 knots ÷ 60), both in the direction 326° True. Plot the intercept for each sight from the advanced AP, but use intercept distance and azimuth based on the actual time of each sight.

Note that the three LOPs form a small triangle where they cross. Usually it is somewhat larger than the one shown here if the sights are taken while you're bouncing around on a small boat in mid-ocean. Your fix, however, is considered to be the center of this triangle. Note also that you don't need the word "fix" as the complete circle at this point denotes a fix; the half circle on the DR course line denotes your DR position at the time of the third sight.

AZIMUTH SEPARATION

When taking sights of two bodies for a fix, try to pick ones that are about 90° in azimuth; one in the southeast and the other in the southwest, for example. That way your LOPs will cross at close to a 90° angle to give you an accurate fix point. If the bodies are too close together or too far apart, the intersecting LOPs will have an indeterminate area in which the point of the fix could lie (see Figs. 10-3, 10-4).

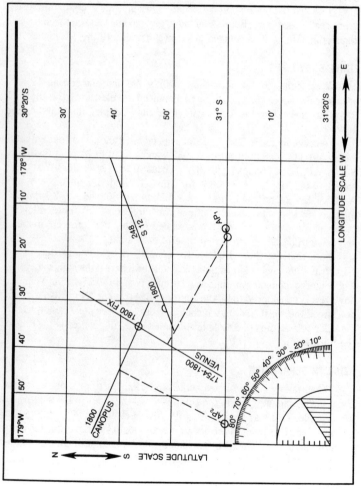

Fig. 10-1. A two-body fix. Here the course line is shown, and the DR position at the time of the fix. Note that AP₁ is advanced in the direction of the course line, and for the distance the boat traveled in the interval between the first and second sight.

138

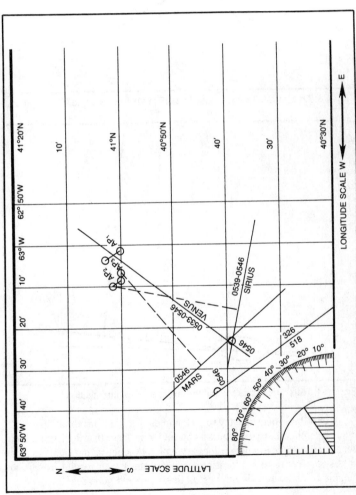

Fig. 10-2. A three-body fix. Both AP₁ and AP₂ are advanced before their intercepts are drawn. The 0546 fix is in the center of the triangle formed where the three LOPs cross.

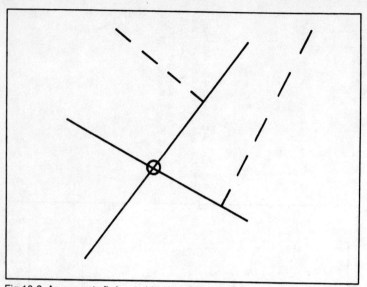

Fig. 10-3. An accurate fix is provided by LOPs that cross at or near 90 degrees.

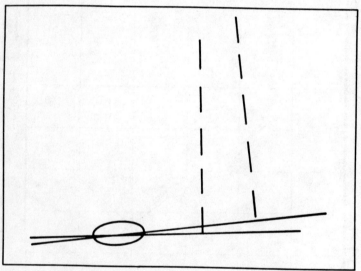

Fig. 10-4. Here it is difficult to tell the exact point of intersection.

For a three-body fix, try to pick bodies that are from about 45° to 60° from each other; certainly you won't want them closer than 30°. You want to get a good triangle in which to locate your fix. Practice taking sights as much as possible from a known location so you can check your results. You'll soon be able to estimate azimuth angles that will yield good results.

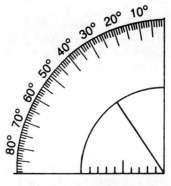

Chapter 11

Polaris

The Pole Star, Polaris, has been the navigators' friend for thousands of years, and it is still a very helpful body on the celestial sphere. While it is not directly above our terrestial North Pole, it is never more than about one degree away. If you can spot Polaris, you know which way is north, at least. More important, however, it can provide you with your exact latitude, without recourse to sight reduction tables such as *Pub. 229*. All the information you need is in the *Nautical Almanac*.

Take your sextant sight in the normal manner. It's a good idea to take several, average the results, and pick the sight closest to the average. Since there's no significant shift in the GHA, you can't do a run of sights and find the slopes as you can with other bodies. Establish the GMT of the chosen sight, and then determine the GHA of Aries based on this time. For example, on 28 September 1980 at watch time 19-40-06 (daylight saving time), you record a sight at Hs 40°52.6'. You're at DR latitude 40°57.9' N, Lo 73°01.7' W; watch error is 0-00; and IC is −1.0'. Your eye height is 12 feet above sea level. For GMT you have:

WT	19-40-06	28 Sept
WE	00-00	
ZT	19-40-06	
ZD	+ 4	
GMT	23-40-06	28 Sept

From the *Almanac* daily page for 28 September you get:

GHA Aries	
23h	352°51.9'
40m 06s	10°03.1'
	362°55.0'

Subtract 360 degrees to get actual GHA:

$$362°55.0'$$
$$- \quad 360°$$
$$\text{GHA Aries} \quad \overline{2°55.0'}$$

Now determine the LHA of Aries from your DR longitude. You don't use an AP longitude in this calculation, and LHA will not be a whole degree. Divide the minutes and tenths of minutes in the answer by 60 to convert these to tenths of degrees:

GHA Aries	2°55.0'	
	(362°55.0')	
DR Lo	73°01.7'	
LHA	289°53.3'	= 289.9°

Note that the answer has been rounded to the nearest tenth.

The last three white pages in the *Almanac* contain the Polaris information, with tables based on LHA Aries at the top of each page. These tables provide an "a_o" correction, and you often need to interpolate to get the a_o figure you need. In this example, the table at the top of the last white page is used. In the LHA 280° to 289° column, an a_o correction of 1°11.3' is shown for 289°, and a correction of 1°10.5' for the next whole degree (290°). You need the correction of 289.9°:

$$289° - 1°11.3'$$
$$290° - \underline{1°10.5'}$$
$$0.8'$$
$$0.8' \times 0.9 = 0.72' \text{ (round to 0.7')}$$
$$289.9° - 1°10.5'$$

Note that the correction is subtracted from the a_o figure for 289°, as a_o is decreasing.

Follow the 280° to 289° column (Table 11-1) down into the tables below for the latitude (a_1) and month (a_2) corrections, as well as the azimuth. No interpolation is needed. For 41° latitude, a_1 is 0.5'; and for September, a_2 is 0.9'. Azimuth for 41° latitude is 1.0°.

Determine the observed altitude of Polaris in the normal manner, subtract one degree from this, then add the a_0, a_1, and a_2 corrections. The result is your actual latitude:

a_0	1°10.6'	Hs	40°52.6'
a_1	0.5'	IC	− 1.0'
a_2	0.9'	Dip	− 3.4'
	1°12.0'	Ha	40°48.2'
		Main	− 1.1'

Table 11-1. Page in the Back of the *Nautical Almanac* Is Used to Work a Polaris Sight.

POLARIS (POLE STAR) TABLES, 1980
FOR DETERMINING LATITUDE FROM SEXTANT ALTITUDE AND FOR AZIMUTH

L.H.A. ARIES	240°–249°	250°–259°	260°–269°	270°–279°	280°–289°	290°–299°	300°–309°	310°–319°	320°–329°	330°–339°	340°–349°	350°–359°
°	a_0	a_0	a_0	a_0	a_0	a_0	a_0	a_0	a_0	a_0	a_0	a_0
0	1 43·0	1 38·6	1 32·9	1 26·2	1 18·6	1 10·5	1 01·9	0 53·3	0 44·8	0 36·8	0 29·4	0 22·9
1	42·7	38·0	32·2	25·4	17·8	09·6	01·1	52·4	44·0	36·0	28·7	22·3
2	42·2	37·5	31·6	24·7	17·0	08·8	1 00·2	51·6	43·2	35·2	28·0	21·7
3	41·8	37·0	31·0	24·0	16·2	07·9	0 59·3	50·7	42·4	34·5	27·3	21·1
4	41·4	36·4	30·3	23·2	15·4	07·1	58·5	49·9	41·5	33·7	26·7	20·6
5	1 41·0	1 35·9	1 29·6	1 22·5	1 14·6	1 06·2	0 57·6	0 49·0	0 40·7	0 33·0	0 26·0	0 20·0
6	40·5	35·3	29·0	21·7	13·8	05·4	56·7	48·2	39·9	32·2	25·4	19·5
7	40·0	34·7	28·3	20·9	12·9	04·5	55·9	47·3	39·1	31·5	24·7	19·0
8	39·6	34·1	27·6	20·2	12·1	03·7	55·0	46·5	38·3	30·8	24·1	18·5
9	39·1	33·5	26·9	19·4	11·3	02·8	54·2	45·7	37·5	30·1	23·5	18·0
10	1 38·6	1 32·9	1 26·2	1 18·6	1 10·5	1 01·9	0 53·3	0 44·8	0 36·8	0 29·4	0 22·9	0 17·5

Lat.	a_1	a_1	a_1	a_1	a_1	a_1	a_1	a_1	a_1	a_1	a_1	a_1
°	′	′	′	′	′	′	′	′	′	′	′	′
0	0·5	0·4	0·3	0·3	0·2	0·2	0·2	0·2	0·2	0·3	0·4	0·4
10	·5	·4	·4	·3	·3	·2	·2	·3	·3	·3	·4	·5
20	·5	·5	·4	·4	·3	·3	·3	·3	·3	·4	·4	·5
30	·5	·5	·5	·4	·4	·4	·4	·4	·4	·4	·5	·5
40	0·6	0·5	0·5	0·5	0·5	0·5	0·5	0·5	0·5	0·5	0·5	0·6
45	·6	·6	·6	·5	·5	·5	·5	·5	·5	·6	·6	·6
50	·6	·6	·6	·6	·6	·6	·6	·6	·6	·6	·6	·6
55	·6	·6	·7	·7	·7	·7	·7	·7	·7	·7	·6	·6
60	·7	·7	·7	·7	·8	·8	·8	·8	·8	·7	·7	·7
62	0·7	0·7	0·8	0·8	0·8	0·8	0·8	0·8	0·8	0·8	0·7	0·7
64	·7	·7	·8	·8	·9	0·9	0·9	0·9	·9	·8	·8	·7
66	·7	·8	·8	0·9	0·9	1·0	1·0	1·0	0·9	·9	·8	·7
68	0·7	0·8	0·9	1·0	1·0	1·0	1·1	1·0	1·0	0·9	0·9	0·8

Month	a_2	a_2	a_2	a_2	a_2	a_2	a_2	a_2	a_2	a_2	a_2	a_2
	′	′	′	′	′	′	′	′	′	′	′	′
Jan.	0·5	0·5	0·5	0·5	0·6	0·6	0·6	0·6	0·6	0·7	0·7	0·7
Feb.	·4	·4	·4	·4	·4	·4	·4	·5	·5	·5	·6	·6
Mar.	·4	·4	·4	·3	·3	·3	·3	·3	·3	·4	·4	·4
Apr.	0·5	0·5	0·4	0·3	0·3	0·3	0·2	0·2	0·2	0·2	0·3	0·3
May	·7	·6	·5	·4	·4	·3	·3	·2	·2	·2	·2	·2
June	·8	·7	·7	·6	·5	·5	·4	·3	·3	·2	·2	·2
July	0·9	0·9	0·8	0·7	0·7	0·6	0·5	0·5	0·4	0·4	0·3	0·3
Aug.	·9	·9	·9	·9	·8	·8	·7	·6	·6	·5	·5	·4
Sept.	·9	·9	·9	·9	·9	·9	·8	·8	·8	·7	·6	·6
Oct.	0·8	0·9	0·9	0·9	0·9	0·9	0·9	0·9	0·9	0·9	0·8	0·8
Nov.	·7	7	·8	·8	·9	·9	1·0	1·0	1·0	1·0	1·0	0·9
Dec.	0·5	0·6	0·6	0·7	0·8	0·8	0·9	0·9	1·0	1·0	1·0	1·0

Lat.	AZIMUTH											
°	°	°	°	°	°	°	°	°	°	°	°	°
0	0·4	0·6	0·6	0·7	0·8	0·8	0·8	0·8	0·8	0·7	0·6	0·5
20	0·5	0·6	0·7	0·8	0·8	0·9	0·9	0·9	0·8	0·7	0·7	0·5
40	0·6	0·7	0·8	0·9	1·0	1·1	1·1	1·1	1·0	0·9	0·8	0·7
50	0·7	0·8	1·0	1·1	1·2	1·3	1·3	1·3	1·2	1·1	1·0	0·8
55	0·7	0·9	1·1	1·3	1·4	1·4	1·4	1·4	1·3	1·2	1·1	0·9
60	0·9	1·1	1·3	1·4	1·6	1·6	1·7	1·6	1·5	1·4	1·2	1·0
65	1·0	1·3	1·5	1·7	1·8	1·9	2·0	1·9	1·8	1·7	1·5	1·2

Latitude = Apparent altitude (corrected for refraction) $- 1° + a_0 + a_1 + a_2$

The table is entered with L.H.A. Aries to determine the column to be used; each column refers to a range of 10°. a_0 is taken, with mental interpolation, from the upper table with the units of L.H.A. Aries in degrees as argument; a_1, a_2 are taken, without interpolation, from the second and third tables with arguments latitude and month respectively. a_0, a_1, a_2 are always positive. The final table gives the azimuth of *Polaris*.

BODY—POLARIS   DR L 40°57.9'N, Lo 73°01.7'W

WT   19-40-06   28 SEPT.   Hs   40°52.6'
WE   0-00   IC   − 1.0'
                                                 DIP   − 3.4'
ZT   19-40-06   Ha   40°48.2'
ZD   +4   Main   − 1.1'
                                                 Ho   40°47.1'
GMT 23-40-06

GHA ϒ   289° — 1°11.3'
23h   352°51.9'   289.9° — 1°10.6'
40m06s 10°03.1'   290° — 1°10.5'

        362°55.0'

GHA ϒ   2°55.0'   Ho 40°47.1'
DR Lo   73°01.7'   − 1°

LHA ϒ 289°53.3' = 289.9°   39°47.1'
                                               + 1°12.0'

a_0 1°10.6'   L 40°59.1'N
a_1 0.5'   DR L 40°57.9'N
a_2 0.9'   l diff.   1.2' = 1.2 mile

       1°12.0'

Zm 1.0°

Table 11-2. The Polaris Sight Worksheet.

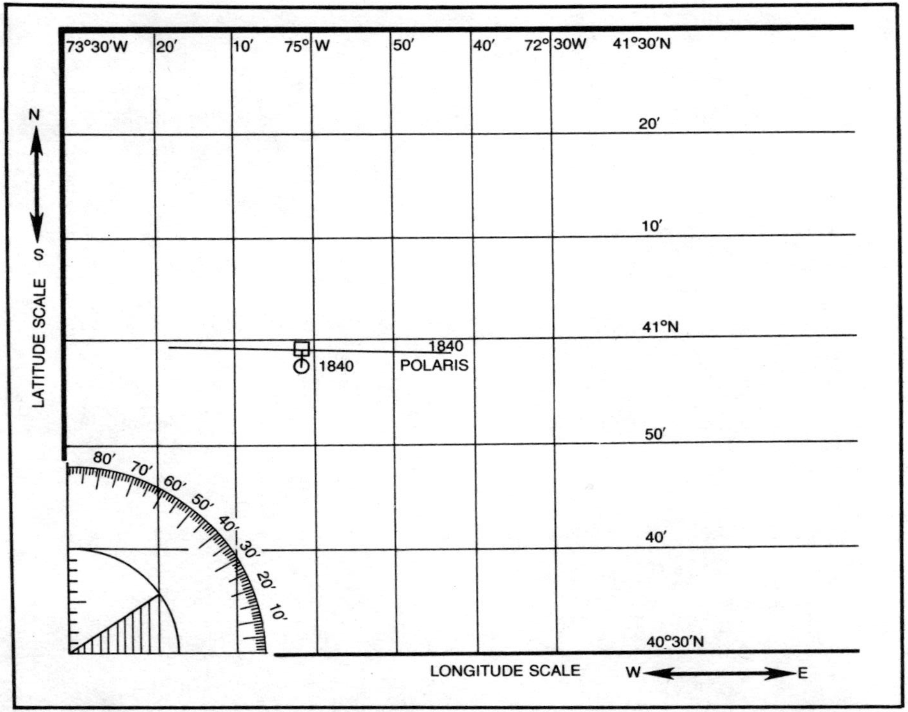

Fig. 11-1. Plot of the Polaris sight. The LOP is at a slight angle because the star is not exactly above the north pole.

$$\begin{array}{ll} \text{Ho} & 40°47.1' \\ & \underline{-\ 1°} \\ & 39°47.1' \\ & \underline{+\ 1°12.0'} \\ \text{L} & \text{N } 40°59.1'\text{N} \end{array}$$

Compare actual latitude to DR latitude:

$$\begin{array}{lll} & \text{L} \quad \text{N} & 40°59.1'\ \text{N} \\ \text{DR} & \text{L} \quad \text{N} & \underline{40°57.9'\ \text{N}} \\ 1 & \text{diff.} & 2.2' = 2.2\ \text{miles} \end{array}$$

Table 11-2 shows the worksheet for this sight and Fig. 11-1 shows the plot. Note that the intercept is drawn from the DR position at a Zn of 001.0° and the LOP is at right angles to the intercept. Your EP, shown by the square, is at the point where the intercept touches the LOP.

Chapter 12

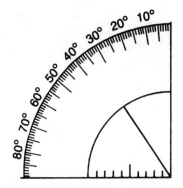

Star Identification

On a clear night, with no moon, you can see thousands of stars. How do you know which are the navigation stars? How can you identify them?

Fortunately, the navigation stars are the brightest ones in the heavens. They are visible during morning and evening twilight. If you can see a star and the horizon, it's a navigation star. There are only 57 of them and about half are below your horizon in the Southern Hemisphere where you'll never see them unless you voyage south of the equator. There are only a couple of dozen to really learn. The only way to do it is to study the night sky at every opportunity.

Star charts are printed in the *Nautical Almanac* and the *Air Almanac*. These are excellent guides. The *Nautical Almanac* has four charts on two pages: Northern Stars, Southern Stars, Equatorial Stars SHA 0° to 180°, and Equatorial Stars SHA 180° to 360° (see Figs. 12-1, 12-2). The *Air Almanac* has a single, fold-out sheet (Fig. 12-3) with the stars in white against a black background more like the actual night sky. Non-navigational stars are also shown on the charts, in small dots, as parts of the constellations in which the brighter stars appear.

Usually, the brightest "stars" that can be seen aren't stars at all; they are planets and are not shown on the star charts. The Almanac given in some daily newspapers or as part of weather broadcasts is a help in identifying these. Mars may be distinguished by its reddish color in comparison to the blue-white of other planets and stars. Also, you can get from the Almanac the approximate GP of each planet at the time you plan to take sights and establish the LHA. If this is between 90° and 180°, the planet will be below your horizon at that time.

There is a Starfinder, once produced by the U.S. Naval Oceanographic Office and now offered by suppliers such as Simex (Coast Navigation, 1934

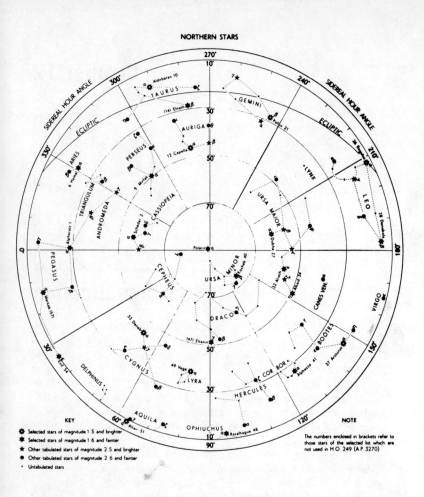

NORTHERN STARS

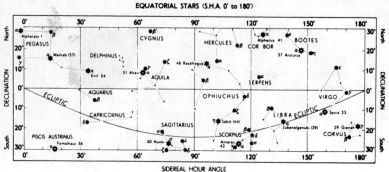

EQUATORIAL STARS (S.H.A. 0° to 180°)

Fig. 12-1. Northern stars as shown in the *Nautical Almanac*. The lower section shows the stars within 30° north and south of the celestial equator.

148

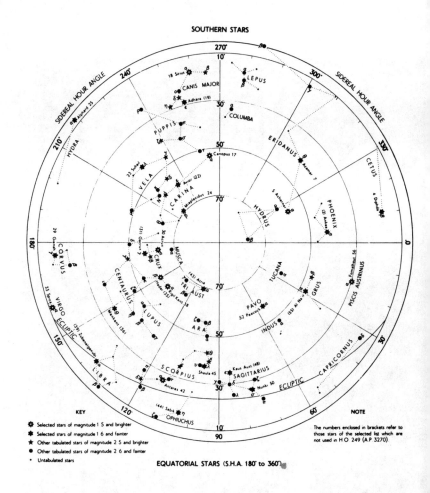

SOUTHERN STARS

EQUATORIAL STARS (S.H.A. 180° to 360°)

KEY

✪ Selected stars of magnitude 1 5 and brighter
✸ Selected stars of magnitude 1 6 and fainter
★ Other tabulated stars of magnitude 2 5 and brighter
● Other tabulated stars of magnitude 2 6 and fainter
· Untabulated stars

NOTE

The numbers enclosed in brackets refer to those stars of the selected list which are not used in H O 249 (A.P. 3270)

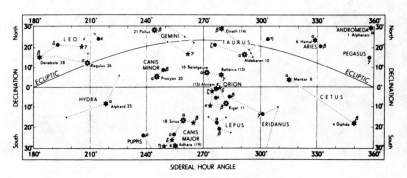

Fig. 12-2. Southern stars shown in the *Nautical Almanac*.

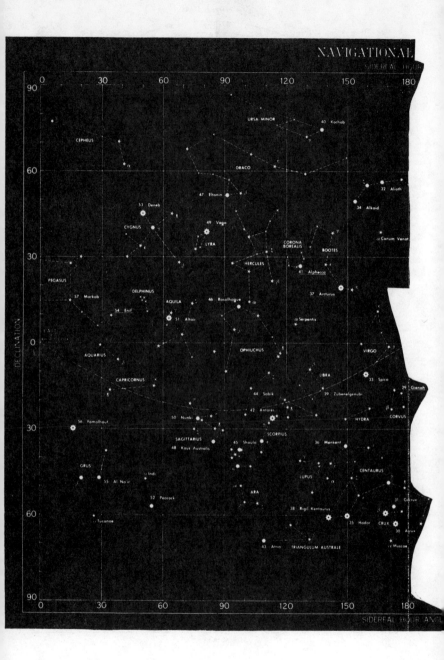

Fig. 12-3. Star chart in the *Air Almanac*.

150

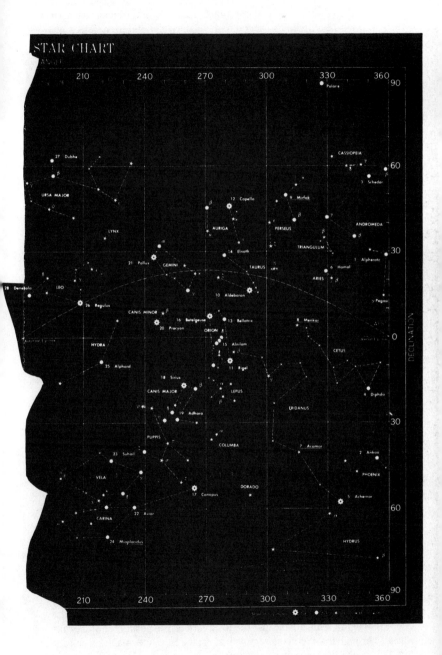

STAR CHART

151

Lincoln Drive, Annapolis, MD 21401) that helps you identify stars on which you have taken sights. Alternatively, you can use it to determine in advance the approximate altitude and azimuth of the stars that will be visible when you plan to take sights.

Suppose you're at a latitude of about N 35°, with scattered clouds in the evening sky. You manage to spot a very bright star about 20° to the west of your meridian, and you get a sextant altitude of 40°32.8′ on it. Your handy Starfinder quickly shows you that you have taken a sight on Spica.

The main component is an opaque, white circular disc, about 8½″ in diameter. On one side are the navigation stars in the Northern Hemisphere, and the southern stars are on the other side. A small pin extends out of the center on each side. A series of nine templates, each representing a latitude increment of 10 degrees, is supplied. By placing the template for the latitude closest to your DR latitude on the appropriate side of the disc, and turning it to line up with the appropriate GHA Aries for the time of the sights, you can quickly note the altitude and azimuth of all the stars that will be visible—or were visible—at that time.

Because the template intervals are 10 degrees, the chances are that you are not at the exact latitude of the template in use. Simply add, or subtract, the difference between DR latitude and template latitude to the altitude of each selected star. Work in whole degrees; don't worry about the minutes and seconds or tenths of minutes and seconds or tenths of minutes because you can't get that sort of precision with this device. The results are so accurate, however, that you'll have no difficulty in finding and identifying the selected stars if sky conditions are favorable (no clouds).

A separate template is supplied that allows you to actually plot the locations of the moon and planets right on the opaque disc (using pencil with a soft lead so that the marks can be erased). The first step is to find the *right ascension* (RA) of the body. The RA is simply the GHA of Aries less the GHA of the body. For example, you want to plot Venus on the Starfinder for GMT 22-30-00 on 4 October 1980:

GHA Aries		**GHA Venus**	
22h	343°44.3′	22h	191°37.0′
30m 00s	7°31.2′	30m 00 s	7°30.0′
GHA Aries	351°15.5′	GHA Venus	199°07.0′
GHA Venus	−199°07.0′		
RA	152°08.5′	Dec.	N 12°02.7′

In practice you can use just the figure for the whole hour, the results will be close enough for practical purposes. Note that declination of the body is recorded.

Now set the special template with the magenta lines on the star base, and make sure both conform to your hemisphere (north or south). Set the zero line of the template against the RA figure on the perimeter of the star base. Through the slot in the template along the zero line, make a small dot on the base at the appropriate declination; in this case at N 12°. The slot is scaled to declinations 30 degrees north and south of the celestial equator, at 2-degree intervals. Identify your dot with a small letter or the symbol for the planet.

Once you have the planets plotted, you can leave them on the card for a couple of weeks or so. Their change in position is gradual enough so the Starfinder will continue to indicate their approximate locations.

The moon can be plotted in the same manner to see if it will be visible at the time of your intended sights. It may be in a position where its light will overpower that of nearby navigational stars. The sun also can be plotted. That might not seem necessary, but it will give you a good indication of its LHA at the time of the sights.

Chapter 13

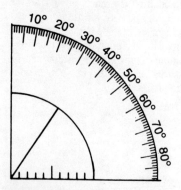

Special Problems

The sights I have discussed thus far are normal ones where the bodies are not low near the horizon, nor nearly overhead, nor on the other side of the celestial pole of your hemisphere. For sights involving these conditions, special reduction techniques are required.

STARS NEAR THE POLES

Earlier I noted that an LHA between 90 and 180 degrees indicates that the body is below your horizon, and the chances are that a mistake was made in recording the sight or in your calculations. However, this is not the case with stars near the North Pole or South Pole, in some circumstances.

The farther north or south of the equator that you are, the more stars you can observe that never set as they revolve around the celestial pole. A sight on such a star, when it is on the far side of its respective pole, will result in an LHA between 90 and 180 degrees.

This brings us to the bottom of the right-hand pages in *Pub. 229* in the section below the short lines in each column. These are C-S (contrary-same) lines. If declination and latitude are contrary names, you won't have a situation where you'll work below the C-S line because this would involve bodies in the hemisphere below your horizon. However, if declination and latitude are the same name, you can end up with an LHA in the 90° to 180° range and use the figures below the C-S lines.

For example, you're at DR L 44°22.2'N, Lo 37°19.1'W, and you get a sight of a star to the northwest of your position. Its declination is 64°05.2'N, and GHA works out to 151°28.2', so LHA is 114°. You calculate an Ho of 30°07.4'.

Using LHA 114°, tabulated declination 64° and AP Lat. 44°, *Pub. 229* Table 13-1 shows:

Table 13-1. *Pub. 229* Data Used for a Star with an LHA Between 90° and 270°.

	38°	39°	40°	41°	42°	43°	44°	45°	
60	21 53.8 + 37.7	22 45.9 + 38.2	23 38.0 + 38.6	24 29.9 + 39.1	25 21.8 + 39.4	26 13.5 + 39.9	27 05.1 + 40.3	27 56.5 + 40.8	60
61	22 31.5 37.4	23 24.1 37.9	24 16.6 38.3	25 09.0 38.7	26 01.2 39.2	26 53.4 39.6	27 45.4 40.0	28 37.3 40.4	61
62	23 08.9 37.2	24 02.0 37.6	24 54.9 38.0	25 47.7 38.4	26 40.4 38.9	27 33.0 39.3	28 25.4 39.7	29 17.7 40.2	62
63	23 46.1 36.9	24 39.6 37.3	25 32.9 37.7	26 26.1 38.2	27 19.3 38.5	28 12.3 38.9	29 05.1 39.4	29 57.9 39.8	63
64	24 23.0 36.6	25 16.9 37.0	26 10.6 37.4	27 04.3 37.8	27 57.8 38.2	28 51.2 38.6	29 44.5 39.1	30 37.7 39.5	64
65	24 59.6 + 36.3	25 53.9 + 36.7	26 48.0 + 37.1	27 42.1 + 37.4	28 36.0 + 37.9	29 29.8 + 38.3	30 23.6 + 38.6	31 17.2 + 39.1	65
66	25 35.9 36.0	26 30.6 36.3	27 25.1 36.7	28 19.5 37.2	29 13.9 37.5	30 08.1 37.9	31 02.2 38.4	31 56.3 38.7	66
67	26 11.9 35.7	27 06.9 36.0	28 01.8 36.4	28 56.7 36.7	29 51.4 37.2	30 46.0 37.6	31 40.6 37.9	32 35.0 38.3	67
68	26 47.6 35.3	27 42.9 35.7	28 38.2 36.1	29 33.4 36.4	30 28.6 36.7	31 23.6 37.1	32 18.5 37.5	33 13.3 37.9	68
69	27 22.9 34.9	28 18.6 35.3	29 14.3 35.6	30 09.8 36.0	31 05.3 36.4	32 00.7 36.8	32 56.0 37.1	33 51.2 37.5	69
70	27 57.8 + 34.6	28 53.9 + 34.9	29 49.9 + 35.3	30 45.8 + 35.7	31 41.7 + 36.0	32 37.5 + 36.3	33 33.1 + 36.7	34 28.7 + 37.1	70
71	28 32.4 34.2	29 28.8 34.5	30 25.2 34.8	31 21.5 35.1	32 17.7 35.5	33 13.8 35.9	34 09.8 36.2	35 05.8 36.5	71
72	29 06.6 33.8	30 03.3 34.2	31 00.0 34.5	31 56.6 34.8	32 53.2 35.1	33 49.7 35.4	34 46.0 35.8	35 42.3 36.2	72
73	29 40.4 33.4	30 37.5 33.6	31 34.5 34.0	32 31.4 34.3	33 28.3 34.6	34 25.1 34.9	35 21.8 35.3	36 18.5 35.6	73
74	30 13.8 33.0	31 11.1 33.3	32 08.5 33.9	33 05.7 33.9	34 02.9 34.2	35 00.0 34.5	35 57.1 34.8	36 54.1 35.1	74
75	30 46.7 + 32.6	31 44.4 + 32.8	32 42.0 + 33.1	33 39.6 + 33.3	34 37.1 + 33.6	35 34.5 + 34.0	36 31.9 + 34.2	37 29.2 + 34.6	75
76	31 19.3 32.0	32 17.2 32.3	33 15.1 32.6	34 12.9 32.9	35 10.7 33.2	36 08.5 33.4	37 06.1 33.8	38 03.8 34.0	76
77	31 51.3 31.7	32 49.5 31.9	33 47.7 32.1	34 45.8 32.4	35 43.9 32.6	36 41.9 32.9	37 39.9 33.1	38 37.8 33.4	77
78	32 23.0 31.1	33 21.4 31.4	34 19.8 31.6	35 18.2 31.8	36 16.5 32.1	37 14.8 32.3	38 13.0 32.6	39 11.2 32.9	78
79	32 54.1 30.6	33 52.8 30.8	34 51.4 31.1	35 50.0 31.3	36 48.6 31.5	37 47.1 31.8	38 45.6 32.0	39 44.1 32.2	79
80	33 24.7 + 30.1	34 23.6 + 30.3	35 22.5 + 30.5	36 21.3 + 30.7	37 20.1 + 31.0	38 18.9 + 31.1	39 17.6 + 31.4	40 16.3 + 31.6	80
81	33 54.8 29.6	34 53.9 29.8	35 53.0 29.9	36 52.0 30.1	37 51.1 30.3	38 50.0 30.6	40 49.0 30.8	40 47.9 31.0	81
82	34 24.4 29.0	35 23.7 29.2	36 22.9 29.4	37 22.2 29.5	38 21.4 29.7	39 20.6 29.9	40 19.8 30.0	41 18.9 30.3	82
83	34 53.4 28.5	35 52.9 28.6	36 52.3 28.8	37 51.7 28.9	39 11.1 29.1	39 50.5 29.4	40 49.8 29.4	41 49.2 29.5	83
84	35 21.9 27.9	36 21.5 28.0	37 21.1 28.1	38 20.6 28.3	39 20.2 28.4	40 19.7 28.6	41 19.2 28.7	42 18.7 28.9	84
85	35 49.8 + 27.3	36 49.5 + 27.4	37 49.2 + 27.5	38 48.9 + 27.6	39 48.6 + 27.7	40 48.3 + 27.8	41 47.9 + 28.0	42 47.6 + 28.1	85
86	36 17.1 26.7	37 16.9 26.8	38 16.7 26.9	39 16.5 27.0	40 16.3 27.0	41 16.1 27.1	42 15.9 27.1	43 15.7 27.3	86
87	36 43.8 26.0	37 43.7 26.1	38 43.6 26.2	39 43.5 26.2	41 43.3 26.3	41 43.2 26.4	42 43.1 26.4	43 43.0 26.5	87
88	37 09.8 25.5	38 09.8 25.4	39 09.8 25.4	40 09.7 25.5	41 09.6 25.6	42 09.6 25.6	43 09.5 25.7	44 09.5 25.7	88
89	37 35.3 24.7	38 35.2 24.8	39 35.2 24.8	40 35.2 24.8	41 35.2 24.8	42 35.2 24.8	43 35.2 24.8	44 35.2 24.8	89
90	38 00.0 + 24.1	39 00.0 + 24.0	40 00.0 + 24.0	41 00.0 + 24.0	42 00.0 + 24.0	43 00.0 + 24.0	44 00.0 + 24.0	45 00.0 + 24.0	90

S. Lat. { L.H.A. greater than 180° ....Zn=180° − Z
         L.H.A. less than 180° ....Zn=180° + Z

LATITUDE SAME NAME AS DECLINATION

L.H.A. 114°, 246°

	Tab Hc	**d**	**Tab Z**
	29°44.5′	+39.1′	27.5° (26.6°@ Dec. 65°)

For a declination increment of 05.2′ you get a d correction of + 3.4′, and a Z correction of −0.1′, so:

	Tab Hc	29°44.5′		Tab Z	27.5°
	d corr.	+3.4′		Z corr.	−0.1′
	Hc	29°47.9′		Z	27.4°

Since LHA is less than 180 degrees and you are in the Northern Hemisphere, Zn is 360° − Z:

$$
\begin{array}{r}
360° \\
-27.4° \\
\hline
\text{Zn} \quad 332.6°
\end{array}
$$

Compare Hc and Ho to get intercept distance, and direction in relation to the body:

	Hc	29°47.9′
	Ho	30°07.4′
	Intercept	19.5 miles towards the body

Figure 13-1 shows the plot for this. Since it's not based on an actual star or time, the LOP is not labeled.

HORIZON SIGHTS

A sight within about 10 degrees of the horizon is not recommended. The *Nautical Almanac* does not provide main corrections for the sun, stars, and planets much below this figure; the *Air Almanac* does give refraction corrections, but for the sun you need to work in the semi-diameter as well. The *Nautical Almanac* does have low altitude corrections for the moon, however, and it is conceivable that you might require such a sight.

Whenever apparent altitude of a body is below 10 degrees, an extra correction should be added to the main correction figure. This is based on temperature and barometric pressure conditions, both of which affect the amount of refraction at these low altitudes. The table on page A4 of the *Nautical Almanac* provides the corrections, and you can see that only in extreme cases are corrections needed above altitudes of 20°40′. For altitudes between 10° and 20°40′, the corrections in normal conditions are so small they can be ignored safely (Table 13-2).

The table is simple to use. In the top section find the *zone letter* that includes your temperature and pressure (which is given in both millibars and inches). Under this letter, in the lower section, the correction is on the line that most closely matches your Ha. The correction is always added.

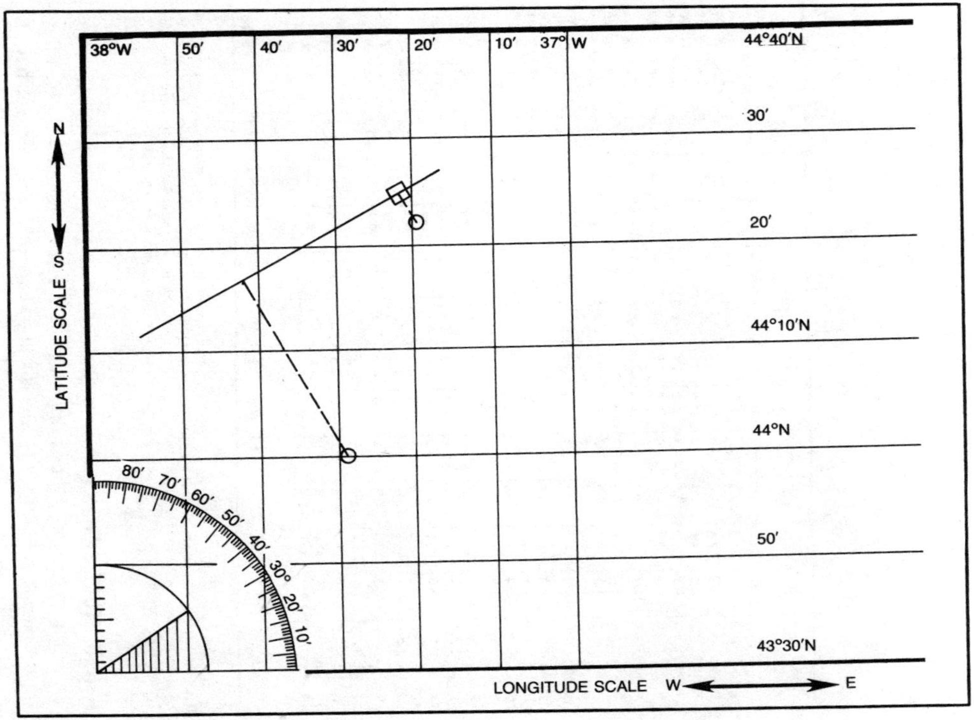

Fig. 13-1. Sight is plotted in the normal manner.

ADDITIONAL REFRACTION CORRECTIONS FOR NON-STANDARD CONDITIONS

Temperature

−20°F. −10° 0° +10° 20° 30° 40° 50° 60° 70° 80° 90° 100°F.

−30°C. −20° −10° 0° +10° 20° 30° 40°C.

Pressure in millibars (left): 1050, 1030, 1010, 990, 970

Pressure in inches (right): 31·0, 30·5, 30·0, 29·5, 29·0

Zone letters: A B C D E F G H J K L M N

App. Alt.	A	B	C	D	E	F	G	H	J	K	L	M	N	App. Alt.
0 00	−6·9	−5·7	−4·6	−3·4	−2·3	−1·1	0·0	+1·1	+2·3	+3·4	+4·6	+5·7	+6·9	0 00
0 30	5·2	4·4	3·5	2·6	1·7	0·9	0·0	0·9	1·7	2·6	3·5	4·4	5·2	0 30
1 00	4·3	3·5	2·8	2·1	1·4	0·7	0·0	0·7	1·4	2·1	2·8	3·5	4·3	1 00
1 30	3·5	2·9	2·4	1·8	1·2	0·6	0·0	0·6	1·2	1·8	2·4	2·9	3·5	1 30
2 00	3·0	2·5	2·0	1·5	1·0	0·5	0·0	0·5	1·0	1·5	2·0	2·5	3·0	2 00
2 30	−2·5	−2·1	−1·6	−1·2	−0·8	−0·4	0·0	+0·4	+0·8	+1·2	+1·6	+2·1	+2·5	2 30
3 00	2·2	1·8	1·5	1·1	0·7	0·4	0·0	0·4	0·7	1·1	1·5	1·8	2·2	3 00
3 30	2·0	1·6	1·3	1·0	0·7	0·3	0·0	0·3	0·7	1·0	1·3	1·6	2·0	3 30
4 00	1·8	1·5	1·2	0·9	0·6	0·3	0·0	0·3	0·6	0·9	1·2	1·6	1·8	4 00
4 30	1·6	1·4	1·1	0·8	0·5	0·3	0·0	0·3	0·5	0·8	1·2	1·5	1·6	4 30
5 00	−1·5	−1·3	−1·0	−0·8	−0·5	−0·2	0·0	+0·2	+0·5	+0·8	+1·0	+1·3	+1·5	5 00
6	1·3	1·1	0·9	0·6	0·4	0·2	0·0	0·2	0·4	0·6	0·9	1·1	1·3	6
7	1·1	0·9	0·7	0·6	0·4	0·2	0·0	0·2	0·4	0·6	0·7	0·9	1·1	7
8	1·0	0·8	0·7	0·5	0·3	0·2	0·0	0·2	0·3	0·5	0·7	0·8	1·0	8
9	0·9	0·7	0·6	0·4	0·3	0·1	0·0	0·1	0·3	0·4	0·6	0·7	0·9	9
10 00	−0·8	−0·7	−0·5	−0·4	−0·3	−0·1	0·0	+0·1	+0·3	+0·4	+0·6	+0·7	+0·8	10 00
12	0·7	0·6	0·5	0·3	0·2	0·1	0·0	0·1	0·2	0·3	0·5	0·6	0·7	12
14	0·6	0·5	0·4	0·3	0·2	0·1	0·0	0·1	0·2	0·3	0·4	0·5	0·6	14
16	0·5	0·4	0·3	0·3	0·2	0·1	0·0	0·1	0·2	0·3	0·3	0·5	0·6	16
18	0·4	0·4	0·3	0·2	0·2	0·1	0·0	0·1	0·2	0·3	0·3	0·4	0·5	18
20 00	−0·4	−0·3	−0·3	−0·2	−0·1	−0·1	0·0	+0·1	+0·1	+0·2	+0·3	+0·3	+0·4	20 00
25	0·3	0·3	0·2	0·2	0·1	−0·1	0·0	+0·1	0·1	0·2	0·2	0·3	0·3	25
30	0·3	0·2	0·2	0·1	0·1	0·0	0·0	0·0	0·1	0·1	0·2	0·3	0·3	30
35	0·2	0·2	0·1	0·1	0·1	0·0	0·0	0·0	0·1	0·1	0·1	0·2	0·3	35
40	0·2	0·1	0·1	0·1	−0·1	0·0	0·0	0·0	+0·1	0·1	0·1	0·2	0·2	40
50 00	−0·1	−0·1	−0·1	−0·1	0·0	0·0	0·0	0·0	0·0	+0·1	+0·1	+0·1	+0·1	50 00

The graph is entered with arguments temperature and pressure to find a zone letter; using as arguments this zone letter and apparent altitude (sextant altitude corrected for dip), a correction is taken from the table. This correction is to be applied to the sextant altitude in addition to the corrections for standard conditions (for the Sun, stars and planets from page A2 and for the Moon from pages xxxiv and xxxv).

HIGH ALTITUDE SIGHTS

A high altitude sight is one that often can be taken with excellent results. At sextant altitudes above about 85 degrees, however, you have to be very careful when taking the sight because it seems as if you can bring the body down to any point on the horizon. Indeed, if the body is directly overhead at the zenith, altitude is 90° and any part of the horizon can be used. In most cases, once you get the body down, pivot left and right and adjust the sextant as necessary until you are certain the body just touches the horizon at the right point.

With a high altitude sight, the GP of the body is very close to your actual position. If it's close enough to be plotted on your plotting sheet, you can set a drawing compass to the distance represented by the observed altitude and draw a circle of position. You don't even have to reduce the sight. Your boat lies somewhere on this circle with your EP at the point closest to your DR position (see Fig. 13-2). Usually, however, it is necessary to determine intercept distance and plot this from an AP.

DOUBLE SECOND DIFFERENCE

In correcting Tab Hc to Hc, an additional factor is added: the *double second difference* (DSD). It is the algebraic difference between the d figure given for the Hc *above* the one you are using, and the one *below* it. For example, at LHA 354°, tabulated declination 46°N at AP Lat. 45°N. *Pub. 229* shows:

Tab Dec	Tab Hc	d	Tab Z
45°		+ 9.1'	
46°	85°40.7'	− 4.8'	87.9°
47°		−18.0'	

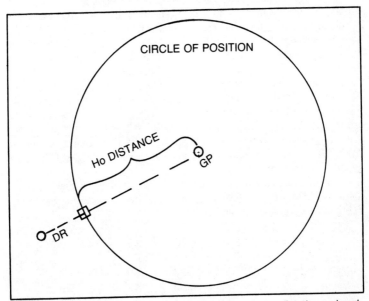

Fig. 13-2. Circle of position can be drawn if body is almost directly overhead.

159

You'll note that a little dot is shown alongside the d figures (Table 13-3). This indicates that the DSD calculation should be made. Here the algebraic difference between d at 45° and d at 47° is:

$$
\begin{array}{lll}
45° & (+)\ 9.1' & 13.9' \\
46° & (-)\ 4.8' & 13.2' \\
47° & (-)18.0' & \overline{27.1'} \\
\end{array}
$$

The DSD factor 27.1' is used with the interpolation table for the appropriate declination increment to find the DSD correction. The factors and corrections are shown to the right of the blocks used for standard d and Z corrections, the DSD factors are in large type; the corrections, to the right, are in smaller type. Pick the correction that lies between the DSD factors that bracket the one you have calculated.

With the above example, at a declination increment of 32.6', you have a DSD correction of 1.7' (Table 13-4). The DSD correction is always added to the regular d correction found by the standard method. It is always positive, regardless of the sign for the standard d correction. The resulting Hc, compared to Ho, gives intercept distance and direction in relation to the body in the normal manner.

TABLE OF OFFSETS

When the GP of the body is so close to your actual position (even though it may not be close enough to put on your plotting sheet), the LOP drawn at right angles to the intercept really should be an arc. In normal sights, the LOP actually represents a segment of an arc, but the radius involved is so long that the arc can be drawn as a straight line.

With a high altitude sight, the straight line LOP does not provide the necessary degree of accuracy, so this must be adjusted by means of a Table of Offsets that appears on page XVI of *Pub. 229*.

Measure the distance from your DR position to the intercept, parallel to the LOP. For the distances in the table that bracket this, pick the offset figures at the Hc of the body. It may be necessary to interpolate to get greatest accuracy. This is particularly true at the highest altitudes and greatest distances. Often you can "eyeball" the interpolation right from the table. Note that the distances are shown as minutes of arc, but these are equal to nautical miles (Table 13-5).

In our example, the distances 20' and 25' bracket our measured distance of 21.0 miles, and Hc is 87.4° (note that Hc is rounded to the nearest tenth of a degree). Interpolation shows:

Hc	20'	25'
87.0°	1.1'	1.7'
87.4°	1.3'	2.0'
87.5°	1.3'	2.1'

Table 13-3. *Pub. 229* Information Used to Establish Double-Second Difference.

6°, 354° L.H.A. — **LATITUDE SAME NAME AS DECLINATION**

Dec.	38° Hc	d	Z	39° Hc	d	Z	40° Hc	d	Z	41° Hc	d	Z	42° Hc	d	Z	43° Hc	d	Z	44° Hc	d	Z	45° Hc	d	N	Dec.
0	51 36.0	+59.5	170.3	50 36.8	+59.5	170.5	49 37.6	+59.6	170.7	48 38.4	+59.6	170.9	47 39.2	+59.5	171.1	46 39.9	+59.6	171.2	45 40.6	+59.6	171.4	44 41.2	+59.7	171.5	0
1	52 35.5	59.4	170.1	51 36.3	59.5	170.3	50 37.2	59.5	170.5	49 38.0	59.5	170.7	48 38.7	59.6	170.9	47 39.5	59.6	171.1	46 40.2	59.6	171.2	45 40.9	59.6	171.4	1
2	53 34.9	59.4	169.9	52 35.8	59.5	170.1	51 36.7	59.5	170.3	50 37.5	59.5	170.5	49 38.3	59.6	170.7	48 39.1	59.6	170.9	47 39.8	59.7	171.1	46 40.5	59.7	171.2	2
3	54 34.3	59.4	169.6	53 35.3	59.4	169.9	52 36.2	59.5	170.1	51 37.0	59.5	170.3	50 37.9	59.5	170.5	49 38.7	59.6	170.7	48 39.5	59.6	170.9	47 40.2	59.6	171.1	3
4	55 33.7	59.3	169.4	54 34.7	59.3	169.6	53 35.6	59.5	169.9	52 36.5	59.5	170.1	51 37.4	59.5	170.3	50 38.3	59.5	170.5	49 39.1	59.5	170.7	48 39.8	59.6	170.9	4
36	84 48.5	19.7	110.8	84 22.5	26.7	120.4	83 48.6	37.2	128.4	83 08.7	42.0	134.9	82 24.3	47.1	140.2	81 36.8	50.0	144.6	80 46.8	52.2	148.1	79 55.0	53.8	151.1	36
37	85 08.2	+8.2	100.0	84 52.1	19.9	111.1	84 25.8	29.0	120.7	83 51.6	37.5	128.7	83 11.4	43.2	135.3	82 26.8	47.3	140.6	81 39.0	50.3	144.9	80 48.8	52.4	148.5	37
38	85 16.4	-4.4	88.2	85 12.0	+8.3	100.1	84 55.7	20.2	111.6	84 29.1	30.2	121.0	83 54.6	37.9	129.1	83 09.1	43.6	135.6	82 29.3	47.6	140.9	81 41.2	50.6	145.3	38
39	85 12.0	16.3	76.1	85 20.3	-4.4	88.1	85 15.9	+8.4	100.3	84 59.3	20.5	111.6	84 32.5	30.6	121.3	83 57.7	38.2	129.5	83 16.9	43.9	136.0	82 31.8	47.9	141.3	39
40	84 55.7	-26.6	64.9	85 15.9	-16.6	75.9	85 24.3	-4.5	88.1	85 19.8	+8.6	100.4	85 03.1	+20.7	111.8	84 35.9	+31.0	121.7	84 00.8	+38.6	129.9	83 19.7	+44.2	136.4	40
41	84 29.1	34.5	55.2	84 59.3	26.8	64.6	85 19.8	16.7	75.7	85 28.4	-4.6	88.0	85 22.8	+8.7	100.5	85 06.9	21.0	112.1	84 39.4	31.4	122.1	84 03.9	39.1	130.3	41
42	83 54.6	40.5	47.1	84 32.5	34.8	54.8	85 03.1	27.2	64.2	85 23.8	16.9	75.5	85 32.5	-4.6	88.0	85 27.9	+8.9	100.7	85 10.8	21.3	112.4	84 43.0	31.7	122.5	42
43	83 14.1	44.8	40.5	83 57.7	40.8	46.6	84 35.9	35.1	54.3	85 06.9	27.5	63.9	85 27.9	17.1	75.2	85 36.8	-4.7	88.0	85 32.1	+9.0	100.9	85 14.7	21.7	112.8	43
44	82 29.3	48.1	35.1	83 16.9	45.1	40.0	84 00.8	41.1	46.1	84 39.4	35.5	53.9	85 10.8	27.8	63.5	85 32.1	17.4	75.0	85 41.1	-4.7	87.9	85 36.4	9.1	101.0	44
45	81 41.2	-50.3	30.7	82 31.8	-48.3	34.6	83 19.7	-45.4	39.5	84 03.9	-41.4	45.6	84 43.0	-35.9	53.4	85 14.7	-28.1	63.1	85 36.4	-17.7	74.8	85 45.5	-4.8	87.9	45
46	80 50.9	52.2	27.2	81 43.5	50.6	30.3	82 34.3	48.5	34.2	83 22.5	45.7	39.0	84 07.1	41.7	45.1	84 46.6	36.3	52.9	85 18.7	28.5	62.7	85 40.7	18.0	74.5	46
47	79 58.7	53.4	24.2	80 52.9	52.3	26.7	81 45.8	50.8	29.8	82 36.8	48.7	33.7	83 25.4	46.0	38.5	84 10.3	42.0	44.6	84 50.2	36.6	52.4	85 22.7	28.8	62.2	47
48	79 05.3	54.5	21.7	80 00.6	53.6	23.8	80 55.0	52.5	26.3	81 48.1	51.0	29.4	82 39.4	49.0	33.2	83 28.2	46.3	38.0	84 13.6	42.5	44.0	84 53.9	37.0	51.9	48
49	78 10.8	55.3	19.6	79 07.0	54.6	21.3	80 02.5	53.7	23.4	80 57.1	52.7	25.9	81 50.4	51.2	28.9	82 42.0	49.3	32.7	83 31.1	46.5	37.4	84 16.9	42.8	43.5	49
80	47 56.2	-59.6	1.6	48 56.1	-59.5	1.6	49 56.1	-59.5	1.6	50 56.1	-59.5	1.7	51 56.1	-59.5	1.7	52 56.0	-59.5	1.7	53 56.0	-59.5	1.8	54 56.0	-59.5	1.8	80
81	46 56.6	59.6	1.4	47 56.6	59.5	1.4	48 56.6	59.5	1.4	49 56.5	59.5	1.5	50 56.5	59.5	1.5	51 56.5	59.5	1.5	52 56.5	59.5	1.6	53 56.5	59.5	1.6	81
82	45 57.0	59.6	1.2	46 57.0	59.6	1.2	47 57.0	59.6	1.2	48 57.0	59.6	1.3	49 57.0	59.6	1.3	50 57.0	59.6	1.3	51 57.0	59.6	1.4	52 57.0	59.6	1.4	82
83	44 57.4	59.6	1.0	45 57.4	59.6	1.0	46 57.4	59.6	1.1	47 57.4	59.6	1.1	48 57.4	59.6	1.1	49 57.4	59.6	1.1	50 57.4	59.6	1.2	51 57.4	59.6	1.2	83
84	43 57.8	59.6	0.9	44 57.8	59.6	0.9	45 57.8	59.6	0.9	46 57.8	59.6	0.9	47 57.8	59.6	0.9	48 57.8	59.6	1.0	49 57.8	59.6	1.0	50 57.8	59.6	1.0	84
85	42 58.2	-59.6	0.7	43 58.2	-59.6	0.7	44 58.2	-59.6	0.7	45 58.2	-59.6	0.8	46 58.2	-59.6	0.8	47 58.2	-59.6	0.8	48 58.2	-59.6	0.8	49 58.2	-59.6	0.8	85
86	41 58.6	59.6	0.6	42 58.6	59.6	0.6	43 58.6	59.6	0.6	44 58.6	59.6	0.6	45 58.6	59.6	0.6	46 58.6	59.6	0.6	47 58.6	59.6	0.5	48 58.6	59.6	0.6	86
87	40 59.0	59.7	0.4	41 59.0	59.7	0.4	42 59.0	59.7	0.4	43 59.0	59.7	0.5	44 59.0	59.7	0.5	45 59.0	59.7	0.5	46 59.0	59.7	0.4	47 59.0	59.7	0.5	87
88	39 59.3	59.7	0.3	40 59.3	59.7	0.3	41 59.3	59.7	0.3	42 59.3	59.7	0.3	43 59.3	59.7	0.3	44 59.3	59.7	0.3	45 59.3	59.7	0.3	46 59.3	59.7	0.3	88
89	38 59.7	59.7	0.1	39 59.7	59.7	0.1	40 59.7	59.7	0.1	41 59.7	59.7	0.1	42 59.7	59.7	0.1	43 59.7	59.7	0.1	44 59.7	59.7	0.1	45 59.7	59.7	0.2	89
90	38 00.0	-59.7	0.0	39 00.0	-59.7	0.0	40 00.0	-59.7	0.0	41 00.0	-59.7	0.0	42 00.0	-59.7	0.0	43 00.0	-59.7	0.0	44 00.0	-59.7	0.0	45 00.0	-59.7	0.0	90

Table 13-4. Double-Second Difference for a Declination Increment of 32.6′.

INTERPOLATION TABLE

Left block

Dec. Inc.	10′	20′	30′	40′	50′	.	0′	1′	2′	3′	4′	5′	6′	7′	8′	9′
28.0	4.6	9.3	14.0	18.6	23.3	.0	0.0	0.5	0.9	1.4	1.9	2.5	3.0	3.5	4.0	4.6
28.1	4.7	9.3	14.0	18.7	23.4	.1	0.0	0.5	1.0	1.5	1.9	2.5	3.0	3.5	4.0	4.6
31.7	5.3	10.6	15.9	21.2	26.4	.7	0.4	0.9	1.4	1.9	2.5	3.0	3.5	4.0	4.6	5.1
31.8	5.3	10.6	15.9	21.2	26.5	.8	0.4	1.0	1.5	2.0	2.5	3.1	3.6	4.1	4.6	5.1
31.9	5.4	10.7	16.0	21.3	26.6	.9	0.5	1.0	1.5	2.0	2.6	3.1	3.6	4.1	4.7	5.2
32.0	5.3	10.6	16.0	21.3	26.6	.0	0.0	0.5	1.1	1.6	2.2	2.7	3.2	3.8	4.3	4.9
32.1	5.3	10.7	16.0	21.4	26.7	.1	0.1	0.6	1.1	1.7	2.2	2.8	3.3	3.8	4.4	4.9
32.2	5.3	10.7	16.1	21.4	26.8	.2	0.1	0.7	1.2	1.7	2.3	2.8	3.4	3.9	4.5	5.0
32.3	5.4	10.8	16.1	21.5	26.9	.3	0.2	0.7	1.2	1.8	2.3	2.9	3.4	4.0	4.5	5.0
32.4	5.4	10.8	16.2	21.6	27.0	.4	0.2	0.8	1.3	1.8	2.4	2.9	3.5	4.0	4.6	5.1
32.5	5.4	10.8	16.3	21.7	27.1	.5	0.3	0.8	1.4	1.9	2.5	3.0	3.5	4.1	4.6	5.1
32.6	5.4	10.9	16.3	21.7	27.2	.6	0.3	0.9	1.4	1.9	2.5	3.0	3.6	4.1	4.7	5.2
32.7	5.5	11.0	16.4	21.8	27.3	.7	0.4	0.9	1.5	2.0	2.6	3.1	3.7	4.2	4.7	5.3
32.8	5.5	11.0	16.4	21.9	27.4	.8	0.4	1.0	1.5	2.1	2.6	3.2	3.7	4.3	4.8	5.3
32.9	5.5	11.0	16.5	22.0	27.5	.9	0.5	1.0	1.6	2.1	2.7	3.2	3.7	4.3	4.8	5.4
33.0	5.5	11.0	16.5	22.0	27.5	.0	0.0	0.6	1.1	1.7	2.2	2.8	3.3	3.9	4.4	5.0
33.1	5.5	11.0	16.5	22.1	27.6	.1	0.1	0.6	1.2	1.7	2.3	2.8	3.4	3.9	4.5	5.0
33.2	5.5	11.0	16.6	22.1	27.6	.2	0.1	0.7	1.2	1.8	2.3	2.9	3.4	4.0	4.5	5.1
33.3	5.6	11.1	16.6	22.2	27.7	.3	0.2	0.7	1.3	1.8	2.4	2.9	3.5	4.1	4.6	5.2
33.4	5.6	11.1	16.7	22.3	27.8	.4	0.2	0.8	1.3	1.9	2.4	3.0	3.6	4.1	4.7	5.2
33.5	5.6	11.2	16.8	22.3	27.9	.5	0.3	0.8	1.4	2.0	2.5	3.1	3.6	4.2	4.7	5.3
33.6	5.6	11.2	16.8	22.4	28.0	.6	0.3	0.9	1.5	2.0	2.6	3.1	3.7	4.3	4.8	5.4
33.7	5.6	11.3	16.9	22.5	28.1	.7	0.4	1.0	1.5	2.1	2.6	3.2	3.8	4.3	4.9	5.5
33.8	5.7	11.3	16.9	22.6	28.2	.8	0.4	1.0	1.6	2.1	2.7	3.3	3.8	4.4	5.0	5.5
33.9	5.7	11.3	17.0	22.6	28.3	.9	0.5	1.1	1.6	2.2	2.8	3.3	3.9	4.4	5.0	5.6

Double-Second Diff. and Corr. (left block):

28.0–28.1 group: 0.8 → 0.1; 2.4 → 0.2

31.7–31.9 group: 31.2 → 2.0; 32.8 → 2.1; 34.4 → 2.1

32.0–33.9 group: 0.8 → 0.1; 2.4 → 0.2; 4.0 → 0.2; 5.7 → 0.3; 7.3 → 0.4; 8.9 → 0.5; 10.5 → 0.7; 12.1 → 0.7; 13.7 → 0.8; 15.4 → 0.9; 17.0 → 1.1; 18.6 → 1.2; 20.2 → 1.3; 21.8 → 1.4; 23.4 → 1.5; 25.1 → 1.6; 26.7 → 1.6; 28.3 → 1.7; 29.9 → 1.9; 31.5 → 2.0; 33.1 → 2.1; 34.7 → 2.2

Right block

Dec. Inc.	10′	20′	30′	40′	50′	.	0′	1′	2′	3′	4′	5′	6′	7′	8′	9′
36.0	6.0	12.0	18.0	24.0	30.0	.7	0.0	0.6	1.2	1.8	2.4	3.0	3.6	4.3	4.9	5.5
36.1	6.0	12.0	18.0	24.0	30.1	.8	0.1	0.7	1.3	1.9	2.5	3.1	3.7	4.3	4.9	5.5
39.7	6.6	13.3	19.9	26.5	33.1	.7	0.5	1.1	1.8	2.4	3.1	3.8	4.4	5.1	5.7	6.4
39.8	6.7	13.3	19.9	26.6	33.2	.8	0.5	1.2	1.9	2.5	3.2	3.8	4.5	5.2	5.8	6.5
39.9	6.7	13.3	20.0	26.6	33.3	.9	0.6	1.3	1.9	2.6	3.3	3.9	4.6	5.3	5.9	6.5
40.0	6.6	13.3	20.0	26.6	33.3	.0	0.0	0.7	1.3	2.0	2.7	3.3	4.0	4.7	5.4	6.0
40.1	6.7	13.3	20.0	26.7	33.4	.1	0.1	0.7	1.4	2.1	2.8	3.4	4.1	4.8	5.5	6.1
40.2	6.7	13.4	20.1	26.8	33.5	.2	0.1	0.8	1.5	2.2	2.8	3.5	4.2	4.9	5.6	6.2
40.3	6.7	13.4	20.1	26.9	33.6	.3	0.2	0.9	1.6	2.3	2.9	3.6	4.3	5.0	5.6	6.3
40.4	6.7	13.5	20.2	26.9	33.7	.4	0.3	0.9	1.6	2.3	3.0	3.6	4.3	5.0	5.7	6.3
40.5	6.8	13.5	20.3	27.0	33.8	.5	0.3	1.0	1.7	2.4	3.0	3.7	4.4	5.1	5.7	6.4
40.6	6.8	13.5	20.3	27.1	33.8	.6	0.4	1.1	1.8	2.4	3.1	3.8	4.5	5.2	5.8	6.5
40.7	6.8	13.6	20.4	27.2	33.9	.7	0.5	1.2	1.8	2.5	3.2	3.9	4.5	5.2	5.9	6.6
40.8	6.8	13.6	20.4	27.2	34.0	.8	0.5	1.2	1.9	2.6	3.3	3.9	4.6	5.3	6.0	6.7
40.9	6.9	13.7	20.5	27.3	34.1	.9	0.6	1.3	2.0	2.7	3.3	4.0	4.7	5.4	6.1	6.7
41.0	6.8	13.6	20.5	27.3	34.1	.0	0.0	0.7	1.4	2.1	2.8	3.5	4.1	4.8	5.5	6.2
41.1	6.8	13.7	20.5	27.4	34.2	.1	0.1	0.8	1.5	2.1	2.8	3.5	4.2	4.9	5.6	6.3
41.2	6.8	13.7	20.6	27.4	34.3	.2	0.1	0.8	1.5	2.2	2.9	3.6	4.3	5.0	5.7	6.4
41.3	6.9	13.8	20.6	27.5	34.4	.3	0.2	0.9	1.6	2.3	3.0	3.7	4.4	5.1	5.8	6.5
41.4	6.9	13.8	20.7	27.5	34.5	.4	0.3	1.0	1.7	2.4	3.1	3.8	4.5	5.2	5.9	6.6
41.5	6.9	13.8	20.8	27.7	34.6	.5	0.3	1.0	1.7	2.4	3.1	3.8	4.5	5.2	5.9	6.6
41.6	6.9	13.9	20.8	27.7	34.7	.6	0.4	1.1	1.8	2.5	3.2	3.9	4.6	5.3	6.0	6.6
41.7	7.0	13.9	20.9	27.8	34.8	.7	0.5	1.2	1.9	2.6	3.3	4.0	4.6	5.3	6.0	6.8
41.8	7.0	14.0	20.9	27.9	34.9	.8	0.6	1.2	1.9	2.6	3.3	4.0	4.7	5.4	6.1	6.8
41.9	7.0	14.0	21.0	28.0	35.0	.9	0.6	1.3	2.0	2.7	3.4	4.1	4.8	5.5	6.2	6.8

Double-Second Diff. and Corr. (right block):

36.0–36.1 group: 0.8 → 0.1

39.7–39.9 group: 32.5 → 1.8; 34.3 → 1.9

40.0–41.9 group: 0.9 → 0.1; 2.8 → 0.2; 4.6 → 0.3; 6.5 → 0.4; 8.3 → 0.5; 10.2 → 0.6; 12.0 → 0.6; 13.9 → 0.7; 15.7 → 0.8; 17.6 → 0.9; 19.4 → 1.1; 21.3 → 1.2; 23.1 → 1.2; 25.0 → 1.3; 26.8 → 1.4; 28.7 → 1.5; 30.5 → 1.7; 32.3 → 1.8; 34.2 → 1.8

Table 13-5. Table of Offsets Used with High Altitude Vega Sight. See Text for Interpolation Explanation.

TABLE OF OFFSETS

DISTANCE ALONG LINE OF POSITION FROM INTERCEPT

ALT.	00′	05′	10′	15′	20′	25′	30′	35′	40′	45′	ALT.
					OFFSETS						
0°	0.0	0.0	0.0	0.0	0.0	0.0	0.0	0.0	0.0	0.0	0°
30	0.0	0.0	0.0	0.0	0.0	0.1	0.1	0.1	0.1	0.2	30
40	0.0	0.0	0.0	0.0	0.1	0.1	0.1	0.2	0.2	0.3	40
50	0.0	0.0	0.0	0.0	0.1	0.1	0.2	0.2	0.3	0.3	50
55	0.0	0.0	0.0	0.0	0.1	0.1	0.2	0.3	0.3	0.4	55
60	0.0	0.0	0.0	0.1	0.1	0.2	0.2	0.3	0.4	0.5	60
62	0.0	0.0	0.0	0.1	0.1	0.2	0.2	0.3	0.4	0.5	62
64	0.0	0.0	0.0	0.1	0.1	0.2	0.3	0.4	0.5	0.6	64
66	0.0	0.0	0.0	0.1	0.1	0.2	0.3	0.4	0.5	0.7	66
68	0.0	0.0	0.0	0.1	0.1	0.2	0.3	0.4	0.6	0.7	68
70	0.0	0.0	0.0	0.1	0.2	0.2	0.4	0.5	0.6	0.8	70
71	0.0	0.0	0.0	0.1	0.2	0.3	0.4	0.5	0.7	0.9	71
72	0.0	0.0	0.0	0.1	0.2	0.3	0.4	0.5	0.7	0.9	72
73	0.0	0.0	0.0	0.1	0.2	0.3	0.4	0.6	0.8	1.0	73
74	0.0	0.0	0.1	0.1	0.2	0.3	0.5	0.6	0.8	1.0	74
75	0.0	0.0	0.1	0.1	0.2	0.3	0.5	0.7	0.9	1.1	75
76	0.0	0.0	0.1	0.1	0.2	0.4	0.5	0.7	0.9	1.2	76
77	0.0	0.0	0.1	0.1	0.3	0.4	0.6	0.8	1.0	1.3	77
78	0.0	0.0	0.1	0.2	0.3	0.4	0.6	0.8	1.1	1.4	78
79	0.0	0.0	0.1	0.2	0.3	0.5	0.7	0.9	1.2	1.5	79
80.0	0.0	0.0	0.1	0.2	0.3	0.5	0.7	1.0	1.3	1.7	80.0
80.5	0.0	0.0	0.1	0.2	0.3	0.5	0.8	1.1	1.4	1.8	80.5
81.0	0.0	0.0	0.1	0.2	0.4	0.6	0.8	1.1	1.5	1.9	81.0
81.5	0.0	0.0	0.1	0.2	0.4	0.6	0.9	1.2	1.6	2.0	81.5
82.0	0.0	0.0	0.1	0.2	0.4	0.6	0.9	1.3	1.7	2.1	82.0
82.5	0.0	0.0	0.1	0.2	0.4	0.7	1.0	1.4	1.8	2.2	82.5
83.0	0.0	0.0	0.1	0.3	0.5	0.7	1.1	1.5	1.9	2.4	83.0
83.5	0.0	0.0	0.1	0.3	0.5	0.8	1.2	1.6	2.0	2.6	83.5
84.0	0.0	0.0	0.1	0.3	0.5	0.9	1.2	1.7	2.2	2.8	84.0
84.5	0.0	0.0	0.2	0.3	0.6	1.0	1.4	1.9	2.4	3.1	84.5
85.0	0.0	0.0	0.2	0.4	0.7	1.0	1.5	2.1	2.7	3.4	85.0
85.5	0.0	0.0	0.2	0.4	0.7	1.2	1.7	2.3	3.0	3.8	85.5
86.0	0.0	0.1	0.2	0.5	0.8	1.3	1.9	2.6	3.4	4.3	86.0
86.5	0.0	0.1	0.2	0.5	1.0	1.5	2.2	2.9	3.8	4.9	86.5
87.0	0.0	0.1	0.3	0.6	1.1	1.7	2.5	3.4	4.5	5.7	87.0
87.5	0.0	0.1	0.3	0.8	1.3	2.1	3.0	4.1	5.4	6.9	87.5
88.0	0.0	0.1	0.4	0.9	1.7	2.7	3.8	5.2	6.9	8.8	88.0
88.5	0.0	0.2	0.6	1.3	2.3	3.5	5.1	7.1	9.4	12.1	88.5
89.0	0.0	0.3	0.8	1.9	3.4	5.5	8.0	11.3	15.3	20.3	89.0

In adjusting the straight LOP to obtain a closer approximation of the arc of the circle of equal altitude, points on the LOP are offset at right angles to the LOP in the direction of the celestial body. The arguments for entering the table are the distance from the intercept to the point on the LOP to be offset and the altitude of the body.

In the use of the table with the graphical method for interpolating altitude for latitude and LHA increments, the offset of the foot of the perpendicular is along the azimuth line in a direction away from the body. The arguments for entering the table are the distance from the DR to the foot of the perpendicular and the altitude of the body.

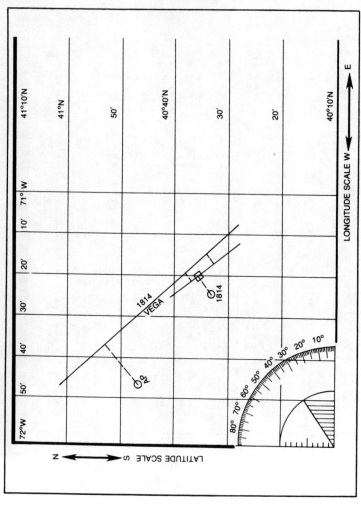

Fig. 13-3. Offset distances indicated by short dashed lines are plotted at right angles to the LOP, in the direction of the body. The EP is on the short line that connects these offsets.

164

Table 13-6. More *Pub. 229* Information for a High Altitude Sight.

2°, 358° L.H.A. **LATITUDE SAME NAME AS DECLINATION** N. Lat. { L.H.A. greater than 180°......Zn=Z / L.H.A. less than 180°......Zn=360°−Z }

Dec.	38° Hc	d	Z	39° Hc	d	Z	40° Hc	d	Z	41° Hc	d	Z	42° Hc	d	Z	43° Hc	d	Z	44° Hc	d	Z	45° Hc	d	Z	Dec.
0	51 57.3	+60.0	176.8	50 57.4	+60.0	176.8	49 57.5	+60.0	176.8	48 57.5	60.0	176.8	47 57.7	+59.9	177.0	46 57.8	+59.9	177.1	45 57.9	+60.0	177.1	44 57.9	+60.0	177.2	0
1	52 57.2	59.9	176.6	51 57.4	59.9	176.8	50 57.5	59.9	176.8	49 57.5	60.0	176.8	48 57.6	59.9	176.9	47 57.7	59.9	177.0	46 57.8	60.0	177.1	45 57.9	60.0	177.1	1
2	53 57.2	59.9	176.6	52 57.3	59.9	176.7	51 57.4	59.9	176.8	50 57.5	60.0	176.8	49 57.6	60.0	176.8	48 57.7	59.9	177.0	47 57.8	59.9	177.0	46 57.8	60.0	177.1	2
3	54 57.2	60.0	176.5	53 57.3	59.9	176.6	52 57.3	59.9	176.6	51 57.4	59.9	176.7	50 57.5	59.9	176.8	49 57.6	59.9	176.8	48 57.7	60.0	176.9	47 57.8	59.9	177.0	3
4	55 57.1	59.9	176.4	54 57.2	59.9	176.4	53 57.3	59.9	176.6	52 57.4	59.9	176.7	51 57.5	59.9	176.8	50 57.6	59.9	176.8	49 57.7	59.9	176.9	48 57.7	60.0	177.0	4
5	56 56.8	+59.9	176.3	55 57.1	+59.9	176.4	54 57.2	+59.9	176.4	53 57.3	+59.9	176.5	52 57.4	+59.9	176.6	51 57.5	+59.9	176.8	50 57.6	+60.0	176.8	49 57.7	+60.0	176.8	5
6	57 56.8	60.0	176.2	56 57.0	60.0	176.4	55 57.1	60.0	176.4	54 57.2	59.9	176.5	53 57.4	59.9	176.6	52 57.5	59.9	176.7	51 57.6	59.9	176.7	50 57.7	59.9	176.8	6
7	58 56.8	59.9	176.1	57 57.0	59.9	176.1	56 57.1	59.9	176.3	55 57.2	59.9	176.4	54 57.3	59.9	176.5	53 57.4	59.9	176.6	52 57.5	60.0	176.7	51 57.6	60.0	176.7	7
8	59 56.7	59.9	176.0	58 57.0	59.9	176.1	57 57.0	59.9	176.3	56 57.1	59.9	176.3	55 57.2	59.9	176.4	54 57.4	59.9	176.5	53 57.5	60.0	176.6	52 57.6	59.9	176.7	8
9	60 56.6	59.9	176.1	59 56.9	59.9	176.1	58 56.9	59.9	176.2	57 57.1	59.9	176.3	56 57.2	59.9	176.3	55 57.3	59.9	176.4	54 57.4	59.9	176.6	53 57.5	60.0	176.6	9
10	61 56.5	+59.9	175.8	60 56.7	+59.9	175.9	59 56.8	+60.0	176.1	58 57.0	+59.9	176.1	57 57.1	+59.9	176.3	56 57.2	+60.0	176.4	55 57.3	+57.4	176.5	54 57.5	+59.9	176.6	10
11	62 56.4	59.9	175.7	61 56.6	59.9	175.8	60 56.8	59.9	175.8	59 56.9	59.9	176.0	58 57.0	59.9	176.1	57 57.2	59.9	176.3	56 57.3	59.9	176.4	55 57.4	59.9	176.5	11
12	63 56.3	59.9	175.5	62 56.5	59.9	175.7	61 56.7	59.9	175.8	60 56.8	59.9	176.0	59 57.0	59.9	176.1	58 57.1	59.9	176.2	57 57.2	60.0	176.3	56 57.3	59.9	176.4	12
13	64 56.2	59.9	175.4	63 56.4	59.9	175.5	62 56.6	59.9	175.7	61 56.8	59.9	175.9	60 56.9	59.9	176.0	59 57.0	59.9	176.1	58 57.2	59.9	176.2	57 57.3	59.9	176.3	13
14	65 56.1	59.9	175.2	64 56.3	59.9	175.4	63 56.5	59.8	175.6	62 56.6	59.9	175.7	61 56.8	59.9	175.9	60 56.9	60.0	176.0	59 57.1	59.9	176.1	58 57.2	59.9	176.2	14
15	66 55.9	+59.9	175.1	65 56.1	+59.9	175.2	64 56.3	+59.9	175.3	63 56.5	+59.9	175.5	62 56.6	+59.9	175.6	61 56.8	+59.9	175.7	60 57.0	+59.9	175.8	59 57.1	+60.0	176.1	15
16	67 55.8	59.8	174.9	66 56.0	59.9	175.1	65 56.2	59.8	175.1	64 56.4	59.9	175.3	63 56.6	59.8	175.5	62 56.8	59.9	175.6	61 56.9	59.9	175.8	60 57.1	59.9	176.0	16
17	68 55.6	59.8	174.7	67 55.9	59.8	174.9	66 56.1	59.8	175.1	65 56.4	59.9	175.1	64 56.5	59.8	175.3	63 56.7	59.9	175.5	62 56.8	59.9	175.8	61 57.0	59.9	176.0	17
18	69 55.4	59.8	174.5	68 55.7	59.8	174.7	67 55.9	59.9	174.9	66 56.2	59.8	175.1	65 56.4	59.8	175.1	64 56.6	59.8	175.3	63 56.7	59.9	175.5	62 56.8	59.9	175.8	18
19	70 55.5	59.6	174.5	69 55.5	59.8	174.5	68 55.8	59.8	174.7	67 56.0	59.8	174.9	66 56.2	59.9	175.0	65 56.4	59.8	175.4	64 56.6	59.9	175.5	63 56.8	59.9	175.7	19
20	71 55.0	+59.7	173.9	70 55.3	+59.8	174.2	69 55.6	+59.8	174.5	68 55.9	+59.8	174.6	67 56.1	+59.7	175.0	66 56.3	+59.9	175.2	65 56.5	+59.9	175.4	64 56.7	+59.9	175.6	20
21	72 54.7	59.8	173.6	71 55.1	59.7	174.0	70 55.5	59.7	174.2	69 55.7	59.8	174.4	68 56.0	59.8	174.6	67 56.2	59.8	174.9	66 56.4	59.9	175.1	65 56.6	59.9	175.4	21
22	73 54.5	59.6	173.3	72 54.9	59.7	173.7	71 55.2	59.8	174.0	70 55.5	59.8	174.3	69 55.8	59.6	174.5	68 56.0	59.8	174.8	67 56.3	59.8	175.0	66 56.5	59.9	175.3	22
23	74 54.1	59.7	173.2	73 54.6	59.7	173.3	72 55.0	59.7	173.7	71 55.3	59.7	174.0	70 55.6	59.7	174.3	69 55.9	59.8	174.6	68 56.1	59.8	174.8	67 56.3	59.9	175.2	23
24	75 53.8	59.6	173.1	74 54.3	59.6	173.0	73 54.7	59.9	173.4	72 55.1	59.7	173.8	71 55.4	59.7	174.1	70 55.7	59.8	174.4	69 56.0	59.8	174.6	68 56.2	59.9	175.1	24
25	76 53.4	+59.5	172.5	75 53.9	+59.5	172.9	74 54.4	+59.7	173.5	73 54.7	+59.8	173.0	72 55.2	+59.7	173.8	71 55.5	+59.8	174.1	70 55.8	+59.8	174.4	69 56.1	+59.8	174.7	25
26	77 52.9	59.4	172.1	76 53.5	59.6	172.0	75 54.1	59.5	172.9	74 54.5	59.7	173.1	73 54.9	59.8	173.5	72 55.3	59.8	173.9	71 55.6	59.8	174.2	70 55.9	59.8	174.5	26
27	78 52.3	59.4	171.4	77 53.3	59.5	171.6	76 53.7	59.5	172.1	75 54.3	59.5	172.6	74 54.7	59.7	173.1	73 55.1	59.8	173.6	72 55.5	59.7	173.9	71 55.8	59.8	174.2	27
28	79 51.7	59.2	169.9	78 52.5	59.2	170.8	77 53.2	59.4	171.6	76 53.8	59.5	172.2	75 54.3	59.7	172.7	74 54.8	59.7	173.2	73 55.3	59.7	173.6	72 55.5	59.7	173.9	28
29	80 50.9	59.1	168.9	79 51.9	59.2	170.0	78 52.7	59.4	170.8	77 53.4	59.5	171.6	76 54.0	59.6	172.3	73 54.5	59.6	172.8	74 54.9	59.6	173.2	73 55.3	59.6	173.7	29
30	81 49.8	+58.7	167.7	80 51.1	+59.1	169.0	79 52.1	+59.3	170.1	78 52.9	+59.3	170.2	75 53.5	+59.5	171.7	76 54.1	+59.5	172.3	75 54.6	+59.7	172.9	74 55.1	+59.7	173.3	30
31	82 48.5	58.7	166.2	81 50.1	58.7	167.8	80 51.3	59.0	169.2	79 52.1	59.3	170.1	78 53.0	59.3	170.3	77 53.7	59.5	171.2	76 54.3	59.6	172.4	75 54.8	59.6	172.5	31
32	83 46.9	57.4	164.1	82 48.8	58.4	166.2	81 50.3	58.4	168.0	80 51.5	59.1	169.3	79 52.4	59.3	170.3	78 53.2	59.3	170.4	77 53.9	59.5	171.4	76 54.4	59.5	171.4	32
33	84 44.5	56.6	161.4	83 47.2	57.7	164.3	82 49.1	58.4	166.5	81 50.3	59.1	169.3	80 51.7	59.1	169.4	79 52.6	59.3	170.4	78 53.4	59.3	170.5	77 53.6	59.5	171.1	33
34	85 41.1	54.7	157.4	84 44.9	56.7	161.6	83 47.5	57.8	164.5	82 49.4	58.4	166.6	81 50.8	58.9	168.2	80 51.9	58.9	169.5	79 52.9	59.3	170.4	78 53.6	59.5	170.5	34
35	86 35.8	+50.6	151.2	85 41.6	+54.8	157.6	84 45.3	+56.8	161.8	83 47.8	+57.9	164.7	82 49.7	+58.4	166.8	81 51.1	+58.9	168.4	81 51.1	+59.1	169.6	79 53.1	+59.1	170.6	35
36	87 28.6	41.1*	140.3	86 36.4	50.8*	151.5	85 42.1	54.9*	157.9	84 45.7	56.8	162.0	83 48.1	58.0	164.8	82 50.0	58.5	166.9	76 54.3	59.6	172.4	80 52.4	59.2	169.7	36
37	88 20.6	41.1*–120.6		87 27.2	41.1*	140.6	86 37.0	51.0*	151.8	85 42.5	54.9*	158.1	84 46.1	56.9	162.2	83 48.5	58.0	165.0	82 50.2	58.6	167.1	81 51.6	58.9	168.8	37
38	88 08.6	–18.1*	121.9	88 06.6	–17.0*	89.4	87 28.0	41.1*	140.8	86 37.6	51.2*	152.1	85 43.0	55.2*	158.4	84 46.5	57.0	162.4	83 48.8	58.1	165.2	82 50.5	58.7	167.5	38
39	88 08.0	40.6*	56.8	88 06.7	–17.0*	89.4	88 09.7	+18.4*–122.3		87 28.8	42.0*–141.9		86 38.2	55.2*–152.5		85 43.5	53.1*–158.7		84 46.9	57.1	162.7	83 49.2	58.1	165.4	39

Z=Z ..... Zn=Z / Zn=360°−Z

So our offsets are 1.3' and 2.0' respectively. Measure the 20 mile and 25 mile distances along the LOP, in the direction of the DR position. From each of these points drop a perpendicular *toward* the sighted body for the offset distance required. Connect the ends of the offsets with a straight line. You will notice that this forms an angle with the LOP, and it represents a short section of the LOP adjusted to more closely indicate the circle of position. A perpendicular dropped from your DR position to this line provides your EP (see Fig. 13-3).

AP LATITUDE ADJUSTMENTS

Note in the plot that latitude of the AP is not on a whole degree. When altitudes greater than about 85 degrees are recorded, the minutes and tenths of minutes of AP latitude are selected to match the declination increment minutes and tenths of minutes. To determine Hc and Z, you need to work with two adjacent columns in *Pub. 229*, with one degree difference in latitude and one degree difference in declination. Your AP latitude will be between the two tabulated latitudes, and declination will be between the two selected tabulated declinations.

In the example illustrated by Fig. 13-3, the GHA of Vega is 73°46.3' and its declination is N 38°46.9'. The DR position is 41°33.5'N, Lo 71°24.4' W. Your AP Lo is established in the usual way; here it is 2°. You want the AP latitude to be within 30' of DR latitude, which means that in this case it is at 41°46.9'N. If DR latitude had been 41°13.5'N, the AP Lat. would be 40°46.9'N.

On the *Pub. 229* page for LHA 2°, 358° (Table 13-6), note the figures for declination 38° at Lat. 40°, and for declination at Lat. 41°. This gives you:

	Lat. 40°		Lat. 41°	
	Hc	**Z**	**Hc**	**Z**
Dec. 38°	87°28.0'	141.5°		
Dec. 39°			87°28.8'	141.9°

Determine the difference for both Hc and Z:

	87°28.8'	141.9°
	87°28.0'	141.5°
d	+ 0.8'	Z diff. +0.4°

Use this d figure to find the d correction in the usual way. Do *not* use the "di" figure that is tabulated; the one you have calculated replaces it. Here the d correction is + 0.6' and it is added to Tab Hc to give you a final Hc of 87°28.6'. Azimuth is corrected in the normal way, using the Z difference factor found above.

It might seem as though a lot of extra work is involved with high altitude sights, but it only takes a few minutes more to do the actual calculations and the plot. Because high altitude sights are often easy to obtain, it's worth the extra effort needed for their reduction.

Chapter 14

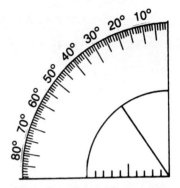

Other Methods of Sight Reduction

Although *Pub. 229* provides a highly accurate and relatively easy method of sight reduction, some skippers prefer other methods, either because they find them easier, or because they require tables that are less bulky. Of course, with a handy pocket calculator, you can dispense with tables entirely.

SIGHT REDUCTION TABLES FOR AIR NAVIGATION

A most popular alternative to *Pub. 229* is the three-volume series *Sight Reduction Tables for Air Navigation (Pub. 249)*. Volume 1 is for 41 selected stars; Volumes 2 and 3 are for all bodies. Volume 2 covers latitudes 0° to 40° and Volume 3 covers latitudes 40° to 89°. Nevertheless, these two books are for declinations less than 30 degrees. In many cases, this eliminates high altitude sights. Also, the data is given to the nearest full minute of arc—not quite as precise as *Pub. 229*, but perfectly adequate for small boat navigation. Volume 1 of *Pub. 249* covers a five year period (epoch) and needs replacement at the end of each epoch.

Pub. 249 is easy to use once you're familiar with its layout (Volume 1 doesn't even require interpolation for the declination increment). In Volumes 2 and 3, data are arranged by whole degrees of latitude (AP latitude). You look up the whole degree of tabulated declination at the top or bottom of the appropriate latitude page and then go down or up the column to the required LHA. LHA 0° to 90° is on the left of each page; LHA 270° to 360° is on the right. Arrangement of "same name" and "contrary name" information is not uniform. Sometimes both are on one page and sometimes they are on separate pages.

In Volumes 2 and 3, you find Hc, d, and Z. Correct Hc with the d correction found using Table 5 in the back of the book (it's also on the "bookmark" sheet). Here's an example of using *Pub. 249*, based on the Sirius sight from Chapter 9 (see Figs. 9-1 and 9-2).

As in Chapter 9, the Ho is 31°42.5′, and Sirius has a GHA of 42°51.9′ and declination of S 16°41.2′. The assumed position is again L 38° N, Lo 63°51.9′W, which gives us LHA 339°. On the page for Lat. 38°, Declination 15° to 29°, contrary name to latitude (Table 14-1), we have at LHA 339°:

Tab Hc	d	Z
32°31′	−57′	156°

The d corr., from Table 5 (Fig. 10-9) is −39′, so:

Tab Hc	32°31′
d corr.	− 39′
Hc	31°52′

Compare this to Ho for intercept distance and direction in relation to Sirius:

Hc	31°52.0′
Ho	31°42.5′
Intercept	9.5 miles away

There is no correction made to the Z figure, and since LHA is greater than 180°, Z = Zn, and Zn is 156°. Figure 14-1 is the plot of this sight.

In Volume 1 of *Pub. 249*, information is again based on latitude. The front half of the book is for northern latitudes and the rear half is for southern latitudes. LHA of Aries is shown at 1-degree intervals along the left side of the page, in groups of 15 (above Lat. 70° LHA is at 2-degree intervals). Seven columns across the page provide Hc and Z for the stars, with seven specific stars listed for each 15 degree grouping of LHA Aries. As an added bonus, the three stars of the seven that make a good three-body fix carry an asterisk.

With this arrangement you don't have to determine the SHA and GHA of the individual stars. You only need to know LHA Aries for the time of the sight, based on an AP longitude that gives LHA in whole degrees. The Hc and Z given in the tables are based on the exact declination of each star, so no Dec. Inc. correction is needed.

H.O. 214

Previous to *Pub. 229*, the government-printed sight reduction tables for marine use was *H.O. 214*, a seven-volume series with each volume covering nine degrees of latitude. While it was last offered by the government in 1946, copies still can be obtained from suppliers such as Coast Navigation. Its use now is largely limited to those who started with it and don't want to make the switch to *Pub. 229*.

Table 14-1. *Pub 249* Data for Lat. 38°, LHA 340°, and Dec. 16°, with Declination Contrary in Name to Latitude.

LAT 38°

	16°	17°	18°	19°	20°	21°	22°	23°	24°	25°	26°	27°	28°	29°	
	Hc d Z	Hc d Z	Hc d Z	Hc d Z	Hc d Z	Hc d Z	Hc d Z	Hc d Z	Hc d Z	Hc d Z	Hc d Z	Hc d Z	Hc d Z	Hc d Z	Zn
	31 29 56 153	30 34 56 153	29 38 56 154	28 42 56 154	27 46 56 154	26 50 56 155	25 54 55 155	24 58 56 156	24 02 56 156	23 06 57 156	22 09 56 157	21 13 56 157	20 17 57 158	19 20 56 158	336
	31 51 56 154	30 55 56 154	29 59 56 155	29 02 56 155	28 06 56 155	27 10 57 155	26 14 57 156	25 17 56 156	24 21 57 157	23 24 56 157	22 28 56 157	21 31 57 158	20 35 57 158	19 38 57 159	337
	32 11 56 155	31 15 56 155	30 18 56 156	29 22 56 156	28 26 56 156	27 29 57 156	26 32 57 157	25 36 56 157	24 39 57 158	23 42 57 158	22 45 56 158	21 49 57 159	20 52 57 159	19 55 57 160	338
	32 11 56 156	31 35 56 156	30 38 57 157	29 41 56 157	28 44 57 157	27 47 58 157	26 50 58 158	25 53 57 158	24 56 58 158	23 59 57 159	23 02 58 159	22 05 57 160	21 08 57 160	20 11 57 161	339
	32 50 -57 157	31 53 -57 157	30 56 -57 158	29 59 -57 158	29 02 -57 158	28 05 -57 159	27 08 -58 159	26 10 -57 160	25 13 -57 160	24 16 -58 160	23 18 -57 161	22 21 -57 161	21 24 -58 161	20 26 -57 162	340
	33 08 58 158	32 11 58 158	31 13 59 159	30 16 59 159	29 19 58 159	28 21 60 160	27 24 58 160	26 27 58 160	25 29 57 161	24 32 58 161	23 34 58 161	22 36 58 162	21 39 58 162	20 41 58 162	341
	33 25 59 159	32 28 59 159	31 30 59 160	30 33 60 160	29 35 60 161	28 37 61 161	27 40 60 161	26 42 59 161	25 45 59 162	24 46 59 162	23 49 58 162	22 51 58 163	21 53 58 163	20 55 58 163	342
	33 41 60 160	32 44 60 160	31 46 60 161	30 48 61 161	29 50 61 161	28 53 60 162	27 55 61 162	26 57 59 162	25 59 60 163	25 01 60 163	24 03 58 163	23 05 58 164	22 06 58 164	21 08 60 164	343
	33 57 58 161	32 59 58 161	32 01 61 162	31 03 60 162	30 05 60 162	29 07 62 163	28 09 59 163	27 11 59 163	26 12 58 164	25 14 58 164	24 16 58 164	23 18 58 165	22 16 58 165	21 21 59 165	344
	34 12 -59 163	33 13 -53 163	32 15 -58 163	31 17 -58 163	30 19 -59 164	29 20 -58 164	28 22 -58 164	27 24 -59 164	26 25 -58 165	25 27 -59 165	24 28 -58 165	23 30 -58 165	22 31 -58 166	21 33 -59 166	345
	34 25 58 164	33 27 58 164	32 29 59 164	31 30 59 164	30 32 58 165	29 33 58 165	28 35 59 165	27 36 58 166	26 37 58 166	25 39 58 166	24 40 58 166	23 41 59 166	22 43 59 167	21 44 59 167	346
	34 38 65 165	33 40 65 165	32 41 65 165	31 42 66 166	30 44 65 166	29 45 66 166	28 46 67 166	27 47 67 167	26 47 67 167	25 50 67 167	24 51 67 167	23 52 67 167	22 53 68 168	21 54 68 168	347
	34 50 66 166	33 51 66 166	32 53 66 167	31 54 66 167	30 55 67 167	29 56 67 167	28 57 67 167	27 57 68 168	26 58 67 168	26 00 68 168	25 01 68 168	24 02 68 168	23 03 69 169	22 04 69 169	348
	35 01 68 168	34 02 68 167	33 03 67 168	32 04 68 168	31 05 68 168	30 06 68 168	29 06 69 168	28 08 68 169	27 09 69 169	26 10 69 169	25 10 69 169	24 11 69 169	23 12 69 170	22 13 69 170	349
	35 11 -59 169	34 12 -59 168	33 13 -59 169	32 14 -59 169	31 15 -60 169	30 15 -59 169	29 16 -59 169	28 17 -59 170	27 18 -60 170	26 18 -59 170	25 19 -59 170	24 20 -60 170	23 20 -59 171	22 21 -59 171	350
	35 21 60 170	34 21 60 170	33 23 59 170	32 23 60 170	31 23 60 170	30 24 60 170	29 24 59 170	28 25 60 170	27 26 60 171	26 26 60 171	25 27 60 171	24 27 60 171	23 28 60 171	22 28 60 172	351
	35 29 60 171	34 29 60 171	33 30 60 171	32 30 60 171	31 31 60 171	30 31 60 171	29 32 60 171	28 32 60 172	27 33 60 172	26 34 60 172	25 34 60 172	24 34 60 172	23 35 60 172	22 35 60 172	352
	35 36 59 172	34 37 59 172	33 37 60 172	32 37 60 172	31 38 60 172	30 38 60 172	29 38 59 173	28 39 59 173	27 40 60 173	26 40 60 173	25 40 60 173	24 40 60 173	23 41 60 173	22 41 60 173	353
	35 42 59 173	34 43 60 173	33 43 60 173	32 43 60 173	31 44 60 173	30 44 60 173	29 44 59 174	28 44 60 174	27 45 60 174	26 45 60 174	25 45 60 174	24 45 60 174	23 46 60 174	22 46 60 174	354
	35 48 -59 174	34 48 -60 174	33 48 -59 174	32 48 -59 174	31 49 -60 175	30 49 -60 175	29 49 -60 175	28 49 -60 175	27 49 -59 175	26 50 -60 175	25 50 -60 175	24 50 -60 175	23 50 -60 175	22 50 -60 175	355
	35 52 60 175	34 52 60 175	33 52 60 175	32 53 60 175	31 53 60 176	30 53 60 176	29 53 60 176	28 53 60 176	27 53 60 176	26 53 60 176	25 53 60 176	24 54 60 176	23 54 60 176	22 54 60 176	356
	35 56 60 176	34 56 60 176	33 56 60 177	32 56 60 177	31 56 60 177	30 56 60 177	29 56 60 177	28 56 60 177	27 56 60 177	26 56 60 177	25 56 60 177	24 56 60 178	23 56 60 177	22 57 60 177	357
	35 58 60 178	34 58 60 178	33 58 60 178	32 58 60 178	31 58 60 178	30 58 60 178	29 58 60 178	28 58 60 178	27 58 60 178	26 58 60 178	25 58 60 178	24 58 60 178	23 58 60 178	23 00 60 178	358
	36 00 60 179	35 00 60 179	34 00 60 180	33 00 60 180	32 00 60 179	31 00 60 179	30 00 60 179	29 00 60 179	28 00 60 179	27 00 60 179	26 00 60 179	25 00 60 179	24 00 60 179	23 00 60 179	359
	36 00 -60 180	35 00 -60 180	34 00 -60 180	33 00 -60 180	32 00 -60 180	31 00 -60 180	30 00 -60 180	29 00 -60 180	28 00 -60 180	27 00 -60 180	26 00 -60 180	25 00 -60 180	24 00 -60 180	23 00 -60 180	360

DECLINATION (15°–29°) CONTRARY NAME TO LATITUDE

LAT 38°

S. Lat. { LHA greater than 180°........ Zn=180−Z } { LHA less than 180°........ Zn=180+Z }

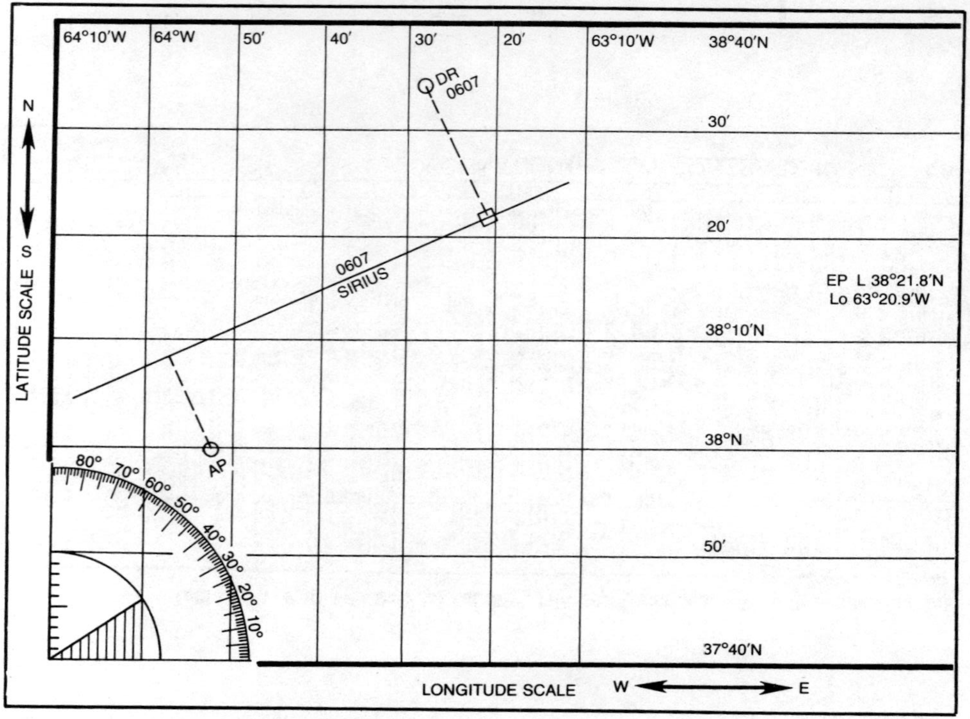

Fig. 14-1. The Sirius plot based on *Pub. 249* sight reduction. Compare this to Fig. 9-2.

EP L 38°21.8'N
Lo 63°20.9'W

AGETON METHOD

A sight reduction method that uses just 36 pages of tables was devised by Lt. Arthur A. Ageton, USN, about 50 years ago. This was published as *H.O. 211*. These tables now appear in Volume 2 of "Bowditch" (Defense Mapping Agency Hydrographic/Topographic Center *Pub. 9*), and they also are offered by some nautical suppliers as a booklet of about 50 pages, including explanation of their use. If you use the booklet, or photocopies of the pages from "Bowditch," you can find your position anywhere without such bulky references as *Pub. 229* or *Pub. 249*.

In this method, you have to look up and record many numbers. Some numbers are added and others are subtracted. It's like a Chinese menu with choices from Column A and Column B. A great deal of care must be taken in locating and recording the numbers. It's best to use prepared sight reduction forms such as those from Simex. Simex also has forms for use with *Pub. 229* and *Pub. 249*.

To use this method, you need to establish *meridian angle t*; this is LHA west or east of your DR longitude. For example, if your DR Lo is 63°W and GHA of the body is 93°, t is 30°W. If GHA of the body is 33°, t is 30°E (see Fig. 14-2).

To demonstrate the method, I will again use the Sirius sight from Chapter 9 as an example. In this case t is:

DR Lo	63°27.8′W
GHA Sirius	42°51.9′
"t"	20°35.9′E

With t and declination you can enter the tables. In the first column of your worksheet (Table 14-3), put in the A number that corresponds to t (Table 14-4) and below it the B number that corresponds to declination

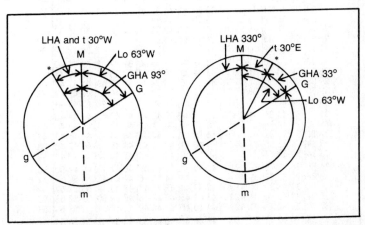

Fig. 14-2. For the Ageton method of sight reduction, the t angle, east or west, replaces LHA.

Table 14-2. *Pub. 249* Shows a Correction of 39′ for a d Factor Of 57 and a Declination Increment of 41′.

41	42	43	44	45	46	47	48	49	50	51	52	53	54	55	56	57	58	59	60	d/′
0	0	0	0	0	0	0	0	0	0	0	0	0	0	0	0	0	0	0	0	0
1	1	1	1	1	1	1	1	1	1	1	1	1	1	1	1	1	1	1	1	1
1	1	1	1	2	2	2	2	2	2	2	2	2	2	2	2	2	2	2	2	2
2	2	2	2	2	2	2	2	2	2	3	3	3	3	3	3	3	3	3	3	3
3	3	3	3	3	3	3	3	3	3	3	3	4	4	4	4	4	4	4	4	4
3	4	4	4	4	4	4	4	4	4	4	4	4	4	5	5	5	5	5	5	5
4	4	4	4	4	5	5	5	5	5	5	5	5	5	6	6	6	6	6	6	6
5	5	5	5	5	5	5	6	6	6	6	6	6	6	6	7	7	7	7	7	7
5	6	6	6	6	6	6	6	7	7	7	7	7	7	7	7	8	8	8	8	8
6	6	6	7	7	7	7	7	7	8	8	8	8	8	8	8	9	9	9	9	9
7	7	7	7	8	8	8	8	8	8	8	9	9	9	9	9	10	10	10	10	10
8	8	8	8	8	8	9	9	9	9	9	10	10	10	10	10	10	11	11	11	11
8	8	9	9	9	9	9	10	10	10	10	10	11	11	11	11	11	12	12	12	12
9	9	9	10	10	10	10	10	11	11	11	11	11	12	12	12	12	13	13	13	13
10	10	10	10	10	11	11	11	11	12	12	12	12	13	13	13	13	14	14	14	14
10	10	11	11	11	12	12	12	12	12	13	13	13	14	14	14	14	14	15	15	15
11	11	11	12	12	12	13	13	13	13	14	14	14	14	15	15	15	15	16	16	16
12	12	12	12	13	13	13	14	14	14	14	15	15	15	16	16	16	16	17	17	17
12	13	13	13	14	14	14	14	15	15	15	16	16	16	16	17	17	17	18	18	18
13	13	14	14	14	15	15	15	16	16	16	16	17	17	17	18	18	18	19	19	19
14	14	14	15	15	15	16	16	16	17	17	17	18	18	18	19	19	19	20	20	20
14	15	15	15	16	16	16	17	17	18	18	18	19	19	19	20	20	20	21	21	21
15	15	16	16	16	17	17	18	18	18	19	19	19	20	20	21	21	21	22	22	22
16	16	16	17	17	18	18	18	19	19	20	20	20	21	21	21	22	22	23	23	23
16	17	17	18	18	18	19	19	20	20	20	21	21	22	22	22	23	23	24	24	24
17	18	18	18	19	19	20	20	20	21	21	22	22	22	23	23	24	24	25	25	25
18	18	19	19	20	20	20	21	21	22	22	23	23	23	24	24	25	25	26	26	26
18	19	19	20	20	21	21	22	22	22	23	23	24	24	25	25	26	26	27	27	27
19	20	20	21	21	21	22	22	23	23	24	24	25	25	26	26	27	27	28	28	28
20	20	21	21	22	22	23	23	24	24	25	25	26	26	27	27	28	28	29	29	29
20	21	22	22	22	23	24	24	24	25	26	26	26	27	28	28	28	29	30	30	30
21	22	22	23	23	24	24	25	25	26	26	27	27	28	28	29	29	30	30	31	31
22	22	23	23	24	25	25	26	26	27	27	28	28	29	29	30	30	31	31	32	32
23	23	24	24	25	25	26	26	27	28	28	29	29	30	30	31	31	32	32	33	33
23	24	24	25	26	26	27	27	28	28	29	29	30	31	31	32	32	33	33	34	34
24	24	25	26	26	27	27	28	29	29	30	30	31	32	32	33	33	34	34	35	35
25	25	26	26	27	28	28	29	29	30	31	31	32	32	33	34	34	35	35	36	36
25	26	27	27	28	28	29	30	30	31	31	32	33	33	34	35	35	36	36	37	37
26	27	27	28	28	29	30	30	31	32	32	33	34	34	35	35	36	37	37	38	38
27	27	28	29	29	30	31	31	32	32	33	34	34	35	36	36	37	38	38	39	39
27	28	29	29	30	31	31	32	33	33	34	35	35	36	37	37	38	39	39	40	40
28	29	29	30	31	31	32	33	33	34	35	36	36	37	38	38	39	40	40	41	41
29	29	30	31	32	32	33	34	34	35	36	36	37	38	38	39	40	41	41	42	42
29	30	31	32	32	33	34	34	35	36	37	37	38	39	39	40	41	42	42	43	43
30	31	32	32	33	34	34	35	36	37	37	38	39	40	40	41	42	43	43	44	44
31	32	32	33	34	34	35	36	37	38	38	39	40	40	41	42	43	44	44	45	45
31	32	33	34	34	35	36	37	38	38	39	40	41	41	42	43	44	44	45	46	46
32	33	34	34	35	36	37	38	38	39	40	41	42	42	43	44	45	45	46	47	47
33	34	34	35	36	37	38	38	39	40	41	42	42	43	44	45	46	46	47	48	48
33	34	35	36	37	38	38	39	40	41	42	42	43	44	45	46	47	47	48	49	49
34	35	36	37	38	38	39	40	41	42	42	43	44	45	46	47	48	48	49	50	50
35	36	37	37	38	39	40	41	42	42	43	44	45	46	47	48	48	49	50	51	51
36	36	37	38	39	40	41	42	42	43	44	45	46	47	48	49	49	50	51	52	52
36	37	38	39	40	41	42	42	43	44	45	46	47	48	49	49	50	51	52	53	53
37	38	39	40	40	41	42	43	44	45	46	47	48	49	50	50	51	52	53	54	54
38	38	39	40	41	42	43	44	45	46	47	48	49	50	50	51	52	53	54	55	55
38	39	40	41	42	43	44	45	46	47	48	49	49	50	51	52	53	54	55	56	56
39	40	41	42	43	44	45	46	47	48	48	49	50	51	52	53	54	55	56	57	57

Table 14-3. Worksheet for Reduction of the Sirius Sight by the Ageton Method.

BODY - SIRIUS   DR L 38°34.4'N, Lo l63°27.8'W

Hs 31°46.7'
1C   + 1.1'
DIP   − 3.7'
Ha 31°44.1'
MAIN − 1.6'
Ho 31°42.5'

WT |06|− '05 −|13 − 28 SEPT
WE (S):+ 01'− '21

ZT |06|− |06 − 34
ZD|+3

GMT 09 − .06 − .34   28 SEPT.

DEC. S 16°41.2

DR Lo 63°27.8'W
GHA * 42°51.9'

t 20°35.9'E

SHA * 258°55.7'
GHA Y
09h   142°17.4'
06m 34s 1°38.8'

402°51.9'
−360°

GHA* 42°51.9'

DR L 38°34.4'N	COL. 1 ADD	COL. 2 SUBTRACT	COL. 3 ADD	COL. 4 SUBTRACT
K 17°45.0'S	A 45382			
K~L 56°19.4'	B 1868	A 54199	B 2617	A 47250
	A 47250	B 2617		
		A 51582 (K 17°45.0)	B 25602	
Ho 31°42.5'			A 28219	B 6912
Hc 31°28.5'			(HC 31°28.5')	A 40348 (Z 156°44.5')
diff. 14.0 MILES TOWARD				
Z - 157°				

(Table 14-5). Note that you are taking the numbers that present angles closest to actual t and declination; you don't need to interpolate to get highly accurate results. In Column 2, put the A number that corresponds to declination alongside its B number in Column 1.

Add the figures in Column 1. The total is an A number (47250), which can be recorded in Column 4 (see Table 14-3). Look up this number, and put the corresponding B number (Table 14-6) adjacent to it in both Column 2 and Column 3. Note that the number you take may not match exactly the one you enter the table with; it is the one closest to it. Here 47242 is the closest number to 47250, so the B figure you record is 2617.

Now *subtract* the B number in Column 2 from the A number above it. The result, 51582, is an A number. Look this up, and from the top of its column take the angle given and add to this the number of minutes shown to the left of the A figure (Table 14-7). Here the angle total is 17°45'.

This angle is the *K factor*. When t is greater than 90°, as it may be in extreme northern or southern latitudes, use the angle given at the bottom of the column and add to it the minutes shown to the right of the A figure.

Give *K* the same name as declination, then find the angle difference between K and your DR latitude. If latitude and K are the same name, just subtract the smaller from the larger. If K and DR latitude are contrary names, as in this case, add the angles (Table 14-3).

Find the B number that is closest to the resulting angle (Table 14-8) and record this in Column 3. Add it to the B number already there. The total

WHEN LHA (E OR W) IS GREATER
THAN 90°, TAKE "K" FROM BOTTOM OF TABLE

'	20° 00'		20° 30'		21° 00'		21° 30'	
	A	B	A	B	A	B	A	B
0	46595	2701	45567	2841	44567	2985	43592	3132
	46577	2704	45551	2844	44551	2988	43576	3135
1	46560	2706	45534	2846	44534	2990	43560	3137
	46543	2708	45517	2848	44518	2992	43544	3140
2	46525	2711	45500	2851	44501	2994	43528	3142
	46508	2713	45483	2853	44485	2997	43512	3145
3	46491	2715	45466	2855	44468	2999	43496	3147
	46473	2717	45449	2858	44452	3002	43480	3150
4	46456	2720	45433	2860	44436	3004	43464	3152
	46439	2722	45416	2862	44419	3007	43448	3155
5	46422	2724	45399	2865	44403	3009	43432	3157
	46404	2727	45382	2867	44386	3012	43416	3160
6	46387	2729	45365	2870	44370	3014	43400	3162
	46370	2731	45348	2872	44354	3016	43385	3165
7	46353	2734	45332	2874	44337	3019	43369	3167
	46335	2736	45315	2877	44321	3021	43353	3170
8	46318	2738	45298	2879	44305	3024	43337	3172
	46301	2741	45281	2881	44288	3026	43321	3175
9	46284	2743	45265	2884	44272	3029	43305	3177
	46266	2745	45248	2886	44256	3031	43289	3180
10	46249	2748	45231	2889	44239	3033	43273	3182
	46232	2750	45214	2891	44223	3036	43257	3185
11	46215	2752	45198	2893	44207	3038	43241	3187
	46198	2755	45181	2896	44190	3041	43225	3190
12	46181	2757	45164	2898	44174	3043	43210	3192
	46163	2759	45147	2901	44158	3046	43194	3195
13	46146	2761	45131	2903	44142	3048	43178	3197
	46129	2764	45114	2905	44125	3051	43162	3200
14	46112	2766	45097	2908	44109	3053	43146	3202
	46095	2768	45081	2910	44093	3056	43130	3205
15	46078	2771	45064	2913	44077	3058	43114	3207
	46061	2773	45047	2915	44060	3060	43099	3210
16	46043	2775	45031	2917	44044	3063	43083	3212
	46026	2778	45014	2920	44028	3065	43067	3215
17	46009	2780	44997	2922	44012	3068	43051	3217
	45992	2782	44981	2924	43995	3070	43035	3220
18	45975	2785	44964	2927	43979	3073	43020	3222
	45958	2787	44947	2929	43963	3075	43004	3225
19	45941	2789	44931	2932	43947	3078	42988	3227
	45924	2792	44914	2934	43931	3080	42972	3230
20	45907	2794	44898	2936	43914	3083	42956	3233
	45890	2797	44881	2939	43898	3085	42941	3235
21	45873	2799	44864	2941	43882	3088	42925	3238
	45856	2801	44848	2944	43866	3090	42909	3240
22	45839	2804	44831	2946	43850	3092	42893	3243
	45822	2806	44815	2949	43834	3095	42878	3245
23	45805	2808	44798	2951	43818	3097	42862	3248
	45788	2811	44782	2953	43801	3100	42846	3250
24	45771	2813	44765	2956	43785	3102	42830	3253
	45754	2815	44748	2958	43769	3105	42815	3255
25	45737	2818	44732	2961	43753	3107	42799	3258
	45720	2820	44715	2963	43737	3110	42783	3260

Table 14-5. The "B" Figure for Column 1, and the "A" Figure for Column 2.

′	15° 00′		15° 30′		16° 00′		16° 30′	
	A	B	A	B	A	B	A	B
0	58700	1506	57310	1609	55966	1716	54666	1826
	58677	1507	57287	1611	55944	1718	54644	1828
1	58653	1509	57265	1612	55922	1719	54623	1830
	58630	1511	57242	1614	55900	1721	54602	1832
2	58606	1512	57219	1616	55878	1723	54581	1834
	58583	1514	57196	1618	55856	1725	54559	1836
3	58559	1516	57174	1619	55834	1727	54538	1837
	58536	1517	57151	1621	55812	1728	54517	1839
4	58512	1519	57128	1623	55790	1730	54496	1841
	58489	1521	57106	1625	55768	1732	54474	1843
5	58465	1523	57083	1627	55746	1734	54453	1845
	58442	1524	57060	1628	55725	1736	54432	1847
6	58418	1526	57038	1630	55703	1738	54411	1849
	58395	1528	57015	1632	55681	1739	54390	1851
7	58372	1529	56992	1634	55659	1741	54368	1853
	58348	1531	56970	1635	55637	1743	54347	1854
8	58325	1533	56947	1637	55615	1745	54326	1856
	58302	1534	56925	1639	55593	1747	54305	1858
9	58278	1536	56902	1641	55572	1749	54284	1860
	58255	1538	56880	1642	55550	1750	54263	1862
10	58232	1540	56857	1644	55528	1752	54242	1864
	58208	1541	56835	1646	55506	1754	54220	1866
11	58185	1543	56812	1648	55484	1756	54199	1868
	58162	1545	56790	1649	55463	1758	54178	1870
12	58138	1546	56767	1651	55441	1760	54157	1871
	58115	1548	56745	1653	55419	1761	54136	1873
13	58092	1550	56722	1655	55397	1763	54115	1875
	58069	1552	56700	1657	55376	1765	54094	1877
14	58046	1553	56677	1658	55354	1767	54078	1879
	58022	1555	56655	1660	55332	1769	54052	1881
15	57999	1557	56632	1662	55311	1771	54031	1883
	57976	1559	56610	1664	55289	1772	54010	1885
16	57953	1560	56588	1665	55267	1774	53989	1887
	57930	1562	56565	1667	55246	1776	53968	1889
17	57907	1564	56543	1669	55224	1778	53947	1890
	57884	1565	56521	1671	55202	1780	53926	1892
18	57860	1567	56498	1673	55181	1782	53905	1894
	57837	1569	56476	1674	55159	1783	53884	1896
19	57814	1571	56454	1676	55138	1785	53864	1898
	57791	1572	56431	1678	55116	1787	53843	1900
20	57768	1574	56409	1680	55095	1789	53822	1902
	57745	1576	56387	1682	55073	1791	53801	1904
21	57722	1578	56365	1683	55051	1793	53780	1906
	57699	1579	56342	1685	55030	1795	53759	1908
22	57676	1581	56320	1687	55008	1796	53738	1910
	57653	1583	56298	1689	54987	1798	53718	1911
23	57630	1584	56276	1691	54965	1800	53697	1913
	57607	1586	56254	1692	54944	1802	53676	1915
24	57584	1588	56231	1694	54922	1804	53655	1917
	57561	1590	56209	1696	54901	1806	53634	1919
25	57538	1591	56187	1698	54880	1808	53614	1921
	57516	1593	56165	1700	54858	1809	53593	1923
26	57493	1595	56143	1701	54837	1811	53572	1925
	57470	1597	56121	1703	54815	1813	53551	1927
27	57447	1598	56099	1705	54794	1815	53531	1929
	57424	1600	56076	1707	54773	1817	53510	1931

175

'	17° 30' A	17° 30' B	18° 00' A	18° 00' B	18° 30' A	18° 30' B	19° 00' A	19° 00' B	19° 30' A	19° 30' B	'
0	52186	2058	51002	2179	49852	2304	48736	2433	47650	2565	30
	52166	2060	50982	2181	49833	2306	48717	2435	47633	2568	
1	52146	2062	50963	2183	49815	2309	48699	2437	47615	2570	29
	52126	2064	50943	2185	49796	2311	48681	2439	47597	2572	
2	52106	2066	50924	2188	49777	2313	48662	2442	47579	2574	28
	52086	2068	50905	2190	49758	2315	48644	2444	47561	2576	
3	52066	2070	50885	2192	49739	2317	48626	2446	47544	2579	27
	52046	2072	50866	2194	49720	2319	48608	2448	47526	2581	
4	52026	2074	50846	2196	49702	2321	48589	2450	47508	2583	26
	52006	2076	50827	2198	49683	2323	48571	2453	47490	2585	
5	51986	2078	50808	2200	49664	2325	48553	2455	47472	2588	25
	51966	2080	50788	2202	49645	2328	48534	2457	47455	2590	
6	51946	2082	50769	2204	49626	2330	48516	2459	47437	2592	24
	51926	2084	50750	2206	49608	2332	48498	2461	47419	2594	
7	51906	2086	50730	2208	49589	2334	48480	2463	47402	2597	23
	51886	2088	50711	2210	49570	2336	48462	2466	47384	2599	
8	51867	2090	50692	2212	49551	2338	48443	2468	47366	2601	22
	51847	2092	50673	2214	49533	2340	48425	2470	47348	2603	
9	51827	2094	50653	2216	49514	2343	48407	2472	47331	2606	21
	51807	2096	50634	2218	49495	2345	48389	2474	47313	2608	
10	51787	2098	50615	2221	49477	2347	48371	2477	47295	2610	20
	51767	2100	50596	2223	49458	2349	48352	2479	47278	2613	
11	51747	2102	50576	2225	49439	2351	48334	2481	47260	2615	19
	51728	2104	50557	2227	49421	2353	48316	2483	47242	2617	

Table 14-6. Here 47242 is Closest to the "A" Total of 47250 in Column 1. The "B" Figure 2617 is Placed in Columns 2 and 3 (See Table 14-3).

Table 14-7. The "A" Total of Column 2 Is closest to 51589, Which Gives a "K" of 17°30' From the Top of the Page, Plus 15' from the Left of the Figure.

'	17° 30' A	17° 30' B	18° 00' A	18° 00' B	18° 30' A	18° 30' B	19° 00' A	19° 00' B
0	52186	2058	51002	2179	49852	2304	48736	2433
	52166	2060	50982	2181	49833	2306	48717	2435
1	52146	2062	50963	2183	49815	2309	48699	2437
	52126	2064	50943	2185	49796	2311	48681	2439
2	52106	2066	50924	2188	49777	2313	48662	2442
	52086	2068	50905	2190	49758	2315	48644	2444
3	52066	2070	50885	2192	49739	2317	48626	2446
	52046	2072	50866	2194	49720	2319	48608	2448
4	52026	2074	50846	2196	49702	2321	48589	2450
	52006	2076	50827	2198	49683	2323	48571	2453
5	51986	2078	50808	2200	49664	2325	48553	2455
	51966	2080	50788	2202	49645	2328	48534	2457
6	51946	2082	50769	2204	49626	2330	48516	2459
	51926	2084	50750	2206	49608	2332	48498	2461
7	51906	2086	50730	2208	49589	2334	48480	2463
	51886	2088	50711	2210	49570	2336	48462	2466
8	51867	2090	50692	2212	49551	2338	48443	2468
	51847	2092	50673	2214	49533	2340	48425	2470
9	51827	2094	50653	2216	49514	2343	48407	2472
	51807	2096	50634	2218	49495	2345	48389	2474
10	51787	2098	50615	2221	49477	2347	48371	2477
	51767	2100	50596	2223	49458	2349	48352	2479
11	51747	2102	50576	2225	49439	2351	48334	2481
	51728	2104	50557	2227	49421	2353	48316	2483
12	51708	2106	50538	2229	49402	2355	48298	2485
	51688	2108	50519	2231	49383	2357	48280	2488
13	51668	2110	50499	2233	49365	2360	48262	2490
	51649	2112	50480	2235	49346	2362	48244	2492
14	51629	2114	50461	2237	49327	2364	48225	2494
	51609	2116	50442	2239	49309	2366	48207	2496
15	51589	2118	50423	2241	49290	2368	48189	2499
	51570	2120	50404	2243	49271	2370	48171	2501
16	51550	2122	50385	2246	49253	2372	48153	2503
	51530	2124	50365	2248	49234	2375	48135	2505
17	51510	2126	50346	2250	49216	2377	48117	2507
	51491	2128	50327	2252	49197	2379	48099	2510
18	51471	2130	50308	2254	49179	2381	48081	2512
	51451	2132	50289	2256	49160	2383	48063	2514
19	51432	2134	50270	2258	49141	2385	48045	2516
	51412	2136	50251	2260	49123	2387	48027	2519
20	51392	2138	50232	2262	49104	2390	48009	2521
	51373	2141	50213	2264	49086	2392	47991	2523
21	51353	2143	50194	2266	49067	2394	47973	2525
	51334	2145	50175	2269	49049	2396	47955	2527
22	51314	2147	50156	2271	49030	2398	47937	2530
	51294	2149	50137	2273	49012	2400	47919	2532
23	51275	2151	50117	2275	48993	2403	47901	2534
	51255	2153	50098	2277	48975	2405	47883	2536
24	51236	2155	50080	2279	48957	2407	47865	2539
	51216	2157	50061	2281	48938	2409	47847	2541
25	51197	2159	50042	2283	48920	2411	47829	2543
	51177	2161	50023	2285	48901	2413	47811	2545
26	51158	2163	50004	2287	48883	2416	47793	2547
	51138	2165	49985	2290	48864	2418	47775	2550

Table 14-8. The Difference Between "K" and
L is 56°19.4', Which Provides This "B" Figure of 25602.

'	55° 00'		55° 30'		56° 00'	
	A	B	A	B	A	B
0	8663	24141	8401	24687	8143	25244
	8659	24150	8396	24696	8138	25253
1	8655	24159	8392	24706	8134	25263
	8650	24168	8388	24715	8130	25272
2	8646	24177	8383	24724	8125	25281
	8641	24186	8379	24733	8121	25291
3	8637	24195	8375	24742	8117	25300
	8633	24204	8370	24752	8113	25309
4	8628	24213	8366	24761	8108	25319
	8624	24222	8362	24770	8104	25328
5	8619	24231	8357	24779	8100	25338
	8615	24240	8353	24788	8096	25347
6	8611	24249	8349	24798	8092	25356
	8606	24258	8344	24807	8087	25366
7	8602	24267	8340	24816	8083	25375
	8597	24276	8336	24825	8079	25385
8	8593	24286	8331	24835	8075	25394
	8589	24295	8327	24844	8070	25403
9	8584	24304	8323	24853	8066	25413
	8580	24313	8318	24862	8062	25422
10	8575	24322	8314	24872	8058	25432
	8571	24331	8310	24881	8053	25441
11	8567	24340	8305	24890	8049	25451
	8562	24349	8301	24899	8045	25460
12	8558	24358	8297	24909	8041	25469
	8553	24367	8292	24918	8036	25479
13	8549	24376	8288	24927	8032	25488
	8545	24385	8284	24936	8028	25498
14	8540	24395	8280	24946	8024	25507
	8536	24404	8275	24955	8020	25517
15	8531	24413	8271	24964	8015	25526
	8527	24422	8267	24973	8011	25536
16	8523	24431	8262	24983	8007	25545
	8518	24440	8258	24992	8003	25554
17	8514	24449	8254	25001	7998	25564
	8510	24458	8249	25011	7994	25573
18	8505	24467	8245	25020	7990	25583
	8501	24477	8241	25029	7986	25592
19	8496	24486	8237	25038	7982	25602
	8492	24495	8232	25048	7977	25611
20	8488	24504	8228	25057	7973	25621
	8483	24513	8224	25066	7969	25630
21	8479	24522	8219	25076	7965	25640
	8475	24531	8215	25085	7961	25649
22	8470	24540	8211	25094	7956	25659
	8466	24550	8207	25104	7952	25668
23	8461	24559	8202	25113	7948	25678
	8457	24568	8198	25122	7944	25687
24	8453	24577	8194	25132	7940	25697
	8448	24586	8189	25141	7935	25706
25	8444	24595	8185	25150	7931	25716
	8440	24605	8181	25160	7927	25725
26	8435	24614	8177	25169	7923	25735
	8431	24623	8172	25178	7919	25744

'	30° 00' A	B	30° 30' A	B	31° 00' A	B
0	30103	6247	29453	6468	28816	6693
	30092	6251	29442	6472	28806	6697
1	30081	6254	29432	6475	28795	6701
	30070	6258	29421	6479	28785	6705
2	30059	6262	29410	6483	28774	6709
	30048	6265	29399	6487	28763	6712
3	30037	6269	29389	6490	28753	6716
	30026	6273	29378	6494	28743	6720
4	30016	6276	29367	6498	28732	6724
	30005	6280	29357	6501	28722	6728
5	29994	6284	29346	6505	28711	6731
	29983	6287	29335	6509	28701	6735
6	29972	6291	29325	6513	28690	6739
	29961	6294	29314	6516	28680	6743
7	29950	6298	29303	6520	28669	6747
	29939	6302	29293	6524	28659	6750
8	29928	6305	29282	6528	28648	6754
	29917	6309	29271	6531	28638	6758
9	29907	6313	29261	6535	28627	6762
	29896	6316	29250	6539	28617	6766
10	29885	6320	29239	6543	28606	6770
	29874	6324	29229	6546	28596	6773
11	29863	6328	29218	6550	28586	6777
	29852	6331	29207	6554	28575	6781
12	29841	6335	29197	6558	28565	6785
	29831	6339	29186	6561	28554	6789
13	29820	6342	29175	6565	28544	6793
	29809	6346	29165	6569	28533	6796
14	29798	6350	29154	6573	28523	6800
	29787	6353	29144	6576	28513	6804
15	29776	6357	29133	6580	28502	6808
	29766	6361	29122	6584	28492	6812
16	29755	6364	29112	6588	28481	6815
	29744	6368	29101	6591	28471	6819
17	29733	6372	29091	6595	28461	6823
	29722	6375	29080	6599	28450	6827
18	29711	6379	29069	6603	28440	6831
	29701	6383	29059	6606	28429	6835
19	29690	6386	29048	6610	28419	6839
	29679	6390	29038	6614	28409	6842
20	29668	6394	29027	6618	28398	6846
	29657	6398	29016	6622	28388	6850
21	29647	6401	29006	6625	28378	6854
	29636	6405	28995	6629	28367	6858
22	29625	6409	28985	6633	28357	6862
	29614	6412	28974	6637	28346	6865
23	29604	6416	28964	6640	28336	6869
	29593	6420	28953	6644	28326	6873
24	29582	6423	28942	6648	28315	6877
	29571	6427	28932	6652	28305	6881
25	29560	6431	28921	6655	28295	6885
	29550	6435	28911	6659	28284	6889
26	29539	6438	28900	6663	28274	6893
	29528	6442	28890	6667	28264	6896
27	29517	6446	28879	6671	28253	6900
	29507	6449	28869	6674	28243	6904
28	29496	6453	28858	6678	28233	6908
	29485	6457	28848	6682	28222	6912

Table 14-10. The "A" Total Column 4 Provides a Z of 156°30' from the Bottom of the Page, and 14.5' from the Right Hand Column.

A	B	A	B	A	B	A	B	
40693	3619	39814	3782	38955	3949	38117	4121	26
40678	3622	39799	3785	38941	3952	38103	4124	
40664	3624	39785	3788	38927	3955	38089	4127	25
40649	3627	39771	3790	38913	3958	38075	4129	
40634	3630	39756	3793	38899	3961	38061	4132	24
40619	3632	39742	3796	38885	3964	38048	4135	
40604	3635	39727	3799	38871	3966	38034	4138	23
40590	3638	39713	3801	38856	3969	38020	4141	
40575	3640	39698	3804	38842	3972	38006	4144	22
40560	3643	39684	3807	38828	3975	37992	4147	
40545	3646	39669	3810	38814	3978	37979	4150	21
40530	3648	39655	3813	38800	3981	37965	4153	
40516	3651	39641	3815	38786	3983	37951	4155	20
40501	3654	39626	3818	38772	3986	37937	4158	
40486	3657	39612	3821	38758	3989	37924	4161	19
40471	3659	39597	3824	38744	3992	37910	4164	
40457	3662	39583	3826	38730	3995	37896	4167	18
40442	3665	39569	3829	38716	3998	37882	4170	
40427	3667	39554	3832	38702	4000	37869	4173	17
40413	3670	39540	3835	38688	4003	37855	4176	
40398	3673	39525	3838	38674	4006	37841	4179	16
40383	3676	39511	3840	38660	4009	37828	4182	
40368	3678	39497	3843	38645	4012	37814	4185	15
40354	3681	39482	3846	38631	4015	37800	4187	
40339	3684	39468	3849	38617	4017	37786	4190	14
40324	3686	39454	3851	38603	4020	37773	4193	
40310	3689	39439	3854	38589	4023	37759	4196	13
40295	3692	39425	3857	38575	4026	37745	4199	
40280	3695	39411	3860	38561	4029	37732	4202	12
40266	3697	39396	3863	38547	4032	37718	4205	
40251	3700	39382	3865	38533	4035	37704	4208	11
40236	3703	39368	3868	38520	4037	37691	4211	
40222	3705	39353	3871	38506	4040	37677	4214	10
40207	3708	39339	3874	38492	4043	37663	4217	
40192	3711	39325	3876	38478	4046	37650	4220	9
40178	3714	39311	3879	38464	4049	37636	4222	
40163	3716	39296	3882	38450	4052	37623	4225	8
40149	3719	39282	3885	38436	4055	37609	4228	
40134	3722	39268	3888	38422	4057	37595	4231	7
40119	3725	39254	3890	38408	4060	37582	4234	
40105	3727	39239	3893	38394	4063	37568	4237	6
40090	3730	39225	3896	38380	4066	37554	4240	
40076	3733	39211	3899	38366	4069	37541	4243	5
40061	3735	39197	3902	38352	4072	37527	4246	
40046	3738	39182	3904	38338	4075	37514	4249	4
40032	3741	39168	3907	38324	4078	37500	4252	
40017	3744	39154	3910	38311	4080	37486	4255	3
40003	3746	39140	3913	38297	4083	37473	4258	
39988	3749	39125	3916	38283	4086	37459	4261	2
39974	3752	39111	3918	38269	4089	37446	4264	
39959	3755	39097	3921	38255	4092	37432	4266	1
39945	3757	39083	3924	38241	4095	37419	4269	
39930	3760	39069	3927	38227	4098	37405	4272	0
A	B	A	B	A	B	A	B	
156° 30'		156° 00'		155° 30'		155° 00'		'

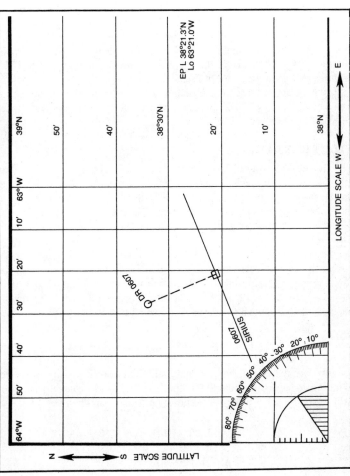

Fig. 14-3. The Sirius sight plotted from the Ageton sight reduction method. Note that the intercept is plotted directly from the DR position. Compare the EP to that of Fig. 9-2 and Fig. 14-1.

181

is again an A number (28219). Look this up and, from the top of its column, again take the angle shown and add to it the minutes to the left of the A figure. Put the corresponding B number in the fourth column. Note in Table 14-9 that the number tabulated (28222) closest to the A number in Column 3 (28219) is about midway between the figures for 28′ and 29′ at the left, so consider the minutes to be 28.5′, and the total angle is 31°28.5′.

This angle is Hc. Compare this to Hc to get intercept distance and direction in relation to the observed body:

$$\begin{array}{ll} \text{Ho} & 31°42.5′ \\ \text{Hc} & \underline{31°28.5′} \\ \text{Intercept} & 14.0 \text{ miles toward} \end{array}$$

To find azimuth, subtract the B number you just entered in Column 4 from the A number already there. Locate the resulting number in the A columns, and take Z angle from the bottom of the page. Add to this the minutes shown to the right of the A number and round to the nearest whole degree (Table 14-10). Here the angle is 156°45.5′, which rounds to 157°. Figure 14-3 shows the plot of this sight.

If K and latitude are the same name, and K is greater than latitude, take the Z angle from the top of the page and add to it the minutes shown to the left of the A number.

POCKET CALCULATOR

A battery-powered pocket calculator with trigonometric functions provides by far the quickest and easiest method of sight reduction—as long as the battery has not run down. The first rule of calculator navigation is to have an adequate supply of fresh batteries. The rechargeable cells are good if you have an on-board generator that you can use for charging—that can take several hours.

From the simplest calculator with trigonometric functions, you can go on up to programmable units designed specially for navigation with the Almanac information already stored in their memory systems. Here we will demonstrate the two basic formulae that are needed: one is for finding calculated altitude and the other is for azimuth. For Hc, you have:

$$\sin^{-1} = (\sin L \sin d) + (\cos L \cos d \cos t)$$

L is DR latitude, d is declination, and t is LHA. If L and d are contrary name, d must be a negative number if it is south, and L must be a negative number if it is south. Again using the Sirius sight from Chapter 9, you have:

$$\begin{array}{ll} \text{L } 38°34.4′N & = 38.573° \\ \text{d S } 16°41.2′ & = -16.686° \\ \text{(LHA)} \quad \text{t } 339°24.1′ & = 339.402° \end{array}$$

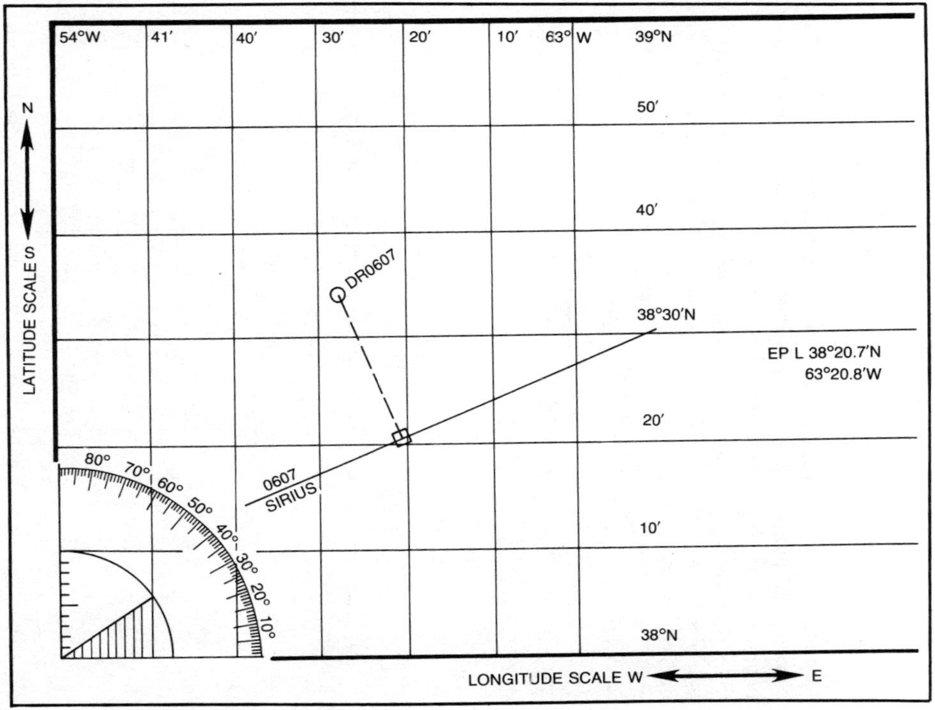
54°W
41′
40′
30′
20′
10′
63° W
39°N
50′
40′
38°30′N
20′
10′
38°N
DR0607
0607
SIRIUS
EP L 38°20.7′N
63°20.8′W
80°
70°
60°
50°
40°
30°
20°
10°
LATITUDE SCALE S N
LONGITUDE SCALE W E

Fig. 14-4. Plot of the Sirius sight based on calculator computation.

Note that the figures are converted to degrees plus decimal places. Divide the minutes and tenths of minutes by 60 on the calculator. The formula then gives you:

$$\sin^{-1} = \sin 38.573° \times \sin -16.686° + \cos 38.573°$$
$$\times \cos -16.686° \times \cos 339.402°$$

With a calculator such as the Texas Instrument TI-30, the sequence is:

Step	Press	Display
1.	38.573 sin	.62351125
2.	× 16.686 +/− sin	−.28712647
3.	+ 38.573 cos	.78181438
4.	× 16.686 +/− cos	.95789268
5.	× 339.402 cos	.93607182
6.	= inv sin	31.465983

This is the calculated altitude, in degrees. If your calculator has a storage capacity, put this figure in storage as it is used in the formula for finding azimuth. To convert this Hc to degrees and minutes, round the decimal portion to three places (.466) and multiply by 60. In this case you get 27.9′ arc, and Hc is 31°27.9′. Ho, as found in Chapter 9, is 31°42.5′, so intercept distance is:

$$
\begin{array}{ll}
\text{Ho} & 31°42.5' \\
\text{Hc} & \underline{31°27.9'} \\
\text{Intercept} & \quad 14.6 \text{ miles toward}
\end{array}
$$

For azimuth, the formula is:

$$\sin^{-1} = \frac{\cos d \, \sin t}{\cos H}$$

H, of course, is the calculated altitude you have just determined. In this case we have:

$$\sin^{-1} = \frac{\cos -16.686° \times \sin 339.402°}{\cos 31.466°}$$

Step	Press	Display
1.	16.686 +/− cos	.95789268
2.	× 339.402 sin	−.35180897
3.	÷ 31.466 cos	.85295007
4.	= inv sin	−23.271821

Of course step 3 might by "rcl cos" if your calculator has the H figure in storage from the previous formula. Since declination is south of your DR

184

latitude and the body is to the east of your position, it is in the southeast quadrant of a circle in which you are at the center. Subtract the number from 180 degrees to get Zn, after rounding it to the nearest tenth:

$$
\begin{array}{rr}
 & 180° \\
Z & -\ 23.3° \\
\hline
Zn & 156.7°
\end{array}
$$

You can see how close the answers are to each other for every method used. The plot of the sight is shown as Fig. 14-4. Note that the intercept is plotted from the DR position, as with the Ageton method. You don't need to establish an assumed position.

With this calculator formula, Z is always "in quadrant," but it's easy to determine which quadrant. If the GP of the body is northeast of your DR position, $Z = Zn$ (if Z is a negative number, ignore the minus sign). If the GP is southeast of the DR position, $Zn = 180° - Z$. If the GP is to the southwest of your DR, Zn is $180° + Z$ (again, if Z is a negative number, ignore the minus sign). And if the GP is to the northwest, $Zn = 360° - Z$.

Because the calculator is so easy to use and provides such accurate results, there's a tendency to rely on it exclusively. Just don't neglect the tabular method of your choice. If the calculator should fail, you'll still be able to reduce your sights to lines of position.

Chapter 15

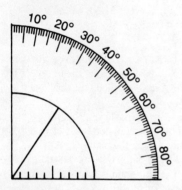

Another Look at the Time

With the information given in Chapter 4, you have all the basics you really need to know for navigation as long as you have an accurate time signal against which you can check your watch prior to taking sights. There is additional information to help you understand time problems, particularly in relation to longitude. This additional information can help you establish your position when certain *Almanac* information is not available.

THE CHRONOMETER

The chronometer is a highly accurate clock that is set to GMT prior to a trip. In practice it is not reset until it gets a specialized cleaning and overhaul. Because this is usually done about every three years, and since not even chronometers (particularly mechanical ones) keep perfect time, considerable *chronometer error* (CE) can develop. This does not affect the usefulness of the instrument as a timepiece because you establish the rate at which it gains or loses time.

Because you want the amount of change per 24 hours in seconds and tenths of seconds, it's easiest to check the amount of error over a 10-day interval, and then divide by 10 to get the daily change rate. Presumably, before radio time signals were available to vessels at sea, this was done before the start of a trip. And if you do not have a radio on board, you will establish the daily change rate before you sail.

Suppose you check chronometer time against a noon, GMT time signal on 17 June 1980 and you have:

```
GMT    12-00-00  17 June
C      12-08-15
CE       (f) 08-15
```

Ten days later, you check again with the noon GMT radio time signal, and you have:

$$
\begin{array}{lll}
\text{GMT} & 12\text{-}00\text{-}00 & 27 \text{ June} \\
\text{C} & \underline{12\text{-}13\text{-}28} & \\
\text{CE (f)} & 13\text{-}28 & 27 \text{ June} \\
\text{CE (f)} & \underline{08\text{-}15} & 17 \text{ June} \\
\text{diff.} & 5\text{-}13 & = 73 \text{ seconds}
\end{array}
$$

Daily change rate: 7.3 seconds, gaining

Now you must establish GMT by chronometer at the time of a sight. To do this, convert elapsed time since GMT was last established to days and tenths of days. Then multiply the result by the daily change rate. This gives you the error to be applied to chronometer time to get GMT. For example, following the time check above, you take a sight on 29 June and chronometer time of the sight is 08-14-26. Elapsed time is:

$$
\begin{array}{ll}
\text{CT } 08\text{-}14\text{-}26 \ 29 \text{ June} = & \\
\quad 56\text{-}14\text{-}26 \ 27 \text{ June} & \\
\text{CT} \underline{-12\text{-}13\text{-}28} \ 27 \text{ June} & \\
\text{ET } 44\text{-}00\text{-}58 = 20\text{-}00\text{-}58 + 1 \text{ day} = 1.83 \text{ days} & \\
\text{ET } 1.83 \text{ days} & \\
\times \quad \underline{7.3} \text{ daily change rate, gaining} & \\
\quad 608 \text{ seconds increase} = 10\text{m } 14\text{s increase} & \\
\text{CE} + \underline{13\text{- }28} \ 27 \text{ June} & \\
\text{CE} \quad 23\text{- }42 \ 29 \text{ June} &
\end{array}
$$

This is the chronometer error with which to correct chronometer time for the sight, so:

$$
\begin{array}{lll}
\text{CT} & 08\text{-}14\text{-}26 & 29 \text{ June} \\
\text{CE (f)} & -\underline{\ \ 23\text{-}42} & \\
\text{GMT} & 07\text{-}50\text{-}44 & 29 \text{ June}
\end{array}
$$

The next time a sight is taken elapsed time is based on the CT of 08-14-26 on 29 June—with the 7.3 second daily change rate applied. If you have a radio time signal to work with, you can periodically verify the chronometer change rate and update it if necessary.

A stopwatch is often used, in conjunction with a chronometer, rather than a standard watch. The stopwatch is started when the second hand of the chronometer is at zero and is stopped by your assistant when you call "mark." Time of the sight is then the chronometer time at which the watch was started, plus the interval on the watch. This must be corrected with the daily change rate, as above, to get GMT. The drawback to this system is that the watch can be stopped only for a single sight. Your assistant can, of course, record elapsed stopwatch times for a run of sights, prior to stopping the watch on a final sight.

Be sure that batteries in a quartz electronic chronometer are fresh at the start of any passage. Winding of a mechanical one should be done at the same time each day to minimize the possibility of a varying daily change rate. Chronometers should be secured where there is the least likelihood of damage to them.

CONVERTING TIME TO ARC

In the *Nautical Almanac,* the time is given for meridian passage of the sun, moon, and planets so that sights on these bodies, when they are on your meridian, provide you with both latitude and longitude—a positive fix. The time given in the *Almanac* is *mean time* at the zone meridian—the mid-point of each time zone (see Chapter 4). If you are exactly at a zone meridian, the body will cross your longitude at the time given in the *Almanac*.

If, however, you take a meridian passage sight and determine that the body crossed your meridian before or after the listed time, you can determine the difference in longitude between your position and the zone meridian by converting the time difference into a measurement of arc. At the rate the earth is revolving about its axis, one hour of time is equal to 15 degrees of arc, so:

$$1h = 15°$$
$$4m = 1°$$
$$1m = 15'$$
$$4s = 1'$$
$$1s = 15''$$

Suppose the sun's meridian passage is listed at 11-47-12, and you determine that it crossed your meridian at 11-44-18. What is the difference in longitude between your position and the zone meridian?

11-47-12	Listed Time
11-44-18	Meridian passage at your position
02-54	Time Difference
2m = 30' arc	
54S = 12'30'' arc	
42'30'' = 42.5' arc of longitude	

Since meridian passage occurred at your position before the sun reached the zone meridian, you are 0°42.5' east of the zone meridian.

Here are some additional problems based on meridian passage of the sun. In each case, determine your longitude difference east or west of the zone meridian. Answers are in the appendix.

31. *Almanac* listed time 12-14-21; actual time 12-12-12
32. *Almanac* listed time 12-08-23; actual time 12-18-15
33. *Almanac* listed time 11-55-52; actual time 12-01-11
34. *Almanac* listed time 11-54-08; actual time 11-41-57
35. *Almanac* listed time 12-00-04; actual time 12-00-08

| | SUN | | Mer. |
| Day | Eqn. of Time | | Pass. |
	00ʰ	12ʰ	
	m s	m s	h m
23	02 04	02 10	12 02
24	02 16	02 23	12 02
25	02 29	02 36	12 03

Table 15-1. Equation of Time Information from the Daily Pages of the *Nautical Almanac*.

EQUATION OF TIME

Of course you have noticed that the *Nautical Almanac* lists meridian passage of the sun as an hour plus whole minutes (see Table 15-1). In the columns adjacent to this listing, figures for the *equation of time* are given. These figures can be used to give exact time of the meridian passage at the zone meridian, not only for the sun, but for the other bodies as well. An examination of the 12h figures and the sun meridian passage figures shows that meridian passage is 12-00-00 plus or minus the equation of time figure, rounded to the nearest minute. In Table 15-1 meridian passage on 23 June is actually 12-02-10; on 24 June it is 12-02-23; and on 25 June it is 12-02-36.

If the earth's orbit about the sun were a perfect circle, meridian passage would be at exactly noon at each zone meridian, every day of the year. Because of the slightly elliptical orbit, this is not the case. Some of the year the earth is slightly ahead of schedule and the remainder of the year it is slightly behind—except at the moments of the Summer and Winter Solstices. The equation of time tells you how far ahead or behind schedule the earth is running.

The equation of time changes relatively slowly. If you keep track of it you can take a meridian passage sight and get a good estimate of your position in an emergency situation when an *Almanac* is not at hand.

APPARENT TIME

Times listed in the *Nautical* or *Air Almanacs* are Greenwich mean time (or "universal time"). If you convert watch time to GMT, you can work out all your navigation problems. Back in the days of time based on local sundials, all time was *apparent time* (time based on position of the sun). Even now, when the sun is directly on your meridian, it is *local apparent noon*. The meridian passage sight of the sun is sometimes called the LAN sight. At the zone meridian of any time zone, the difference in time between ZT 12-00-00 and the time of ALN is the equation of time.

Chapter 16

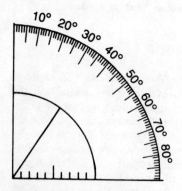

Ocean Charts

All the plotting we have done so far is to record positions on charts that cover a very small area: 1 full degree of latitude and longitude. You need something much larger, however, to plot your intended track from one port to another and to keep a record of your actual passage over the ocean.

Easiest to use is a Mercator projection chart that covers a wide expanse of the earth's surface. On it, a straight line drawn between any two points is a *rhumb line*. A rhumb line course, if followed exactly, will take you from your point of departure to your destination with no change in heading. A rhumb line course, unless it is exactly north-south, is not the shortest distance between two points on a sphere. If time or your fuel supply are crucial factors, and the passage is a long one, a great circle course is your better choice.

GREAT CIRCLE CHARTS

The shortest distance between any two points on the surface of a sphere is along a great circle (see Chapter 1), but you can't plot a great circle course directly on a Mercator projection chart. For this, use the great circle chart that covers the ocean area in which you will be voyaging. These are published by the Defense Mapping Agency Hydrographic/Topographic Center. They are available from major chart suppliers. Figure 16-1 shows a portion of such a chart. On it, a great circle course is plotted as a straight line between any two points.

On a great circle chart, however, you can't easily measure course angles. You must note the points at which your course crosses each line of latitude or longitude and transfer these points to the equivalent Mercator chart. Connect the points on the Mercator chart and you have your great circle course as a series of short, straight lines.

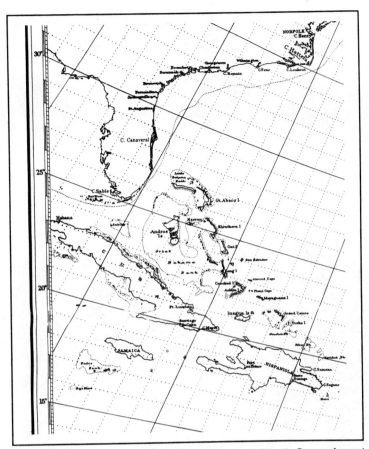

Fig. 16-1. Portion of a Great Circle Chart of the North Atlantic Ocean. A great circle course can be plotted as a straight line.

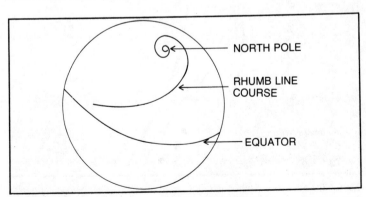

NORTH POLE

RHUMB LINE
COURSE

EQUATOR

Fig. 16-2. A rhumb line course spirals towards the poles.

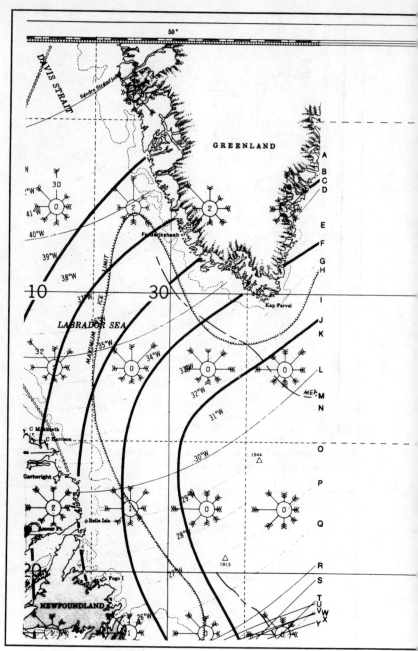

Fig. 16-3. Portion of a Pilot Chart of the North Atlantic Ocean. See text for detail on the weather information this provides.

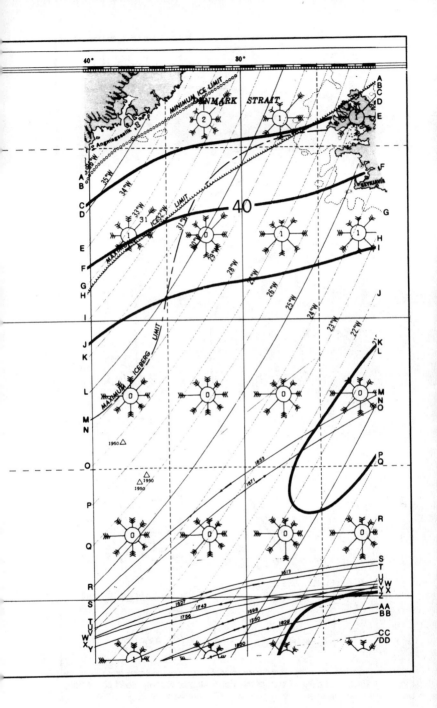

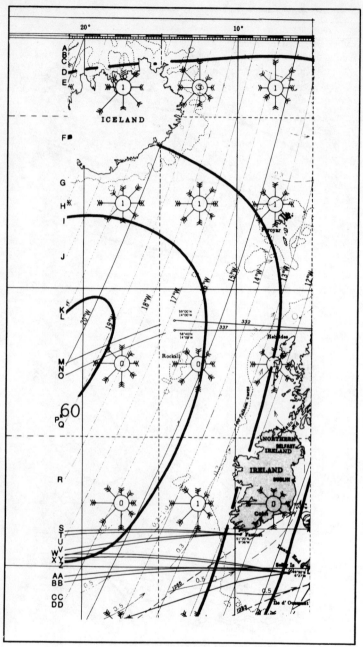

Fig. 16-3. Portion of a Pilot Chart of the North Atlantic Ocean. See text for detail on the weather information this provides. (Continued from page 193.)

Note that your initial course won't take you to your destination—unless you are on the exact north-south heading. Each section of your course as plotted on the Mercator chart is a rhumb line and you have to change heading periodically to maintain your great circle track. Note also that any rhumb line, unless it is exactly east-west—if extended far enough—will spiral about the earth until it reaches a pole. If it's a north-south line, of course, there will be no spiral (see Fig. 16-2).

PILOT CHARTS

Pilot Charts cover wide ocean expanses, such as the North Atlantic. They are published to indicate average meteorological conditions for each month of the year. A single sheet has the charts for a two-month sequence. Figure 16-3 shows a portion of one.

Information includes major tracks between ports for commercial shipping (black lines), ice limits (lines of red circles or curlicues), iceberg limits (thin red lines), percentage of frequency of wave heights above 12 feet (thick red lines), ocean current strength and direction (green arrows), and blue "wind roses" that indicate average duration and strength of wind from each direction—plus the percentage of calms. Very fine black lines show changes in magnetic variation.

While the Pilot Chart is not designed for use in plotting or recording courses, it is the most important form of ocean chart that you should carry.

Chapter 17

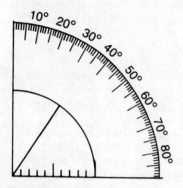

Distances and Courses

It's one thing to know where you are in mid-ocean, but another to know the distance and direction to a destination from your point of departure or from a dead reckoning position of fix at sea. Given a Mercator projection chart that covers a wide enough area, you can plot the course directly on the chart and measure both its distance and direction.

Sometimes, however, this is not possible. You may not have a chart or the destination is not on the same chart as your point of departure or present position. In this case, you have to calculate the needed information. It's easy to do using one of the sight reduction methods described in previous chapters. Latitude and longitude of your destination are the equivalent of the geographic position (declination and GHA) of a celestial body. The difference in longitude between departure and destination is equivalent to the LHA of an observed body. The calculated altitude that you find, subtracted (or added in some cases) from 90 degrees, gives you distance to your objective, and Zn is your initial heading for a great circle course.

PUB. 229

In using *Pub. 229*, you must establish an assumed position close to your departure point, just as for a sight reduction. The departure point, for the purpose of this discussion, may also represent a DR position or fix at sea. Your difference in longitude (DLo) must be a whole number of degrees, just as if it were LHA.

You can consider DLo as the smallest angle west or east of your position rather than convert a longitude to your east to an LHA always measured west from your position. In this respect, it is the equivalent of meridian angle t. When Z is determined from the tables, it is an angle in quadrant so you must put the designation of your departure latitude (N or S) ahead of it and the direction of DLo after it.

When latitude of the point of departure and destination are the same name, and DLo is less than 90 degrees, use the appropriate left-hand page in *Pub. 229* and find Hc, d, and Z. For example, your departure position is L 31°22.7'N, Lo 118°22.5W, and your destination is at L 44°18.6'N, Lo 138°7.42'W. Your AP is then 31° N, Lo 118°47.2W, and DLo is 20°. Destination latitude 44° is tabulated declination, and the latitude increment

Table 17-1. *Pub. 229* **Data Used in Plotting a Great Circle Course.**

Dec.	30° Hc	30° d	30° Z	31° Hc	31° d	31° Z
°	° '	'	°	° '	'	°
0	54 28.1	+ 51.4	143.9	53 39.4	+ 51.9	144.8
1	55 19.5	51.1	143.1	54 31.3	51.6	143.9
2	56 10.6	50.5	142.1	55 22.9	51.2	143.0
3	57 01.1	50.1	141.1	56 14.1	50.8	142.1
4	57 51.2	49.6	140.1	57 04.9	50.2	141.1
5	58 40.8	+ 49.0	139.0	57 55.1	+ 49.8	140.1
6	59 29.8	48.4	137.9	58 44.9	49.2	139.0
7	60 18.2	47.8	136.7	59 34.1	48.6	137.9
8	61 06.0	47.0	135.5	60 22.7	48.0	136.7
9	61 53.0	46.3	134.2	61 10.7	47.3	135.5
10	62 39.3	+ 45.4	132.8	61 58.0	+ 46.5	134.2
11	63 24.7	44.5	131.4	62 44.5	45.7	132.9
12	64 09.2	43.6	129.9	63 30.2	44.7	131.4
13	64 52.8	42.4	128.3	64 14.9	43.8	129.9
14	65 35.2	41.2	126.6	64 58.7	42.7	128.3
15	66 16.4	+ 40.0	124.8	65 41.4	+ 41.5	126.6
16	66 56.4	38.5	122.9	66 22.9	40.2	124.9
17	67 34.9	37.0	120.9	67 03.1	38.9	123.0
18	68 11.9	35.4	118.9	67 42.0	37.3	121.0
19	68 47.3	33.5	116.7	68 19.3	35.6	118.9
20	69 20.8	+ 31.6·	114.3	68 54.9	+ 33.9	116.7
21	69 52.4	29.4·	111.9	69 28.8	31.9·	114.4
22	70 21.8	27.2·	109.3	70 00.7	29.7·	111.9
23	70 49.0	24.7·	106.6	70 30.4	27.5·	109.4
24	71 13.7	22.1·	103.8	70 57.9	25.1·	106.7
25	71 35.8	+ 19.4·	100.9	71 23.0	+ 22.4·	103.8
26	71 55.2	16.4·	97.9	71 45.4	19.6·	100.9
27	72 11.6	13.4·	94.8	72 05.0	16.7·	97.9
28	72 25.0	10.2·	91.6	72 21.7	13.6·	94.7
29	72 35.2	6.9·	88.3	72 35.3	10.4·	91.5
30	72 42.1	+ 3.6·	85.0	72 45.7	+ 7.1·	88.2
31	72 45.7	+ 0.2·	81.6	72 52.8	3.7·	84.8
32	72 45.9	− 3.1·	78.2	72 56.5	+ 0.4·	81.4
33	72 42.8	6.5·	74.9	72 56.9	− 3.1·	78.0
34	72 36.3	9.8·	71.5	72 53.8	6.5·	74.6
35	72 26.5	− 12.9·	68.2	72 47.3	− 9.8·	71.2
36	72 13.6	16.1·	65.0	72 37.5	13.0·	67.9
37	71 57.5	18.9·	61.9	72 24.5	16.1·	64.7
38	71 38.6	21.8·	58.8	72 08.4	19.0·	61.5
39	71 16.8	24.4·	55.9	71 49.4	21.9·	58.4
40	70 52.4	− 26.8·	53.1	71 27.5	− 24.6·	55.5
41	70 25.6	29.2·	50.4	71 02.9	27.0·	52.6
42	69 56.4	31.3·	47.8	70 35.9	29.4·	49.9
43	69 25.1	33.3	45.4	70 06.5	31.5·	47.3
44	68 51.8	35.1	43.0	69 35.0	33.4	44.9
45	68 16.7	− 36.8	40.8	69 01.6	− 35.4	42.5

18.6' is declination increment. In *Pub. 229* (Table 17-1) you find:

Tab Hc	d	Tab Z
69°35.0'	−33.4'	44.9°

Using the Interpolation Table (Table 17-2) you get:

$$
\begin{array}{ll}
\text{d} & - \ 9.3' \\
(+) & - \ 1.3' \\
\text{d corr.} & -10.6'
\end{array}
$$

Since this is negative, subtract it from Tab Hc:

$$
\begin{array}{ll}
\text{Tab Hc} & 69°35.0' \\
\text{d corr.} & - \ \ 10.6' \\
\text{Hc} & 69°24.4'
\end{array}
$$

Now subtract Hc from 90 degrees to get distance from your AP to your destination:

$$
\begin{array}{ll}
 & 90° \\
\text{Hc} - & 63°24.4' \\
 & 20°35.6'
\end{array}
$$

Multiply full degrees by 60 and add the minutes of arc. The result is distance in nautical miles:

$$
\begin{array}{ll}
60 \times 20 = & 1200 \\
+ & \ \ \ 35.6 \\
 & 1235.6 \text{ miles}
\end{array}
$$

It is not necessary to interpolate for Z to get your course direction. Therefore, in this case Tab Z is 44.9°. Round it to a full degree and you have:

Course N 45° W

In other words, the initial segment of a great circle course from your assumed position is 50° west of true north, or 360° − 45° = 315° True. You can plot a segment of this course line as shown in Fig. 17-1. Drop a perpendicular from this line to your departure point. Then measure the distance from this perpendicular to the AP. Add or subtract this as necessary to get actual distance from your point of departure. In this case, the perpendicular meets the course line 1.8 miles ahead of the AP. So the total distance from the point of departure is 1,233.8 nautical miles along a great circle.

LATITUDES CONTRARY NAMES

Working a problem where DLo is within 90 degrees, but latitude of departure and latitude of destination are contrary names is done in exactly

Table 17-2. Interpolation
Data Used When the Latitude
Increment of the Destination Is 18.6'.

Altitude Difference (d)

Dec. Inc.	Tens 10'	20'	30'	40'	50'	Decimals	Units 0'	1'	2'	3'	4'	5'	6'	7'	8'	9'
16.0	2.6	5.3	8.0	10.6	13.3	.0	0.0	0.3	0.5	0.8	1.1	1.4	1.6	1.9	2.2	2.5
16.1	2.7	5.3	8.0	10.7	13.4	.1	0.0	0.3	0.6	0.9	1.1	1.4	1.7	2.0	2.2	2.5
16.2	2.7	5.4	8.1	10.8	13.5	.2	0.1	0.3	0.6	0.9	1.2	1.4	1.7	2.0	2.3	2.5
16.3	2.7	5.4	8.1	10.9	13.6	.3	0.1	0.4	0.6	0.9	1.2	1.5	1.7	2.0	2.3	2.6
16.4	2.7	5.5	8.2	10.9	13.7	.4	0.1	0.4	0.7	0.9	1.2	1.5	1.8	2.0	2.3	2.6
16.5	2.8	5.5	8.3	11.0	13.8	.5	0.1	0.4	0.7	1.0	1.2	1.5	1.8	2.1	2.3	2.6
16.6	2.8	5.5	8.3	11.1	13.8	.6	0.2	0.4	0.7	1.0	1.3	1.5	1.8	2.1	2.4	2.6
16.7	2.8	5.6	8.4	11.2	13.9	.7	0.2	0.5	0.7	1.0	1.3	1.6	1.8	2.1	2.4	2.7
16.8	2.8	5.6	8.4	11.2	14.0	.8	0.2	0.5	0.8	1.0	1.3	1.6	1.9	2.1	2.4	2.7
16.9	2.9	5.7	8.5	11.3	14.1	.9	0.2	0.5	0.8	1.1	1.3	1.6	1.9	2.2	2.4	2.7
17.0	2.8	5.6	8.5	11.3	14.1	.0	0.0	0.3	0.6	0.9	1.2	1.5	1.7	2.0	2.3	2.6
17.1	2.8	5.7	8.5	11.4	14.2	.1	0.0	0.3	0.6	0.9	1.2	1.5	1.8	2.1	2.4	2.7
17.2	2.8	5.7	8.6	11.4	14.3	.2	0.1	0.3	0.6	0.9	1.2	1.5	1.8	2.1	2.4	2.7
17.3	2.9	5.8	8.6	11.5	14.4	.3	0.1	0.4	0.7	1.0	1.3	1.5	1.8	2.1	2.4	2.7
17.4	2.9	5.8	8.7	11.6	14.5	.4	0.1	0.4	0.7	1.0	1.3	1.6	1.9	2.2	2.4	2.7
17.5	2.9	5.8	8.8	11.7	14.6	.5	0.1	0.4	0.7	1.0	1.3	1.6	1.9	2.2	2.5	2.8
17.6	2.9	5.9	8.8	11.8	14.7	.6	0.2	0.5	0.8	1.0	1.3	1.6	1.9	2.2	2.5	2.8
17.7	3.0	5.9	8.9	11.8	14.8	.7	0.2	0.5	0.8	1.1	1.4	1.7	2.0	2.2	2.5	2.8
17.8	3.0	6.0	8.9	11.9	14.9	.8	0.2	0.5	0.8	1.1	1.4	1.7	2.0	2.3	2.6	2.9
17.9	3.0	6.0	9.0	12.0	15.0	.9	0.3	0.6	0.8	1.1	1.4	1.7	2.0	2.3	2.6	2.9
18.0	3.0	6.0	9.0	12.0	15.0	.0	0.0	0.3	0.6	0.9	1.2	1.5	1.8	2.2	2.5	2.8
18.1	3.0	6.0	9.0	12.0	15.1	.1	0.0	0.3	0.6	1.0	1.3	1.6	1.9	2.2	2.5	2.8
18.2	3.0	6.0	9.1	12.1	15.1	.2	0.1	0.4	0.7	1.0	1.3	1.6	1.9	2.2	2.5	2.8
18.3	3.0	6.1	9.1	12.2	15.2	.3	0.1	0.4	0.7	1.0	1.3	1.6	1.9	2.2	2.6	2.9
18.4	3.1	6.1	9.2	12.3	15.3	.4	0.1	0.4	0.7	1.0	1.4	1.7	2.0	2.3	2.6	2.9
18.5	3.1	6.2	9.3	12.3	15.4	.5	0.2	0.5	0.8	1.1	1.4	1.7	2.0	2.3	2.6	2.9
18.6	3.1	6.2	9.3	12.4	15.5	.6	0.2	0.5	0.8	1.1	1.4	1.7	2.0	2.3	2.7	3.0
18.7	3.1	6.3	9.4	12.5	15.6	.7	0.2	0.5	0.8	1.1	1.4	1.8	2.1	2.4	2.7	3.0
18.8	3.2	6.3	9.4	12.6	15.7	.8	0.2	0.6	0.9	1.2	1.5	1.8	2.1	2.4	2.7	3.0
18.9	3.2	6.3	9.5	12.6	15.8	.9	0.3	0.6	0.9	1.2	1.5	1.8	2.1	2.4	2.7	3.1

Double Second Diff. and Corr.

Diff.	Corr.
1.0	0.1
3.0	0.2
4.9	0.3
6.9	0.4
8.9	0.5
10.8	0.6
12.8	0.7
14.8	0.8
16.7	0.9
18.7	1.0
20.7	1.1
22.7	1.2
24.6	1.3
26.6	1.4
28.6	1.5
30.5	1.6
32.5	1.6
34.5	1.7

Diff.	Corr.
0.9	0.1
2.8	0.2
4.6	0.3
6.5	0.4
8.3	0.5
10.2	0.6
12.0	0.7
13.9	0.8
15.7	0.9

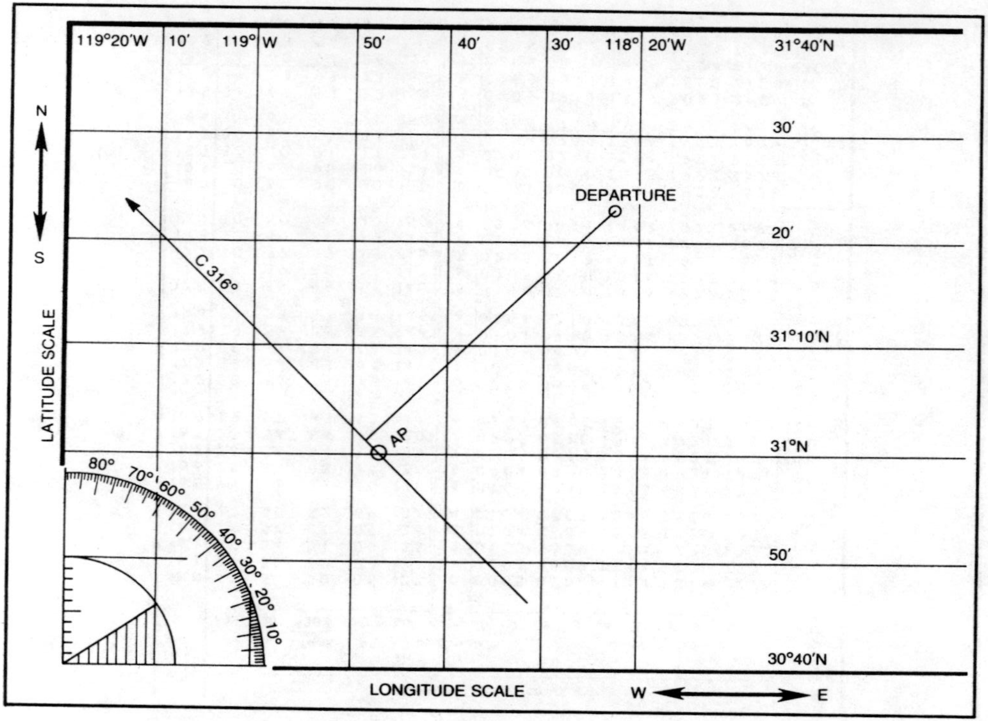

Fig. 17-1. Start of a great circle course is plotted from an assumed position.

the same manner as long as all the information falls above a contrary-same (C-S) line on the right hand page.

If destination latitude is such that the data are below a C-S line, the tabulated numbers are considered to be negative numbers. Calculate Hc in the normal manner, but add the result to 90 degrees to get distance. You're actually subtracting a negative number, but it amounts to the same thing. In a case of this type, subtract the Tab Z from 180 degrees to get Z.

Here's an example. Your point of departure is L 42°07.6′N, Lo 68°19.7′W; destination is L 39°18.2′S, Lo 9°24.6′E. Set AP Lo so that DLo will be in whole degrees:

$$
\begin{array}{ll}
\text{AP Lo} & 68°35.4′\,\text{W} \\
\text{Dest. Lo} & +\ 9°24.6′\,\text{E} \\
\hline
\text{DLo} & 78°\quad\ \ \text{E}
\end{array}
$$

Figure 17-2 shows this relationship, based on longitude of the point of departure. Use DLo 78° as LHA in entering the tables; AP latitude will be 42°N; 39°S is tabulated declination, and 18.2° is declination increment. In *Pub. 229* (Table 17-3) you have:

$$
\begin{array}{ccc}
\textbf{Tab Hc} & \textbf{d} & \textbf{Tab Z} \\
17°31.1′ & +38.8 & 52.9°
\end{array}
$$

The d correction found from the Interpolation Table is +14.8′ so Hc is 17°45.9′. You can consider this a negative number and subtract it from 90 degrees to find distance or simply add it to 90 degrees as a positive number:

$$
\begin{array}{rl}
 & 90° \\
\text{Hc} & +17°45.9′ \\
\hline
 & 107°45.9′ \\
107 \times 60 & = 6420 \\
 & +\quad 45.9 \\
\hline
\text{Distance} & = 6465.9 \quad \text{miles from the AP.}
\end{array}
$$

When this is plotted, you can determine the distance to add or subtract to get actual distance from the point of departure. For course direction, subtract Tab Z from 180 degrees, so:

$$
\begin{array}{rl}
 & 180° \\
\text{Tab Z} & -\ 52.9° \\
\hline
\text{C} & \text{N}127.1°\text{E} = 127°\ \text{True}
\end{array}
$$

Again, this can be plotted as a course segment so that actual distance from the departure point can be determined.

DLo GREATER THAN 90 DEGREES

In many cases your destination may lie more than 90 degrees to the east or west of your point of departure. When both latitudes are the same

Table 17-3. *Pub. 229* Data Used When LHA is 78° and the Two Latitudes Are Contrary Names.

42° Hc	d	Z	43° Hc	d	Z	44° Hc	d	Z	45° Hc	d	Z	Dec
8 53.3	-40.7	98.1	8 44.8	-41.5	98.2	8 36.1	-42.2	98.4	8 27.2	-42.9	98.5	0
8 12.6	40.7	98.8	8 03.3	41.5	99.0	7 53.9	42.3	99.1	7 44.3	43.0	99.3	1
7 31.9	40.9	99.6	7 21.8	41.6	99.7	7 11.6	42.3	99.8	7 01.3	43.0	100.0	2
6 51.0	40.9	100.3	6 40.2	41.6	100.4	6 29.3	42.4	100.5	6 18.3	43.2	100.7	3
6 10.1	40.9	101.1	5 58.6	41.7	101.2	5 46.9	42.4	101.3	5 35.1	43.1	101.4	4
5 29.2	-41.1	101.8	5 16.9	-41.8	101.9	5 04.5	-42.5	102.0	4 52.0	-43.2	102.1	5
4 48.1	41.0	102.5	4 35.1	41.8	102.6	4 22.0	42.6	102.7	4 08.8	43.3	102.8	6
4 07.1	41.1	103.3	3 53.3	41.8	103.3	3 39.4	42.5	103.4	3 25.5	43.3	103.4	7
3 26.0	41.2	104.0	3 11.4	41.8	104.0	2 56.9	42.6	104.1	2 42.2	43.3	104.1	8
2 44.8	41.1	104.7	2 29.6	41.8	104.8	2 14.3	42.6	104.8	1 58.9	43.3	104.8	9
2 03.7	-41.2	105.4	1 47.7	-41.9	105.5	1 31.7	-42.7	105.5	1 15.6	-43.3	105.5	10
1 22.5	41.2	106.2	1 05.8	41.9	106.2	0 49.0	42.6	106.2	0 32.3	-43.3	106.2	11
0 41.3	41.2	106.9	0 23.9	-42.0	106.9	0 06.4	-42.6	106.9	0 11.0	+43.4	73.1	12
0 00.1	-41.2	107.6	0 18.1	+41.9	72.4	0 36.2	+42.7	72.4	0 54.4	43.3	72.4	13
0 41.1	+41.2	71.7	1 00.0	41.9	71.7	1 18.9	42.6	71.7	1 37.7	43.3	71.7	14
1 22.3	+41.2	70.9	1 41.9	+41.9	70.9	2 01.5	+42.6	71.0	2 21.0	+43.3	71.0	15
2 03.5	41.2	70.2	2 23.8	41.9	70.2	2 44.1	42.5	70.3	3 04.3	43.3	70.3	16
2 44.7	41.1	69.5	3 05.7	41.8	69.5	3 26.6	42.6	69.6	3 47.6	43.2	69.6	17
3 25.8	41.1	68.7	3 47.5	41.8	68.8	4 09.2	42.5	68.9	4 30.8	43.2	68.9	18
4 06.9	41.1	68.0	4 29.3	41.8	68.1	4 51.7	42.5	68.2	5 14.0	43.1	68.2	19
4 48.0	+41.0	67.3	5 11.1	+41.7	67.4	5 34.2	+42.4	67.4	5 57.1	+43.1	67.5	20
5 29.0	40.9	66.5	5 52.8	41.7	66.6	6 16.6	42.3	66.7	6 40.2	43.0	66.8	21
6 09.9	40.9	65.8	6 34.5	41.6	65.9	6 58.9	42.3	66.0	7 23.2	43.0	66.1	22
6 50.8	40.9	65.1	7 16.1	41.5	65.2	7 41.2	42.2	65.3	8 06.2	42.9	65.4	23
7 31.7	40.7	64.3	7 57.6	41.5	64.5	8 23.4	42.1	64.6	8 49.1	42.8	64.7	24
8 12.4	+40.7	63.6	8 39.1	+41.3	63.7	9 05.5	+42.1	63.9	9 31.9	+42.7	64.0	25
8 53.1	40.6	62.9	9 20.4	41.3	63.0	9 47.6	42.0	63.1	10 14.6	42.7	63.3	26
9 33.7	40.5	62.1	10 01.7	41.2	62.3	10 29.6	41.8	62.4	10 57.3	42.5	62.6	27
10 14.2	40.4	61.4	10 42.9	41.1	61.5	11 11.4	41.8	61.7	11 39.8	42.4	61.9	28
10 54.6	40.3	60.6	11 24.0	40.9	60.8	11 53.2	41.6	61.0	12 22.2	42.3	61.1	29
11 34.9	+40.1	59.8	12 04.9	+40.9	60.0	12 34.8	+41.5	60.2	13 04.5	+42.2	60.4	30
12 15.0	40.1	59.1	12 45.8	40.9	59.3	13 16.3	41.4	59.5	13 46.7	42.1	59.7	31
12 55.1	39.9	58.3	13 26.5	40.6	58.5	13 57.7	41.3	58.7	14 28.8	41.9	59.0	32
13 35.0	39.7	57.6	14 07.1	40.4	57.8	14 39.0	41.1	58.0	15 10.7	41.8	58.2	33
14 14.7	39.7	56.8	14 47.5	40.3	57.0	15 20.1	40.9	57.2	15 52.5	41.6	57.5	34
14 54.4	+39.4	56.0	15 27.8	+40.1	56.2	16 01.0	+40.8	56.5	16 34.1	+41.4	56.7	35
15 33.8	39.3	55.2	16 07.9	40.0	55.5	16 41.8	40.7	55.7	17 15.5	41.3	56.0	36
16 13.1	39.1	54.4	16 47.9	39.8	54.7	17 22.5	40.4	54.9	17 56.8	41.1	55.2	37
16 52.2	38.9	53.7	17 27.7	39.6	53.9	18 02.9	40.3	54.2	18 37.9	40.9	54.4	38
17 31.1	38.8	52.9	18 07.3	39.4	53.1	18 43.2	40.0	53.4	19 18.8	40.8	53.7	39
18 09.9	+38.5	52.1	18 46.7	+39.2	52.3	19 23.2	+39.9	52.6	19 59.6	+40.5	52.9	40
18 48.4	38.3	51.3	19 25.9	39.0	51.5	20 03.1	39.6	51.8	20 40.1	40.3	52.1	41
19 26.7	38.2	50.4	20 04.9	38.7	50.7	20 42.7	39.4	51.0	21 20.4	40.0	51.2	42

name, use the bottom of the right-hand page for the appropriated DLo and proceed in the normal manner if all the information is below the C-S line in the Latitude column.

If the destination latitude tabulated declination brings you above a C-S line, consider the numbers in the table to be negative. This means you determine Hc in the usual way, and add it to 90 degrees to get course distance. Subtract Tab Z from 180 degrees to get your initial great circle course heading.

The final situation is where departure and destination latitudes are contrary names, and DLo is greater than 90 degrees. The problem is simple

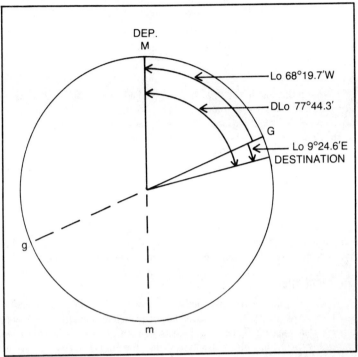

Fig. 17-2. Time diagram shows relation of departure and destination longitudes to the Greenwich meridian.

to work. Look up the appropriate LHA figure on the bottom of a right hand page, but use the left-hand page to look up the information. The tabulated data are considered to be negative numbers; Hc is added to 90 degrees for distance, and Tab Z is subtracted from 180 degrees to get direction.

Suppose your departure point AP latitude is 34°N, DLo is 147°W, and destination latitude is 12°21.4′S. On the left hand page (33°, 327° LHA) equivalent of LHA 147°, you have:

Tab Hc	d	Tab Z
52°47.0′	+39.6′	118.3°

The d correction from the Interpolation Table is +14.1′, so Hc is 53°01.1′. Add this to 90 degrees for distance:

$$
\begin{array}{r}
90° \\
\text{Hc} \quad + \underline{53°01.1′} \\
143°01.1′
\end{array}
$$

$$143 \times 60 = 8580$$
$$+ \quad \underline{1.1}$$
Distance $= \overline{8581.1}$ miles from your AP

Subtract Tab Z from 180°:

$$
\begin{array}{rr}
& 180° \\
\text{Tab Z} & -117.8° \\
\text{C} & \overline{\text{N } 62.2°} \text{ W}
\end{array}
$$

Your initial course is 62° west of the true north, or 360° − 62° = 298° True. You know your destination is well to the south of your point of departure, so how can you have a course that is taking you somewhat to the north? Remember that this is a segment of a great circle and that you will be changing course every few degrees of longitude in order to maintain a great circle route. It is the shortest distance between departure and destination points (see Fig. 17-3).

PUB. 249

You can use *Pub. 249* in the same manner as *Pub. 229*, but you are limited to destination latitudes less than 30 degrees north or south of the equator, and less than 90 degrees east or west of your point of departure.

AGETON METHOD

The procedure with this method is exactly the same as for sight reduction. Just substitute destination latitude and longitude for declination and meridian angle t (LHA west or east of your position). You must be extremely careful in looking up and recording the numbers, as well as your arithmetic. You can plot a course segment directly from the departure point

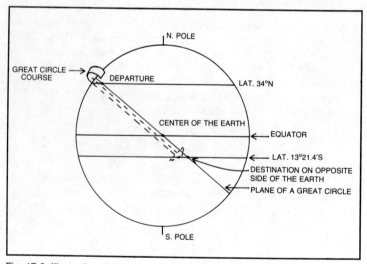

Fig. 17-3. Illustration shows why initial course angle for a great circle route may lead north, even though latitude of the destination is south of the departure latitude.

and the distance that you determine is the actual great circle distance from that point.

Some special points must be noted in using the tables. When departure and destination latitude are the same name:

- Take K from the top of the table when t is less than 90°.
- Take K from the bottom of the table when t is greater than 90°.
- Take Z (Course) from the top of the table when K is greater than departure latitude.
- Take Course from the bottom of the table when K is less than departure latitude.
- Take Distance from the top of the table except when KL and t are both greater than 90°.

When departure latitude and destination latitude are contrary names:

- Take K from the top of the table when t is less than 90°.
- Take K from the bottom of the table when t is greater than 90°.
- Take Course from the bottom of the table except when KL is greater than 180°.
- Take Distance from the bottom of the table except when KL and t are both less than 90°.
- When KL is greater than 180°, subtract 180° before entering the table.

POCKET CALCULATOR

The handy pocket calculator provides the quickest and easiest method to get distance and heading information—as long as its batteries are up to full strength. It must have the basic trigonometric functions as a minimum. You can get programmable models designed specifically for navigation.

With the basic model, use the same formulae as for sight reduction (90° − Hc gives course distance). Azimuth becomes course-in-quadrant, so prefix it with the latitude of your departure point and add the east-west direction after it to determine the quadrant, as with the *Pub. 229* solution. If latitude of departure and latitude of destination are contrary name, make the south latitude a negative number.

For example, you are at latitude 14°18.7′S, Lo 8°53.9′W, and your destination is L 35°39.5′N, Lo 72°44.6′W. Determine LHA (t), and convert the latitude and t figures to degrees and tenths of degrees:

$$
\begin{array}{lr}
\text{Destination Lo} & 72°44.6' \\
\text{Departure Lo} & 8°53.9' \\
\hline
\text{LHA} & 63°50.7'
\end{array}
$$

Departure L 14°18.7′S = L −14.311°
Destination L 35°39.5′N = d 35.658°
LHA 63°50.7′ = t 63.845°

The distance formula is then:

$$\text{Distance} = 90° - \sin^{-1}(\sin L \sin d) + (\cos L \cos d \cos t)$$
$$= \sin^{-1}(-14.311° \sin \times 35.658° \sin) + (-14.311° \cos \times 35.658° \cos \times 63.845° \cos)$$

With a calculator such as the Texas Instrument TI-30, the sequence is:

Step	Press	Display
1.	14.311 +/− sin	−.24718505
2.	× 35.658 sin	.58294577
3.	+ 14.311 +/− cos	.96896829
4.	× 35.658 cos	.81251107
5.	× 63.845 cos	.44080101
6.	= inv sin	11.709289

This is Hc in the formula; subtract it from 90 degrees to get your distance. If your machine has a memory capability, put Hc into storage for use in the second formula. Here 90° − Hc is 78.290711°. Multiply the first three decimal places by 60 to get minutes and tenths of minutes, so Hc = 78°17.4′. Multiply the whole degrees by 60 and add the minutes of arc to get great circle distance:

$$78 \times 60 = 4680$$
$$+ \; 17.4$$
$$\text{Distance} = 4697.4 \text{ nautical miles.}$$

Since you are working from your departure point, and not an assumed position, no adjustment is needed to this distance figure.

For the initial heading, the formula is:

$$C = \sin^{-1} \frac{\cos d \sin t}{\cos H} =$$
$$C = \sin^{-1} \frac{35.658° \cos \times 63.845° \sin}{11.709° \cos}$$

Step	Press	Display
1.	35.658 cos	.81251107
2.	× 63.845 sin	.89760485
3.	+ 11.709 cos	.97919095
4.	= inv sin	48.143016

Since this is course in quadrant, it is S 48.1°W, or 48.1° to the west of due south. This means it is added to 180 degrees to give you the true heading—228°— for the first segment of your great circle course. Note again that a great circle course seems to be heading in the wrong direction—slightly south of west—when you know your destination is to the north. Keep in mind that you will be adjusting the course as you go along. The great circle is really the shortest route to follow.

Chapter 18

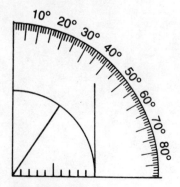

Sailings

A *sailing* is a mathematical method of determining course and distance between two points or the coordinates of a destination when departure point, heading, and distance traveled through the water are known. If you know the latitude and longitude of both the departure and destination points, it's best to find course and distance by one of the methods described in the previous chapter. It is in determining where you will be, after a given time, that a sailing is most useful to the average navigator.

There are three types of sailings that are generally used: midlatitude, great circle, and Mercator. While the first is considered the least accurate of the three, it is the simplest to use and perfectly adequate for all practical purposes.

For midlatitude sailing problems, you need trigonometric and logarithmic tables such as those in "Bowditch," (Defense Mapping Agency Hydrographic/Topographic Center *Pub. 9*), slide rule with these functions, or your handy pocket calculator.

In sailings, the word "departure" is used to denote the distance in nautical miles along the parallel of latitude on which the destination is located. It is indicated in formulae and on diagrams by the letter "p." It is measured between the longitude of the point of departure and longitude of the destination. I use just the letter p to avoid confusion with the departure point. The angular distance between the two points is difference in longitude (DLo) measured from the pole. Difference in latitude between departure point and destination is "l." The angle a course line makes with true north, from the departure point, is "C" (see Fig. 18-1).

The right triangle formed by p, l, and the course line does not take into account the curvature of the earth. One solution to sailing problems is to

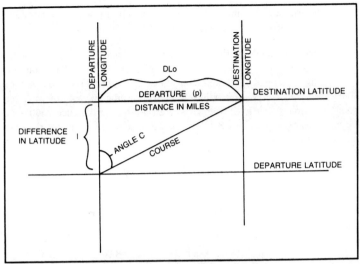

Fig. 18-1. Diagram shows elements of a sailing problem.

base answers on the middle latitude between departure and destination points; hence the term "midlatitude" sailings.

In midlatitude sailings, the course line is a rhumb line course and its distance represents your DR track for the time you plan to remain on that course. Thus your destination will be your DR position at the end of that time. Here are the steps to follow to find the coordinates of that position.

- Determine difference in latitude with the formula:

$$1 = \text{Dist.} \times \cos C$$

Dist. is the length of the course line, in nautical miles.
- Determine distance p along the latitude of the destination with the formula:

$$p = \text{Dist.} \times \sin C$$

- Determine latitude of the destination by adding 1 to the latitude of the departure point if it is to the north (in northern latitudes), or subtracting it if destination is to the south of your departure point. In southern latitudes, of course, the procedure is reversed.
- Determine midlatitude (Lm) by adding half of 1 to departure latitude, or subtracting it, as above.
- Determine DLo with the formula:

$$\text{DLo} = p \div \cos \text{Lm}$$

• Determine longitude of the destination point by adding or subtracting DLo to (or from) the departure longitude, as necessary.

Here's an example, with the solution worked out by pocket calculator. A morning fix places you at L 18°44.6'N, Lo 162°18.7'W. You are proceeding at a speed of 7.8 knots on a heading of 285° True. You want to know what your DR position will be the next morning, in case you can't get noon or evening sights. On the basis of 24 hours, the distance you cover through the water will be 7.8 × 24 = 187.2 nautical miles, so for the first formula you have:

$$l = 187.2 \times \cos 285° = 48.4'N$$

Because your course is north of west, the suffix "N" shows that latitude of destination is north of the latitude of your point of departure. Distance p is:

$$p = 187.2 \times \sin 285° = 180.8'W$$

Note that the answer is in minutes of arc, but these are equal in distance to minutes of arc of latitude, and hence equal to nautical miles.

Latitude of the destination is:

$$\begin{array}{ll} L_1 & 18°44.6'N \\ l \ + & 48.4'N \\ \hline L_2 & 19° \ 3.0'N \end{array}$$

Midlatitude is:

$$\begin{array}{ll} L_1 & 18°44.6'N \\ \tfrac{1}{2}l \ + & 24.2'N \\ \hline Lm & 19°08.8'N \end{array}$$

DLo, then, is:

$$187.2 \div \cos 19.146° = 198.2°W$$

This answer represents minutes of arc of longitude. Divide by 60 to get the whole degrees and you have 2°18.2'. Because you are west bound in west longitude, add the answer to your departure Lo:

$$\begin{array}{ll} Lo_1 & 162°18.7'W \\ DLo & 2°18.2'W \\ \hline Lo_2 & 164°36.9'W \end{array}$$

Your DR position in 24 hours will be L 19°33.0'N, Lo 164°36.9'W.

In using the logarithmic tables, which run from 0° to 180° for angles, you have to convert an angle greater than 180 degrees to an equivalent which is 360 degrees minus the given angle. Thus 285° = 75° for the purposes of the tables. In the above example, then:

$$
\begin{array}{rlr}
\text{Dist. } 187.2 & = & \log 2.272305 \\
\text{C } 285° \, (75°) & = & \log \cos -.587003 \\
\mathrm{l} & = & \log \overline{1.685302} = 48.4'\text{N} \\
\text{Dist. } 187.2 & = & \log 2.272305 \\
\text{C } 285° \, (75°) & = & \log \sin -.015056 \\
\mathrm{p} & = & \log \overline{2.257249} = 180.8'\text{W}
\end{array}
$$

In solving for l and p, the two logarithmic numbers are added. In each case, the second number is negative and adding a negative number is the same as subtracting a positive number.

Latitude of the destination and midlatitude are found as in working the calculator solution. With midlatitude you can solve for DLo:

$$
\begin{array}{rlr}
\text{Dist. } 187.2 & = & \log 2.272305 \\
\text{Lm } \quad 19.146° & = & \log \cos \underline{-.024721} \\
\text{DLo} & = & \log \overline{2.297018} = 198.2\text{W}
\end{array}
$$

Here you are subtracting a negative number, which is the same as adding a positive one. Adding and subtracting logarithms is easier than multiplying or dividing by the actual sines and cosines of the numbers involved.

Chapter 19

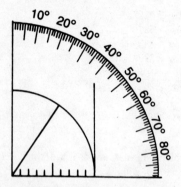

Life Raft Navigation

A question often asked in navigation classes is, "What do you do if you have to abandon your boat so quickly you can't get your sextant, *Almanac*, sight reduction tables, and other navigation gear into your life raft?"

This is a situation that should not arise. Have an inexpensive plastic sextant, a current *Almanac*, and at least the Ageton tables in a waterproof container, along with pencils, protractor, ruler, a Mercator ocean chart, and a Pilot Chart for the ocean in which you are operating. Attach this container to your life raft so that when the raft is launched, the container goes with it and can be hauled aboard.

You can get by without a sextant if you have a watch with a known error and daily change rate, a circular protractor, some sort of straightedge, a piece of string, and a weight. All but the watch are used to make up a substitute for a sextant. Tie the weight to a short length of string, run the other end of the string through the hole in the middle of the protractor, and knot it so it can't slip back through. Place a ruler or other straightedge on the side of the protractor opposite the string, with the upper edge of the ruler aligned with 90° to 270° axis (see Fig. 19-1). When the ruler is aimed at the sun, the string crosses the protractor scale at the observed altitude (you can't correct for eye height or refraction with this method, beyond a rough guess).

You can't sight along the edge of the ruler toward a bright sun without adequate protection for your eyes. You might try the following method. With the plane of the protractor in line with the sun, look at the side of the protractor. If the sun is to your south, you will be facing either east or west. Observe the upper edge of the ruler. If it is in shadow, it is aimed too high. If it is not in shadow it is aimed too low. At the point of transition from shadow to no shadow, take your reading (see Fig. 19-2).

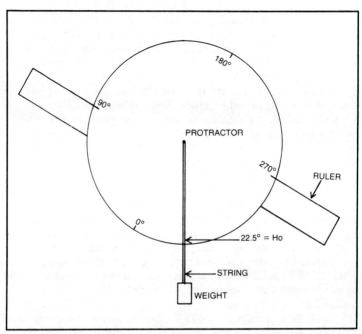

Fig. 19-1. Protractor, ruler, and weight used as a makeshift sextant.

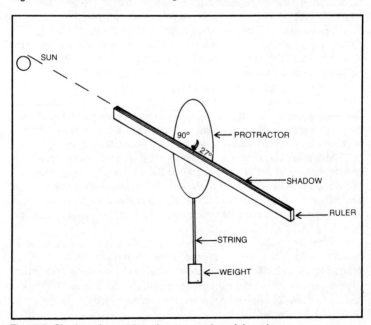

Fig.19-2. Shadow observed on the upper edge of the ruler.

213

If you have an *Almanac* and sight reduction tables, you can time readings of morning and afternoon sights to get a running fix. Since your only movement is the result of current action, this can be reasonably accurate—within 30 miles or so of your actual location, depending on the accuracy of your readings.

Without an *Almanac*, try for a meridian passage sight so you can establish both latitude and longitude. Based on your last known position, you should be able to estimate approximate time of local apparent noon. If you can time a sunset, and the sunrise the next morning, you can also establish the approximate time of LAN. The midpoint between the two times would be local apparent midnight and exactly 12 hours later it would be local apparent noon.

As the time of meridian passage approaches, start taking readings, even as much as 20 minutes before the estimated time of LAN. Take as many as you can until it is evident that altitude of the sun is decreasing. With even a makeshift graph, you can get a good idea of the actual time of meridian passage and the altitude of the sun at that time.

This is where the equation of time is important. Presumably you have kept track of it from day to day as part of your navigational routine, so you know fairly well what it was before you boarded the raft, and you know its approximate rate of change.

Apply the equation of time to 1200 hours to determine time of meridian passage at the zone meridian. The difference between this time and the zone time of your LAN sight can be converted to arc to give you your longitude east or west of the zone meridian (see Chapter 15).

To find your latitude, you need to know the declination of the sun. By subtracting your observed altitude from 90°, you have zenith distance to the sun's GP—directly south (or north) of you. By adding or subtracting declination as necessary, you get your latitude (see Chapter 5).

Of course, your day-to-day sun sights should have given you a good idea of the body's declination at the time of the emergency. If you use the last known declination figure, it will provide an adequate result. The day-to-day noon declination change is less than a full degree. There is another way you can establish approximate declination of the sun.

With your protractor, draw a semicircle (Fig. 19-3) above a base line, with 0° at one end of the line and 180° at the other end of the line, and the 90° mark above the mid-point of the base line. Drop a perpendicular from the 90° mark to the base and divide this perpendicular into eight equal segments. Each segment then represents a three-degree change in declination.

Since the number of degrees in a circle (360°) is almost equal to the number of days in a year, 1 degree of arc on the circumference of the protractor can be considered to equal one day. At the vernal equinox (March 21) the sun is at 0°; at the summer solstice (June 21) it is at 23.5°N; it is again at 0° on 22 September (autumnal equinox) and at 23.5°S on 22 December (winter solstice).

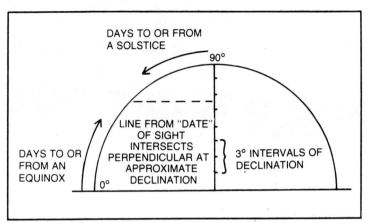

Fig.19-3. Scale used to determine approximate declination of the sun.

Count the days to or from the equinox or solstice closest to your actual date. For days to or from an equinox, measure up the protractor scale toward 90°. For days to or from a solstice, measure down from the 90° mark toward the 0° mark. From the point thus located, draw a line to the center perpendicular, parallel to the base line. The point where this line crosses the perpendicular indicates the sun's approximate declination.

Using this method, along with the protractor "sextant," should give you an estimated position within about 60 miles of your actual location. If seas are rough and you're bobbing about, the accuracy of your sight readings will be considerable downgraded. Still it will give you a good idea of the approximate area of the ocean in which you are located.

Again, it should *never* be necessary to have to rely on such methods. When you board your raft, it should be with the assurance that you have all the navigation gear you need—plus an Electronic Position Indicating Radio Beacon (EPIRB). While it's good to be able to learn your position, it's nice to know there's a good chance that others will be able to locate it as well.

Appendix

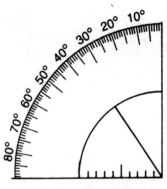

Appendix

The following equations are answers to problems stated in the text.

1.   ZT   0558 18 Jun
              ZD   + 6
              GMT   1158 18 Jun

2.   ZT   1702 12 Apr
              ZD   − 3
              GMT   1402 12 Apr

3.   ZT   1159 22 Sept
              ZD   − 9
              GMT   0259 22 Sept.

4.   ZT   1022 14 Nov.
              ZD   + 5
              GMT   1522 14 Nov.

5.   ZT   0604 8 Aug
                   24
              ZT   3004 7 Aug
              ZD   − 10
              GMT   2004 7 Aug

Note that it was necessary to subtract a day from the date, and add the 24 hours to the ZT.

6.   GMT   1924 6 Nov
       (Rev) ZD   − 8
           ZT   2624 6 Nov
                   − 24

		ZD	0224	7 Nov
		GMT	022	18 Oct
7.	(Rev)	ZD	+ 11	
		ZT	1122	18 Oct
8.		GMT	0022	18 Oct
		GMT	2422	17 Oct
	(Rev)	ZD	− 11	
		ZT	1322	17 Oct
9.		GMT	1313	18 Aug
	(Rev)	ZD	+ 5	
		ZT	0813	18 Aug
10.		GMT	0924	1 Jan
	(Rev)	ZD	+ 8	
		ZT	0124	1 Jan
11.		ZT	1442	6 Apr
		ZD	+ 6	
		GMT	2042	6 Apr
		ET	+ 815	
		GMT	2857	6 Apr
		GMT	0457	7 Apr
		GMT	2857	6 Apr
	(Rev)	ZD	+ 6	
		ZT	2257	6 Apr
12.		ZT	0631	21 Jun
		ZT	3031	20 Jun
		ZD	− 10	
		GMT	2031	20 Jun
		ET	+ 2723	
		GMT	4754	20 Jun
		GMT	2354	21 Jun
	(Rev)	ZD	− 9	
		ZT	3254	21 Jun
		ZT	0854	22 Jun
13.		ZT	1200	16 Jul
		ZD	+ 9	
		GMT	2100	16 Jul
		ET	1200	
		GMT	3300	16 Jul
		GMT	0900	17 Jul
	(Rev)	ZD	+ 8	
		ZT	0100	17 Jul
14.		ZT	1924	24 December
		ZD	+ 12	
		GMT	3124	24 December
		GMT	0724	25 December

```
            ET    + 715
            GMT    1439      25 December
    (Rev) ZD   − 12
            ZT     2639      25 December
            ZT     0239      26 December
```

Note that the International Date Line was crossed in this problem.

```
15.              ZT     0408        6 Feb
                 ZD   −  4
                 GMT    0008        6 Feb
                 ET   +  18
                 GMT    0026        6 Feb
          (Rev) ZD   −  3
                 ZT     0326        6 Feb

16.              WT     06-08-14   23 Apr
                 WE(f)    -0-12
                 ZT     06-08-02
                 ZD   +  5
                 GMT    11-08-02   23 Apr

17.              WT     19-38-00   23 Apr
                 WE(s)    +3-32
                 ZT     19-41-32
                 ZD   +  4
                 GMT    23-41-32   (one zone east of ZD 5)

18.              WT     05-54-41   19 Jul
                 WE(s)    +1-12
                 ZT     05-55-53
                 ZD   −  3
                 GMT    02-55-53

19.              WT     07-23-45   12 Oct
                 WE(f)    −1-51
                 ZT     07-21-54   12 Oct
                 ZT     31-21-54   11 Oct
                 ZD   −  9
                 GMT    22-21-54   11 Oct

20.              WT     16-45-13   28 Sept
                 WE(f)    −0-28
                 ZT     16-44-45
                 ZD   +  6
                 GMT    22-44-45   28 Sept

21.                     90°00.0′
                     −55°12.6′   Ho
                      34°47.4′   Zenith Distance
```

21.

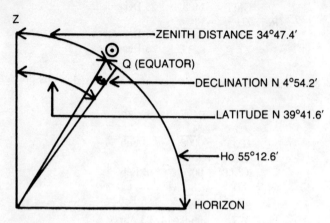

22.

$$
\begin{array}{rl}
+N & \underline{4°54.2'} \quad \text{Declination} \\
N & 39°41.6' \quad \text{Latitude}
\end{array}
$$

$$
\begin{array}{rl}
& 90°00.0' \\
- & \underline{68°17.0'} \quad \text{Ho} \\
& 21°43.0' \quad \text{Zenith Distance} \\
-S & \underline{11°21.4'} \quad \text{Declination} \\
N & 10°21.6' \quad \text{Latitude}
\end{array}
$$

22.

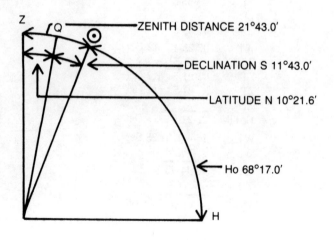

23.

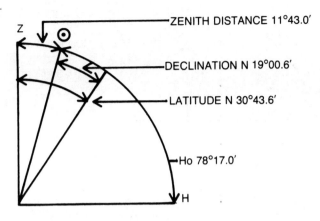

ZENITH DISTANCE 11°43.0′

DECLINATION N 19°00.6′

LATITUDE N 30°43.6′

Ho 78°17.0′

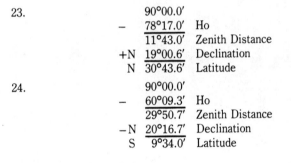

23.
$$
\begin{array}{rll}
& 90°00.0′ & \\
- & 78°17.0′ & \text{Ho} \\
\hline
& 11°43.0′ & \text{Zenith Distance} \\
+\text{N} & 19°00.6′ & \text{Declination} \\
\text{N} & 30°43.6′ & \text{Latitude}
\end{array}
$$

24.
$$
\begin{array}{rll}
& 90°00.0′ & \\
- & 60°09.3′ & \text{Ho} \\
\hline
& 29°50.7′ & \text{Zenith Distance} \\
-\text{N} & 20°16.7′ & \text{Declination} \\
\text{S} & 9°34.0′ & \text{Latitude}
\end{array}
$$

24.

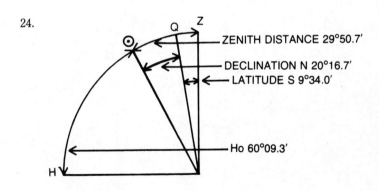

ZENITH DISTANCE 29°50.7′

DECLINATION N 20°16.7′

LATITUDE S 9°34.0′

Ho 60°09.3′

223

25.

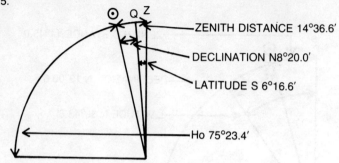

ZENITH DISTANCE 14°36.6'

DECLINATION N8°20.0'

LATITUDE S 6°16.6'

Ho 75°23.4'

25.
$$
\begin{array}{rll}
 & 90°00.0' & \\
- & \underline{75°23.4'} & \text{Ho} \\
 & 14°36.6' & \text{Zenith Distance} \\
-N & \underline{8°20.0'} & \text{Declination} \\
S & 6°16.6' & \text{Latitude}
\end{array}
$$

26.
$$
\begin{array}{rll}
\text{GHA} & 13°17.6' & \\
+ & \underline{360°} & \\
 & 373°17.6' & \\
\text{AP Lo}- & \underline{84°17.6'} & \text{W} \\
\text{LHA} & 289°00.0' &
\end{array}
$$

26.

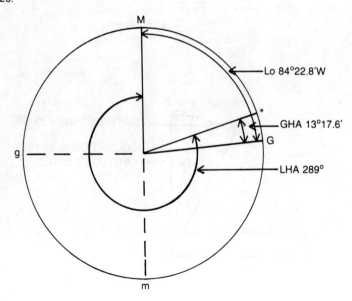

Lo 84°22.8'W

GHA 13°17.6'

G

LHA 289°

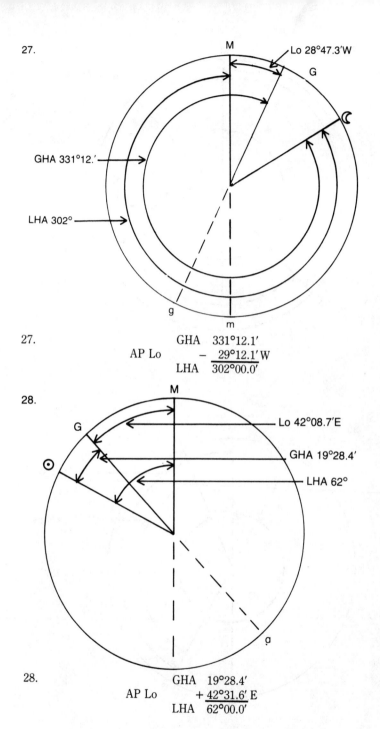

27.

$$\begin{array}{r} \text{GHA} \quad 331°12.1' \\ \text{AP Lo} \quad - \quad \underline{29°12.1'\,\text{W}} \\ \text{LHA} \quad 302°00.0' \end{array}$$

28.

$$\begin{array}{r} \text{GHA} \quad 19°28.4' \\ \text{AP Lo} \quad + \quad \underline{42°31.6'\,\text{E}} \\ \text{LHA} \quad 62°00.0' \end{array}$$

225

29.

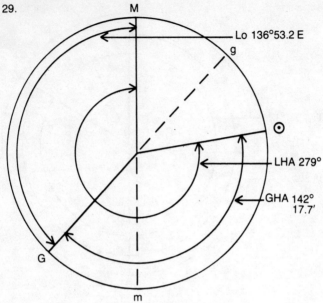

Lo 136°53.2 E

g

☉

LHA 279°

GHA 142°
17.7'

G

m

30.

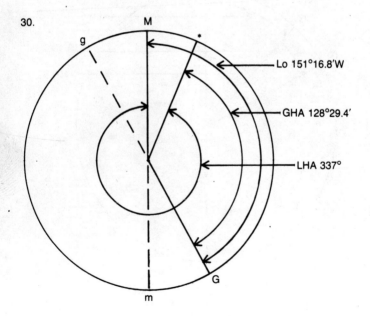

M

g

*

Lo 151°16.8'W

GHA 128°29.4'

LHA 337°

G

m

29.
$$
\begin{array}{lrl}
& \text{GHA} & 142°17.7' \\
\text{AP Lo} & + & 136°42.3' \text{ E} \\
& \text{LHA} & 279°00.0'
\end{array}
$$

30.
$$
\begin{array}{lrl}
& \text{GHA} & 128°29.4' \\
& + & 360° \\
& & 488°29.4' \\
\text{AP Lo} & - & 151°29.4' \text{ W} \\
& \text{LHA} & 337°00.0'
\end{array}
$$

31.
$$
\begin{array}{ll}
12\text{-}14\text{-}21 & \text{Listed Time} \\
\underline{12\text{-}12\text{-}12} & \text{Actual Time} \\
2\text{-}09 & \text{Difference}
\end{array}
$$

$$
\begin{array}{l}
2m = 30' \\
9s = \underline{\quad 2'15''} \\
\quad\quad\; 23'15'' \quad = 32.2' \text{ East of ZM}
\end{array}
$$

32.
$$
\begin{array}{ll}
12\text{-}18\text{-}15 & \text{Actual Time} \\
\underline{12\text{-}08\text{-}23} & \text{Listed Time} \\
9\text{-}52 & \text{Difference}
\end{array}
$$

$$
\begin{array}{l}
9m = 2°15' \\
52s = \underline{\quad 12'00''} \\
\quad\quad 2°27'00'' = 2°27.0' \text{ West of ZM}
\end{array}
$$

33.
$$
\begin{array}{ll}
12\text{-}01\text{-}11 & \text{Actual Time} \\
\underline{11\text{-}55\text{-}52} & \text{Listed Time} \\
5\text{-}19 & \text{Difference}
\end{array}
$$

$$
\begin{array}{l}
5m = 1°15' \\
19s = \underline{\quad 4'45''} \\
\quad\; 1°19'45'' = 1°19.8' \text{ West of ZM}
\end{array}
$$

34.
$$
\begin{array}{ll}
11\text{-}54\text{-}08 & \text{Listed Time} \\
\underline{11\text{-}41\text{-}57} & \text{Actual Time} \\
12\text{-}11 & \text{Difference}
\end{array}
$$

$$
\begin{array}{l}
12m = 3° \\
11s = \underline{\quad 02'45''} \\
\quad 3°02'45'' = 3°02.8' \text{ East of ZM}
\end{array}
$$

35.
$$
\begin{array}{ll}
12\text{-}00\text{-}08 & \text{Actual Time} \\
\underline{12\text{-}00\text{-}04} & \text{Listed Time} \\
00\text{-}04 & \text{Difference} \\
4s = 1' & \text{West of ZM}
\end{array}
$$

Glossary

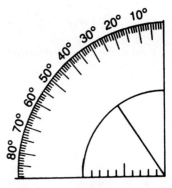

Glossary

Ageton method—A means of converting sextant sight and time information of a line of position using a set of tables by Lt. Arthur A. Ageton, USN.

almanac—A book in which positions of celestial bodies are tabulated on a daily basis for a period of time such as six months or a year. The U.S. government *Air Almanac* and *Nautical Almanac* are the volumes of most use to the navigator.

altitude (H)—The angle above the horizon of a celestial body, measured from the point of the observer.

apparent altitude (Ha)—The altitude of a celestial body above the horizon, as corrected for index (sextant error) and height of eye above the sea.

apparent time—Time in relation to the position of the sun, without regard to any timepiece.

Aries—The point on the celestial sphere at which the apparent path of the sun (the ecliptic) crosses the celestial equator on or about 21 March as the sun changes from south declination to north declination. Sidereal hour angles (SHA) are measured from Aries.

assumed position—A point at which an observer is assumed to be located.

azimuth angle (Z)—The horizontal angle of a celestial body or point from a point on the surface of the earth, measured clockwise or counterclockwise from true north or true south. It is prefixed N or S to agree with the latitude of the point of observation, and E or W to agree with the meridian angle.

azimuth separation—The angle(s) between two or more celestial bodies, selected so that sights will produce lines of position which will result in accurate fixes.

azimuth (Zn)—The horizontal direction of a celestial body or point from a point on the surface of the earth, measured clockwise from true north.

calculated (computed) altitude—The altitude (angle) of a celestial body above the celestial horizon, from a specific point on the surface of the earth, as determined by any tabular or mathematic means.

celestial equator—The circle on the celestial sphere formed where the plane of the earth's equator intersects the celestial sphere. Also called the Equinoctical.

celestial horizon—That circle of the celestial sphere formed by the intersection of the celestial sphere and a plane through the center of the earth perpendicular to the zenith-nadir line at an observer's position.

celestial sphere—An assumed sphere in space, with the earth at its center, on which all the stars are fixed points, and on which the sun, moon, and planets move along their orbits.

chart—A graphic representation of a portion of the earth's surface, drawn to scale, with information needed for navigation. *See* Great Circle Chart, Mercator Projection, and Pilot Chart.

chronometer—A precision time piece carried aboard vessels. Usually set to Greenwich mean time.

chronometer error—The difference between time shown by a chronometer and actual Greenwich mean time.

circle of position—A circle or portion of a circle, drawn on a chart, that represents a known distance from a point or object at the center of the circle. As the distance to the center is known, the boat is presumed to be at some point on this circle.

course—A projected or actual straight-line path of the boat through the water or over the bottom.

course angle—A course measured from 0° at the reference direction clockwise or counterclockwise through 180°, in the same manner as azimuth angle.

course line—The graphic representation of a boat's course, drawn on a chart or plotting sheet.

dead reckoning—Use of the boat's course through the water, speed through the water, and time to determine course or position information.

declination—The angle to a celestial body above or below the celestial horizon. The celestial equivalent of latitude.

declination increment—That portion of total declination expressed as minutes and tenths of minutes of arc.

departure—A boat's position at the start of a run; also distance along a parallel of latitude in some navigation problems.

dip—A sextant correction based on eye height above sea level.
double-second difference—A correction needed for high-altitude sights when using *Pub. 229* for sight reduction.

elapsed time—Time interval between two reference times.
equation of time—Difference in time between apparent time and mean time at the zone meridian.
estimated position—A position plotted in reference to a DR course line and a single line of position.

fix—A position plotted as the result of two or more lines of position.

geographical position—The point where a line drawn from a celestial body to the earth's center passes through the earth's surface.
great circle—Any circle around the surface of a sphere whose plane passes through the center of the sphere.
great circle chart—A navigation chart on which a great circle course can be plotted as a straight line.
great circle course—A course between two points on the earth's circle that lies along a great circle. It represents the shortest distance between the two points.
Greenwich mean time—Mean time at the Greenwich meridian, which is 0° longitude. Sometimes called universal time. It is the standard reference time for all navigation problems and all *Almanac* data.

horizon—The point at which the sky appears to meet the surface of the earth. *See* Chapter 1.
horizontal parallax—A correction figure needed for moon sights due the slight apparent flattening of this body as it goes through its phases.

index error—A sextant error resulting from slight misalignment of index and horizon mirrors; the error may vary as sextant metal or plastic expands or contracts.
intended track—The course over the ground between two points which the navigator plans to maintain.
intercept—The distance, in nautical miles, equal to the difference in arc between calculated altitude and observed altitude.
international date line—A line that lies mostly along the 180° meridian of longitude, but with deflections east or west as necessary to keep geographic units within the same date reference.

latitude—An imaginary line lying along a small circle parallel to the equator. Lines of latitude are marked in a scale of 0° to 90°, with 0° at the equator and 90° at the respective poles.
line of position—A line plotted on a chart as a result of a bearing or a sight; the boat is presumed to lie at some point on this line.

local apparent noon—The point at which the sun reaches maximum altitude; noon by sun time without reference to any timepiece. *See* meridian passage.

local hour angle—Angle in longitude between the position of an observer and the geographical position of a celestial body, measured westward from the observer.

longitude—An imaginary line lying along a great circle that passes through the two poles. Lines of longitude (also called meridians of longitude) are marked in a scale of 0° to 180° both east and west of the 0° Greenwich meridian.

main correction—A sextant correction for refraction for all celestial bodies, plus semi-diameter and parallax for the sun and moon.

mean time—Standard time at the zone meridian of each time zone.

Mercator projection—A chart projection in which all lines of longitude are straight and parallel, and cross lines of latitude at right angles.

meridian—*See* longitude.

meridian passage—The point at which any celestial body passes the meridian of an observer.

mid-latitude—Latitude midway between a departure point and a destination, used in sailing problems. *See* Chapter 18.

nadir—That point on the celestial sphere 180° from an observer's zenith.

name—The labels N and S which are attached to latitude and declination in sight reduction problems. They are the same name when both are north and south, and contrary name when one is north and the other is south.

navigational triangle—The spherical triangle solved in computing altitude and azimuth. It is formed on the celestial sphere by the great circles connecting the pole closest to the observer, zenith of the assumed position, and a celestial body. The terrestrial triangle is formed by the great circles connecting a pole with any two points on the earth's surface, such as assumed position and GP of an observed body.

observed altitude—A sextant sight after all corrections have been made.

parallax—Difference in angle between the sun or moon as measured from the center of the earth and as measured from the surface of the earth.

parallels of latitude—*See* latitude.

Pilot Charts—Charts issued on a monthly basis that provide meteorological and other special information for the charted area.

plotting—Drawing course and position information on a chart or plotting sheet.

point of departure—Position at the start of a run, such as a sea buoy at the start of a trip, or a fix at sea.

prime meridian—The 0° line of longitude at Greenwich, England; also called the Greenwich meridian.

refraction—Bending of light rays as they pass through the earth's atmosphere. Refraction is greatest when a body is near the horizon, and least when it is near the zenith.

rhumb line—A line connecting two points on a Mercator projection chart. Such a line will cross each meridian of longitude at the same angle, and represents a course that can be steered between two points. On a long ocean passage, it may be a longer route than a great circle course.

running fix—A fix determined by advancing a line of position and crossing it with a later line of position.

run of sights—Three or more sights taken in quick succession; a run of five sights is used to establish slope to determine which of the five is the most accurate sight.

sailing—A mathematical means of determining course and distance between two points, or the coordinates of a destination when departure point, heading, and distance travelled through the water are known. *See* Chapter 18.

semidiameter—Half the diameter of the sun or moon; a correction for this is applied to equate an upper or lower limb sight to one taken on the center of the body.

sextant—A navigational instrument used to measure angles, primarily the angle between the horizon and a celestial body. It also can be used to measure horizontal angles.

sextant altitude (Hs)—The sextant reading on a celestial body before any corrections are applied.

sidereal hour angle—Angular distance measured west from Aries.

sight—A measurement of the altitude of a celestial body, using a sextant or some other means of determining angle of the body above the horizon.

sight log—A record of the time and altitude of each sight, plus conditions at the time of the sight such as DR position, time zone, watch error, index error, and height of eye.

sight reduction—Conversion of a sight to a line of position through use of mathematics or tables.

sight reduction tables—Tables of calculated altitude and azimuth from known positions, which may be a boat's DR position or assumed positions near the DR location.

slope—A line drawn on a graph that represents the rate of altitude change of an observed body over a given time period.

small circle—The intersection of a sphere and a plane which does not pass through the center of the sphere.

table of offsets—A table of corrections in *Pub. 229* that are used in conjunction with high altitude sights in order to more properly indicate that the line of position actually represents a small segment of a circle of position.

tabulated altitude—An altitude taken directly from a sight reduction table, without correction for a declination increment or other factor.

tabulated declination—Whole degrees of declination in sight reduction tables.

time diagrams—Drawings used to graphically show the relationship of the observer's position in relation to the Greenwich meridian, the GHA of the sun, other body, or Aries, and the SHA of a start if this is the observed body.

time zones—Zones representing one hour difference in time, east and west of Greenwich, with each zone covering 15° of longitude. *See* Chapter 4.

universal time—*See* Greenwich mean time.

watch error—The difference between watch time and actual zone time at the time of a sight.

zenith—That point on the celestial sphere that is directly above an observer.

zenith distance—Angular distance from the zenith to a point on the celestial sphere. It is equal to the angular distance on the earth's surface from the observer to the GP of this celestial point.

zone meridian—The center meridian of each time zone (*see* Chapter 4, Fig. 4-1).

zone time—Mean time employed in each time zone.

Index

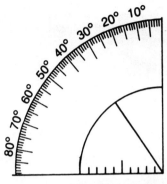

Index